Synergistic Design of Sustainable Built Environments

Synergistic Design of Sustainable Built Environments

Chitrarekha Kabre

CRC Press
Taylor & Francis Group
Boca Raton London New York

CRC Press is an imprint of the
Taylor & Francis Group, an **informa** business

First edition published 2020
by CRC Press
6000 Broken Sound Parkway NW, Suite 300, Boca Raton, FL 33487-2742

and by CRC Press
2 Park Square, Milton Park, Abingdon, Oxon, OX14 4RN

© 2021 Taylor & Francis Group, LLC

CRC Press is an imprint of Taylor & Francis Group, LLC

Library of Congress Cataloging-in-Publication Data

Names: Kabre, Chitrarekha, author.
Title: Synergistic design of sustainable built environments / Chitrarekha Kabre.
Description: First edition. | Boca Raton : CRC Press, 2021. | Includes bibliographical references and index.
Identifiers: LCCN 2020026401 (print) | LCCN 2020026402 (ebook) | ISBN 9780367564834 (hbk) | ISBN 9781003102960 (ebk)
Subjects: LCSH: Sustainable buildings. | Buildings--Energy conservation. | Buildings--Environmental engineering. | Sustainable architecture--Case studies.
Classification: LCC TH880 .K334 2021 (print) | LCC TH880 (ebook) | DDC 720/.47--dc23
LC record available at https://lccn.loc.gov/2020026401
LC ebook record available at https://lccn.loc.gov/2020026402

ISBN: 978-0-367-56483-4 (hbk)
ISBN: 978-1-003-10296-0 (ebk)

Typeset in Times
by Deanta Global Publishing Services, Chennai, India

Visit the eResources: https://www.routledge.com/9780367564834

Cover: A sculpture of daylight above the reconstructed Reichstag (German Parliament), Berlin; design by Pritzker prize laureate architect Sir Norman Foster, 1999. Photo credit Tapan Kumar Ghoshal

Contents

Preface

Nature never breaks her own laws.

Leonardo da Vinci

Over the past half-century, a discourse emphasizing that the environmental health of our earth is profoundly affected by the design of our built environment has emphatically shaped high-performance (green) building practices and associated building regulations (standards and codes), building environmental assessment methods and sustainability assessment systems. The trajectory of environmentally responsive design as applicable to the built environment delineates the transition from technological (high-performance) design paradigm to biocentric (ecological) design paradigm. It is professing a sustainability framework in synergy with nature or ecosystem not only to preserve the environment but also to revitalize and regenerate to have net positive environmental benefits for the living world. This implies a living or whole-systems approach, a more expansive notion of the built environment, one where dynamic relationships exist between a greater number of built and unbuilt elements and where a balance, sustainable relationship between these elements is explored and harnessed. The systems approach in the present context is a much-needed call for building professionals to redefine architecture and adopt principles of regenerative design and examine how it does (or does not) relate to their everyday practice. Mies van der Rohe said that 'less is more'; in the present context a better way of putting it may be, as Alexandro Tombazis says, 'less is beautiful.'

This book explores the theories, principles, and practices of the regenerative design and aims to delineate a novel systems approach, 'synergistic design of sustainable built environment' that fosters transition to both a biocentric (ecological) design paradigm and design excellence. The specific objectives of the book are as follows:

- To articulate parameters of the thermal and luminous environment: climate, sun, occupant comfort, and well-being
- To explain the qualitative and quantitative methods of analysis to make design decisions
- To elucidate the spectrum of thermal environment and luminous environment design strategies
- To explore potentials of renewable energy systems and their integration with the built environment
- To illustrate design case studies for each of the major climate zones
- To present the climatic data and sun-path diagrams in a readily usable format

The manuscript is organized into seven distinct chapters, starting with the introduction of the subject; the second chapter, 'Climate and Comfort,' classifies climate and defines the elements of climate, parameters of thermal comfort, and solar geometry. The third chapter, 'Thermal Environment Design Strategies,' illustrates and explains passive, hybrid (low-energy), and active design strategies of heating, cooling, and ventilation. The fourth chapter, 'Luminous Environment Design Strategies,' describes the fundamentals of daylighting and explains architectural and state-of-the-art technological daylighting design strategies and its integration with electric lighting. The palette of thermal and visual environment design strategies, along with the quantitative and qualitative information included in Chapters 3 and 4, enables designers to generate design solutions in response to climate, occupant comfort, and program during the early phases of the design, when built form can evolve in synergy with natural forces of the sun, wind, sky, water, and earth. The designer can optimize design solutions to ensure the required indoor conditions with little or no use of energy, other than from ambient or renewable sources. The fifth chapter explores potentials of renewable energy from the sun, wind, biomass, geothermal, hydro, hydrogen, and fuel cells and its integration in the built environment. Since carbon neutral buildings can be fully powered by renewable resources, a future of regenerative buildings is not only necessary but also elegantly achievable. The issue of energy storage, smart metering, and smart grid are also covered in Chapter 5. The sixth chapter showcases AIA COTE-awarded contemporary design case studies in each of the five major climatic zones of the United States. Each exemplar design study includes an overview of the design intentions, climate and site responses, thermal strategies, energy systems, sustainability thinking, and a design profile as a snapshot of the project. These building precedent studies are used to illustrate how designers have approached issues related to the sustainable built environment that is ecologically appropriate and meaningful for varied climates, programs, and occupants. The lessons of these precedents should be interpreted for the possibilities they suggest rather than their particular solutions. Chapter 7 presents the climatic data and sun-path diagrams for 50 capital cities, representing each of the states in the United States. Temperature, humidity, sunshine hours, solar radiation, rainfall, and precipitation data are given both numerically and in graphic form; the latter for a quick, visual appreciation, the former for a more detailed analysis. Some single-figure indices are included: an indication of temperature variability; outdoor design conditions recommended as a basis for calculating the required heating or cooling capacity. Wind roses show the direction and frequency of winds, while average speeds are given in numerical form.

This book addresses the quintessential part of the much larger picture of the sustainable built environment; it is just one way of understanding how complex layers of issues need to be integrated into a comprehensive whole that is appropriate for the place, program, and users and promises a regenerative future. Through synergistic design, a vibrant relationship can be established – connections can be made – that weaves together people and built and unbuilt environments into an ecological whole. As a result, the framework of 'synergistic design' should be viewed as something that is inherently broad and comprehensive in scope, dynamic, adaptable, and capable of change and growth.

This book is a practical tool or handbook for architects, building professionals, researchers, and students that will provide them with a theoretical understanding of the physical phenomena to be dealt with and methodology of implementing the synergistic design of sustainable (regenerative) built environment. This book explains and demonstrates how the design wisdom of passive solar architecture can be integrated with the best of modern technological advancement to create sustainable and regenerative yet beautifully designed humane architecture. This book is also an important reference for those architects who are concerned about the aesthetic aspects of sustainability.

Acknowledgments

This book grew out of my past 30 years of academic and professional experiences from four continents: Asia, Australia, Europe, and the USA. These experiences include my research and teaching at leading institutions: School of Planning & Architecture, New Delhi; the Indian Institute of Technology, Kharagpur; Manipal University; the University of Queensland, Australia; North Dakota State University, Fargo; and currently the DCR University of Science & Technology, Murthal. I collaborated with THOWL, Detmold, Germany, under the aegis of CREED (Clime Related Energy Efficient Design); Nanyang Technological University, Singapore; University of Liege, Belgium; TU Delft, Netherlands; ETH Zurich, Switzerland; Yonsei University, Seoul, South Korea and University of Minnesota, Minneapolis, USA for research and development.

I am grateful to all the esteemed organizations and individuals who helped me to collate the information for climate, design case studies, and other topics.

I would like to acknowledge CRC Press, Taylor & Francis, for appreciating the value and need for this book within the architectural profession and for making it a reality.

My father late Mr. Ram Gopal Kabre has been a constant source of motivation to excel in life. My mentor mechanical engineer Mr. Tapan Kumar Ghoshal stood by me during every struggle and all successes. I wish to dedicate this book to the lotus feet of Maa Vajreshwari and my saviour Shri Nityanand Saraswati Maharaj.

About the Author

Prof. Dr. Chitrarekha Kabre earned her doctorate in architecture from the University of Queensland, Australia, in 2008 (recipient of the Australian Development Cooperation Scholarship). In 1989 she received her master's in building engineering and management (recipient of gold medal) from the School of Planning and Architecture, New Delhi (an institute of National Importance). In 1985 she received her bachelor's in architecture from the Maulana Azad National Institute of Technology, Bhopal. Prof. Dr. Chitrarekha Kabre has 30 years of academic and professional experience in the field of computer-aided architectural design, project management, and sustainable architecture. She has developed courses on sustainable architecture at undergraduate, postgraduate, and doctoral levels. She introduced M.Tech. (construction and real estate management), an innovative program awarded by the University Grants Commission, Government of India. Prof. Dr. Chitrarekha Kabre has been the pioneer of sustainable architecture education and research at the eminent institutions like the Indian Institute of Technology, Kharagpur, Manipal University, and presently Deenbandhu Chhotu Ram University of Science and Technology, Murthal (Sonepat). As Fulbright visiting professor, North Dakota State University, Fargo (2012), Prof. Dr. Chitrarekha Kabre contributed in pedagogy of sustainable architecture. She is an active member of the Society of Building Science Educators (SBSE) and recipient of Jeffrey Cook Memorial Scholarship in 2019. She has authored more than 36 research papers in international conferences and journals (*Building & Environment and Architectural Science Review*) and has served as a reviewer for the journals *Building & Environment* (Elsevier Science) and *Indoor & Built Environment* (Sage Publications). Prof. Dr. Chitrarekha Kabre is the author of the book *Sustainable Building Design: Application Using Climatic Data in India*, published by Springer, Germany, and the chief editor of the book *Energy Efficient Design of Buildings and Cities*, published by DCR University of Science & Technology, Murthal, and Hochschule Ostwestfalen-Lippe (HSOWL), Detmold, Germany. She has an extensive citation index in Google Scholar. She is a life member of the International Association of Passive and Low Energy Architecture (PLEA) and an International Associate of American Institute of Architects. She is also LEED® Green Associate of US GBC. Prof. Dr. Chitrarekha Kabre is a certified professional as well as an evaluator (architect and construction management) of Green Rating for Integrated Habitat Assessment (GRIHA), a national green rating of India. She is the master trainer for the Energy Conservation Building Code administered by the Bureau of Energy Efficiency, Ministry of Power, Government of India. Prof. Dr. Chitrarekha Kabre's biography is published in Marquis Who's Who in the World, the United States, as one of the leading achievers.

List of Abbreviations

Δ	change or change in
A/C	Air conditioning
AC	Alternating Current
ADC	Active Downdraft Cooling
AEO	Annual Energy Output
AH	Absolute Humidity
AHU	Air Handling Unit
AIA	American Institute of Architects
ALT	Altitude
AT/FP	Anti Terrorism/Force Protection
ARRA	American Recovery and Reinvestment Act
ANSI	American National Standards Institute
ASHRAE	American Society of Heating, Refrigerating and Air-conditioning Engineers
AZI	Azimuth
BIM	Building Information Modelling
BIPV	Building Integrated Photovoltaics
BRE	Building Research Establishment
BREEAM®	Building Research Establishment Environmental Assessment Method
BS	British Standards
Btu	British thermal unit
C	Centigrade (temperature scale)
CASBEE	Comprehensive Assessment System for Built Environment Efficiency
CAV	Constant Air Volume
CDD	Cooling Degree Days
CFC	chlorofluorocarbon
CFD	Computational Fluid Dynamics
CIE	Commission Internationale de l'Eclaiage (International Commission on Illumination)
CLEAR	Center for Living Environments and Regeneration
CO_2	Carbon dioxide
CoP	Coefficient of Performance
COTE	Committee on the Environment
DBT	Dry Bulb Temperature
DC	Direct Current
DGNB	German Sustainable Building Council
DOAS	Dedicated Oudoor Air System
DOE	Department of Energy
EISA	Energy Independence and Security Act
EPA	Environmental Protection Agency
EUI	Energy Use Intensity
F	Fahrenheit (temperature scale)
FDFA	Federal Department of Foreign Affairs
FSC	Forest Stewardship Council
ft²	square feet
GHG	Green House Gas
HSA	Horizontal Shadow Angle
HDC	Hybrid Downdraft Cooling

HDD	Heating Degree Days
HDH	Heating Degree Hours
HVAC	Heating, Ventilation, and Air Conditioning
IAQ	Indoor Air Quality
ICC	International Code Council
IDP	Integrate Design Process
IEA	International Energy Agency
IEQ	Indoor Environmental Quality
IESNA	Illuminating Engineering Society of North America
IPCC	Intergovernmental Panel on Climate Change
I-P system	inch-pound; English system of units
ISO	International Organization for Standardization
J	Joule
K	Kelvin or absolute (temperature scale)
kWh	kilowatt-hour
LAT	Latitude
LBC	Living Building Challenge
LCA	Life Cycle Assessment
LCC	Life Cycle Cost
LCI	Life Cycle Inventory
LEED	Leadership in Energy and Environmental Design
LID	Low Impact Development
low-E	low emissivity
LPD	Lighting Power Density
lx	Lux
m^2	square meter
MEP	Mechanical, Electrical and Plumbing
MRT	Mean Radiant Temperature
N	Newton
NAVFAC	Naval Facilities Engineering Command
NOAA	National Oceanic and Atmospheric Administration
NREL	National Renewable Energy Laboratory
OECD	Organisation for Economic Co-operation and Development
OPEC	Organization of the Petroleum Exporting Countries
OSHPD	Office of Statewide Health Planning and Development
PPA	Power Purchase Agreement
PDEC	Passive Downdraft Evaporative Cooling
PHIUS	Passive House Institute the US
PV	Photovoltaic
RH	Relative Humidity
RIBA	Royal Institute of British Architects
SDGs	Sustainable Development Goals
SET	Standard Effective Temperature
SHGC	Solar Heat Gain Coefficient
SI	System International (d'unités) International System (of Units)
SPeAR	Sustainable Project Assessment Routine
SPV	Solar Photovoltaic
SWH	Solar Water Heater
TRNSYS	The Transient Energy System Simulation Tool
UIA	Union Internationale des Architects
UNCED	UN Conference on Environment and Development

UNEP	UN Environment Program
USGBC	United States Green Building Council
VAV	Variable Air Volume
VCP	Visual Comfort Probability
VLT	Visual Light Transmission/Transmittance
VOC	Volatile Organic Compound
VSA	Vertical Shadow Angle
VRF	Variable Refrigerant Flow
VRV	Variable Refrigerant Volume
WBT	Wet Bulb Temperature
WMO	World Meteorological Organization
WWR	Window to Wall Ratio
ZNE	Zero Net Energy

1 Introduction

1.1 BACKGROUND

The new millennium is now almost two decades old. That which began with great festive optimism was soon followed by the events of September 11, 2001; the Fukushima Daiichi nuclear disaster of March 11, 2011; a spate of natural disasters; and the more recent unprecedented catastrophe of humanity, COVID-19, which has led to profound environmental, economic, social, and cultural effects on a global scale that are now being more and more dominated by an interconnected set of existential questions with far-reaching consequences for the future existence of mankind.

The global climate strike led by 16-year-old Greta Thunberg and millions of school children from Sydney to Manila, Dhaka to London, and New York echoed the inconvenient truth that the fate of the planet is at stake (*The Guardian*, September 21, 2019). The Decade of Action for the Sustainable Development Goals (SDGs) launched by the United Nations in early 2020 under the rallying cry 'For People, For Planet' urged the world to address the challenges of climate and nature, gender, and inequality.

As a result of this recent discussion, the issue of ecological balance and climate change has risen to prominence worldwide. Among the many problems humanity will have to address in the 21st century, three that ought to be accorded utmost priority, because of an increasing world population that should total 10 billion people by the end of the first century of the new millennium, are the following:

- Securing healthy and sufficient nutrition
- Providing access to clean drinking water
- Assuring disease control and adequate health care

The buildings and construction sector is a key player in the fight against climate change: it accounted for 36% of global final energy use and 39% of energy-related carbon dioxide (CO_2) emissions in 2017 (IEA 2018). In the United States, the construction sectors accounted for $840 billion, or 4.1% of the gross domestic product (GDP), more than many industries, including information, arts and entertainment, utilities, agriculture, and mining (BEA 2019). By 2060, the world is projected to add 230 billion m^2 (2.5 trillion ft^2) of buildings or an area equal to the entire current global building stock (UN Environment and International Energy Agency 2017). This is the equivalent of adding an entire New York City to the planet every 34 days for the next 40 years. This trend points to the questions concerning the securing of a stable and sustainable built environment that is of great importance.

It is no longer just a question of following a particular architectural style or design philosophy; building professionals are urged to transform the global built environment from being a major contributor of greenhouse gas (GHG) emissions to being environmentally sustainable and regenerative. Our best chance is to ensure that the architecture, planning, and development community, the primary agents shaping the built environment through design and construction, has access to the knowledge and tools necessary for the transition to a sustainable and regenerative world. The synergistic design of the sustainable built environment is a much-needed call for building professionals to redefine architecture to help this transition to an environmentally sustainable and regenerative built environment.

This chapter discusses the need for a sustainable built environment and the transition from the technological (high-performance) design paradigm to a biocentric (ecological) design paradigm. This chapter emphasizes the importance of the synergistic design of a sustainable built environment, an innovative systems framework, as the premise of the book.

The next section defines the built environment. The third section presents the climate-responsive architecture of yesteryears. The fourth section discusses sustainable development and sustainability and its relevance to sustainable architecture. Further, the fifth section presents a technological (high-performance) design paradigm, delineating the technical approach, regulatory approach, and rating system approach. The sixth section explains the biocentric design paradigm including ecological theories and life cycle assessment. Finally, the chapter delineates an innovative systems framework for the synergistic design of a sustainable built environment.

1.2   BUILT ENVIRONMENT

As the natural environment with varying climate is not suitable to the lifestyle of man, man is always trying for suitable transformation in the natural surroundings. This transformed environment is known as 'Alan-made' or 'built environment.'

The built environment generally refers to the 'manmade surroundings that provide the setting for human activity, ranging from the large-scale civic surroundings to the personal places' (Moffatt and Kohler 2008). The built environment includes both urban and rural forms.

The built environment intends to provide a comfortable environment for humans to reside and work in and also delivers economic, social, and cultural benefits. The built environment also, however, has wide-ranging negative environmental aspects and impacts, including air quality, water and energy consumption, transport accessibility, materials use, and management of waste (Table 1.1).

1.3   CLIMATE-RESPONSIVE ARCHITECTURE

Even before the first built shelters, humans utilized climate elements to improve thermal comfort. About 2500 years ago, Aeschylus, the Greek playwright, in his play *Prometheus* (the mythological *fire stealer*) observed that ignorant primitives and barbarians 'lacked knowledge of houses built of bricks and turned to face the winter sun, dwelling beneath the ground like swarming ants in sunless caves.'

The evolution of the built environment, with responses to multiple and complex requirements, started by providing the shelter needed for protection from attack by human enemies and wild animals, as well as protection from hostile and unfavorable aspects of the physical environment. At later stages, durability, status, fashion, and improved environmental quality were the motors of development (Rapoport 1969). According to this sequence, the protection from climate was one of the initial factors that have remained a constant preoccupation and priority in the long process of the development of the built environment and the history of architecture (Oliver 1987). From the early huddle of buildings at Catalhöyük in Anatolia, 7000 BC, the indigenous building design

---

TABLE 1.1
Impacts of the Built Environment

Aspects of Built Environment	Consumption	Environmental Effects	Ultimate Effects
Siting	Energy	Waste	Harm to human health
Design	Water	Air pollution	Environment degradation
Construction	Materials	Water pollution	Loss of resources
Operation	Natural resources	Indoor pollution	
Maintenance		Heat islands	
Renovation		Stormwater runoff	
Deconstruction		Noise	

Source: https://archive.epa.gov/greenbuilding/web/html/about.html

---

demonstrated ingenuity for climate amelioration through a basic understanding of the thermal and structural behavior of natural materials. Native American traditional buildings and villages have also utilized passive solar principles for more than 2000 years. In Southern California, the Indians of the Yokut Tule Lodge (Figure 1.1) not only protected their huts but in a generous, direct manner provided for pleasant living and shaded communal areas.

From Aristotle to Montesquieu, many scholars believed that climate had pronounced effects on human physiology and temperament. In book III, Chapter VIII, of Xenophon's *Memorabilia* of the Greek philosopher Socrates (470–399 BC), written a few decades after Aeschylus, and during the Greek wood fuel shortage, Socrates' Megaron house (Figure 1.2) exemplifies the essential, timeless principles of sun-tempered architecture:

Now in houses with a south aspect, the sun's rays penetrate the porticos in winter, but in the summer, the path of the sun is right over our heads and above the roof, so that there is shade. If then this is the best arrangement, we should build the south side loftier to get the winter sun and the north side lower to keep out the winter winds. To put it shortly, the house in which the owner can find a pleasant retreat at all seasons and can store his belongings safely is presumably at once the pleasantest and the most beautiful.

However, the first written documents to explain the functioning of the house in relation to climate impacts are those of the Greeks and Romans. The Roman architect Marcus Vitruvius Pollio (Morgan 1960) wrote 2000 years ago:

If our designs for private houses are to be correct, we must at the outset take note of the countries and climates in which they are built. One style of the house seems appropriate to build in Egypt, another in Spain, a different kind in Pontus, one still different in Rome, and so on with lands and countries of other characteristics. This is because one part the earth is directly under the sun's course, another is far away from it, while another lies midway between these two … it is obvious that designs for houses ought similarly to conform to the nature of the country and diversities of climate.

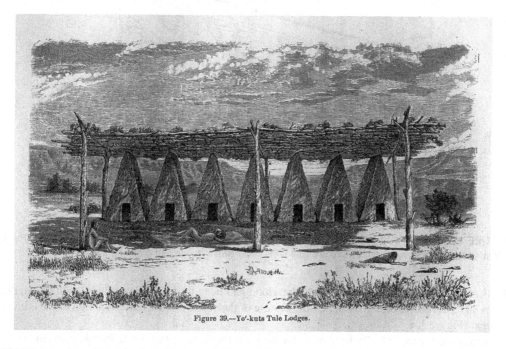

Figure 39.—Yo'-kuts Tule Lodges.

FIGURE 1.1 Yokut Tule Lodge, Southern California. Source: https://missionscalifornia.com/sites/default/files/2019-11/16-Tule_lodges_0f_Yokuts.jpg "Yo'-kuts Tule Lodges" from Contributions to North American Ethnology, Volume III. Washington: Government Printing Office, 1877. California Historical Society, North Baker Research Library Collection, FN-32152.

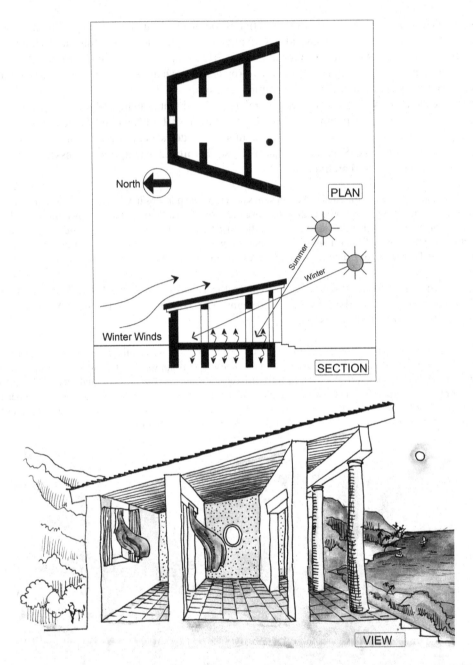

FIGURE 1.2 Socrates' Megaron House (470-399 BC). Source: https://ednovak99.wordpress.com/2016/12/13/passive-solar-design-overview/ Credits: linework by Ar Shiva Bagga and Ar Kapil Grover

Thus, indigenously built habitats across the globe had been an expression of the locally available materials and construction techniques, the culture of the communities, and a function of the climate context (Figure 1.3).

As a consequence of the Industrial Revolution, the instinctive attention to how humankind interacts with the natural environment underwent a brusque inversion and led to the advent of modernity, bringing cultural, territorial, and technological transformations (Frampton 1985). The tone of the international style was set by a few internationally recognized master architects. Most of their

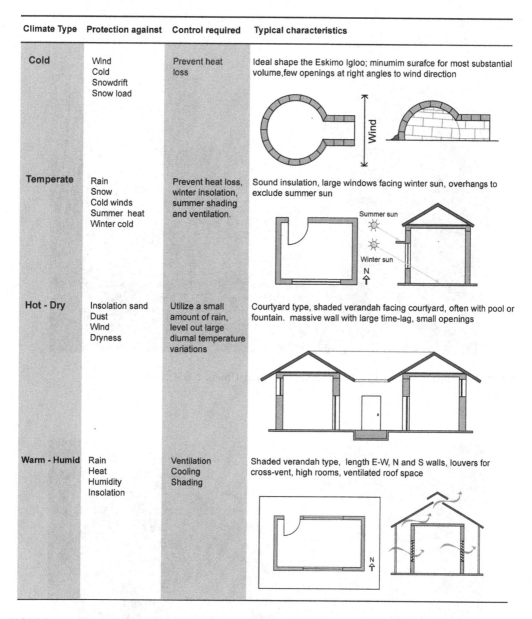

Climate Type	Protection against	Control required	Typical characteristics
Cold	Wind Cold Snowdrift Snow load	Prevent heat loss	Ideal shape the Eskimo Igloo; minumim surafce for most substantial volume,few openings at right angles to wind direction
Temperate	Rain Snow Cold winds Summer heat Winter cold	Prevent heat loss, winter insolation, summer shading and ventilation.	Sound insulation, large windows facing winter sun, overhangs to exclude summer sun
Hot - Dry	Insolation sand Dust Wind Dryness	Utilize a small amount of rain, level out large diurnal temperature variations	Courtyard type, shaded verandah facing courtyard, often with pool or fountain. massive wall with large time-lag, small openings
Warm - Humid	Rain Heat Humidity Insolation	Ventilation Cooling Shading	Shaded verandah type, length E-W, N and S walls, louvers for cross-vent, high rooms, ventilated roof space

FIGURE 1.3   Four primary climate types and indigenously built response. Credits: linework by Vaibhav Ahuja

solutions involve highly transparent glass-steel design that essentially requires the massive inclusion of active building systems, such as air conditioning, to provide the minimal comfort conditions for the building occupants. Figure 1.4 shows the skyline of Chicago, Illinois.

In the post-war years, climate-responsive design of buildings became a concern and realization was evident that since there is no international climate, architecture also cannot be international. In the 1950s and 1960s, modern architects Le Corbusier and Louis I. Kahn designed several buildings driven by a sound response to the climate (Ali and Yannas 1999). One of the physical hallmarks of modern architecture of the tropics was the sun-screen, usually called the *brise-soleil*, located on the facades that faced the sun to prevent its rays penetrating the building's interior in the summer (Figure 1.5).

FIGURE 1.4   Skyline of Chicago. Photo credit: Ms. Cindy Urness, NDSU, Fargo

FIGURE 1.5   Mill Owners' Association Building, Ahmedabad, Le Corbusier 1954.© Tapan Kumar Ghoshal

Walter Gropius (1955), considering regional expression, writes:

true regional character cannot be found through a sentimental or imitative approach by incorporating either old emblems or the newest local fashions which disappear as fast as they appear. But if you take ... the basic difference imposed on architectural design by the climatic condition ... diversity of expression can result ... if the architect will use the utterly contrasting indoor-outdoor relations ... as focus for design conception.

The term 'bioclimatic design' has been coined by Victor Olgyay in 1953 in his research paper 'Bioclimatic Approach to Architecture' and later expanded in his influential book *Design with Climate* with the subtitle *Bioclimatic Approach to Architectural Regionalism* (1963). The term is defined as the architecture that responds to its climatic environment and achieves comfort for the occupants through appropriate design decisions. Olgyay synthesized elements of climatology, human physiology, and building physics, with strong advocacy of architectural regionalism in terms of designing in sympathy with the environment. In many ways, he can be considered an important progenitor of what is now called 'sustainable architecture.'

The first climate and architecture conference was held in February 1979 at the behest of federal sponsors, where more than 50 architects, engineers, home builders, and climatologists convened in Washington. Most importantly, the climate and architecture conferees agreed that designers understand the two fundamentals of climate-conscious architecture. Design that responds to climate – and the research that supports such work – can't be approached as 'solar,' 'geothermal,' or 'underground construction' but as solutions that consider all elements of climate in a holistic approach to energy-conservative design for human comfort (Green 1979).

1.4 SUSTAINABLE DEVELOPMENT AND SUSTAINABILITY

A broad spectrum of concepts and notions of sustainable development or sustainability evolved in different regions/forums, and it is axiomatic that there is no common definition of sustainable development or sustainability. The modern environmental movement is believed to have begun in the United States in 1962, inspired by Rachel Carson's book *Silent Spring*, the publication of which caused a paradigm shift in understanding the environmental impact of pesticide use (IISD 2012). Barbara Ward and Rene Dubos (1972) presented the state of affairs in their book *Only One Earth*.

Environmental degradation was the main concern at the Stockholm UNEP Conference in 1972 (Dodds et al. 2012). The Organization of the Petroleum Exporting Countries' (OPEC) oil embargo and price increases of the 1970s brought the realization of the finite nature of our fossil fuel supplies and spurred significant research and activity to improve energy efficiency and find renewable energy sources. The Organization for Economic Co-operation and Development (OECD) established the International Energy Agency (IEA) in 1974 to help countries co-ordinate a collective response to major disruptions in the supply of oil. In 1987, the United Nations established the World Commission on Environment and Development (WCED), which was later known as the Brundtland Commission after its Chair Gro Harlem Brundtland, the Norwegian prime minister. The commission's report, known as the Brundtland Report (WCED 1987), defined 'sustainable development' as 'development that meets the needs of the present without compromising the ability of future generations to meet their own needs.' It has two intrinsic requirements. First, it entails inter- and intra-generational equity within the constituent domains of sustainability – environmental, social, cultural, and economic, the notion of the 'triple bottom line' (Figure 1.6). Second, sustainability and any reference to sustainable development require thinking long term and assuming responsibility for the future.

The Montreal Protocol on Substances That Deplete the Ozone Layer, an international treaty, was signed in 1987 to phase out organofluorides, which are affecting the ozone layer and admitting more ultraviolet irradiation, causing the greenhouse effect.

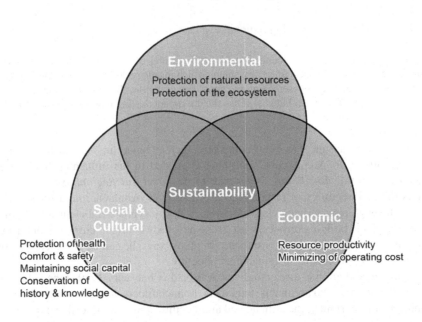

FIGURE 1.6   Triple bottom-line approach to sustainability.

The Intergovernmental Panel on Climate Change (IPCC), in its first assessment report in 1990, firmly established that the climate is changing due to anthropogenic influences caused by the emission of greenhouse gases by human activities (Houghton et al. 1990).

The Agenda 21 (UNSD 1992) prescribed key points for the sustainable construction industry as the utilization of indigenous and local materials and technologies; labor-intensive construction and maintenance technologies; energy-efficient designs and technologies and sustainable utilization of natural resources (i.e., recycling of materials and waste prevention); development of knowledge on the environmental impacts of buildings; and self-help housing for the urban and rural poor.

The energy crisis of the 1970s and the environmental concerns of the 1980s laid the foundation for the contemporary green (sustainable) building movement. The term 'sustainability' was not formally defined then in the context of the built environment, but it echoed in various forums. Sir Alexander John Gordon (1917–1999), then president of the Royal Institute of British Architects, espoused his ideas at the RIBA Conference in 1972 and defined 'good architecture' as buildings that exhibit 'long life, loose fit and low energy,' nicknamed as 3L Principle and which are measurable (Gordon 1972). The philosophical basis of this was that it would be ecologically beneficial to erect buildings that last, which are designed in a way to remain adaptable for changed uses and which use little energy in their operation. In 1973, the American Institute of Architects (AIA) established an energy task force and in 1975 a committee on energy conservation. In 1989, the AIA Committee on the Environment (COTE) was formed. The Environmental Resource Guide was published by AIA, funded by EPA (1992). The first local green building program was introduced in Austin, Texas (1992).The AIA National Convention in 1993 was themed 'Sustainability – Architecture at the Crossroads.' Susan Maxman, FAIA, then president of AIA echoed in her address

> We have the knowledge, we have the riches, we have the power. What is called for is a profound shift in the way we regard this planet and everything on it. Exploitation must be replaced by stewardship. And for stewardship to extend its healing hand, we must act responsibly. (AIA 2007)

At the Chicago Congress, more than 3,000 AIA members joined Maxman and the Union Internationale des Architects (UIA) in signing the Declaration of Interdependence for a Sustainable Future, a document placing 'environmental and social sustainability at the core of our practices and professional responsibilities' (AIA 2007). Its scope and breadth suggested that: 'sustainable design integrates considerations of resource and energy efficiency, healthy buildings and materials, ecologically and socially sensitive land-use, and an aesthetic sensitivity that inspires, affirms, and ennobles.' Many national bodies and institutions of architecture adopted this declaration and developed environmental policies, building energy codes and standards, and green rating systems. 'Greening of the White House' initiative was launched by the Clinton administration in 1993.

The historic United Nations Summit held in September 2015 in Paris adopted the 2030 Agenda for Sustainable Development – the 17 SDGs will universally apply to all over the next 15 years. One of the goals is to make sustainable cities and communities.

The landmark agreement reached at the COP21 UN Forum on Climate Change Conference in December 2015 in Paris insinuates an end to the fossil fuel era. It commits nearly 200 countries – including the United States, China, India, and EU Nations – to keep the global average temperature increase to 'well below 2°C above pre-industrial levels and to drive efforts to limit the temperature increase to 1.5°C above pre-industrial levels.' To meet this target, the world must reach zero fossil fuel and CO_2 emissions in the built environment by about 2050 and zero total global GHG emissions between 2060 and 2080 (Hare et al. 2014, 2030 Challenge).

Sustainable development is favored by the government and the private sector, while the term 'sustainability' has been increasingly used by academics, environmentalists, and nongovernmental organizations (Robinson 2004). Generally, the distinction is drawn between 'anthropocentric-' and 'biocentric'-framed sustainability models (Robinson 2004, Cole 2005). Sustainable development maintains an anthropocentric view and represents the diametrically opposing imperatives of technological advancement and growth, on the one hand, and ecological (and perhaps social and economic) sustainability, on the other. Sustainability, by contrast, promotes a biocentric view that places the human presence within a larger natural context and focuses on constraints and fundamental value and behavioral change.

1.5 TECHNOLOGICAL (HIGH-PERFORMANCE) DESIGN PARADIGM

The terms 'green,' 'sustainable,' and 'high performance,' when used to describe buildings, have different shades of meaning to some. For this book, however, they are used interchangeably. This use is consistent with the U.S. Environmental Protection Agency's definition of green building as

> the practice of creating structures and using processes that are environmentally responsible and resource-efficient throughout a building's life-cycle, from siting to design, construction, operation, maintenance, renovation, and deconstruction. This practice expands and complements the classical building design concerns of economy, utility, durability, and comfort. A green building is also known as a sustainable or high-performance building.

The *ASHRAE Green Guide* (2006) defines green design as 'one that is aware of and respects nature and the natural order of things; it is a design that minimizes the negative human impacts on the natural surroundings, materials, resources, and processes that prevail in nature.'

Within the anthropocentric view to sustainability, the value of a building is still generally defined in terms of human benefit 'most often measured in relatively short-term financial returns and human health' (Mang and Reed 2015). The technological (high-performance) design paradigm has taken three approaches: technical, regulatory, and rating system.

1.5.1 TECHNICAL APPROACH

Globally the use of a different list of indicators in different approaches makes sustainability assessment subjective and causes a kaleidoscope of problems in comparing results from different methods. A research and technical approach to this problem is the development of international standards to regulate these rating systems for the wider green building market (ISO 2008). The ISO TC 59 'Building Construction' and its subcommittee (SC) 17 'Sustainability in Building Construction' is responsible for the following standards (Table 1.2):

i) ISO/CD 21929-1:2011 Sustainability in Building Construction – Sustainability Indicators – Part 1: Framework for the Development of Indicators for Buildings and Core Indicators, (cancels and replaces the ISO/TS 21929-1:2006)

ii) ISO 21930:2007 Sustainability in Building Construction – Environmental Declaration of Building Products

iii) ISO/TS 21931-1:2006 Sustainability in Building Construction – Part 1: Framework for Methods of Assessment for Environmental Performance of Construction Works – Part 1: Buildings, published, stage: 90.92 (2006-06-30)

iv) ISO 15392:2008 Sustainability in Building Construction – General Principles

According to ISO 15392 (2008), construction sustainability includes 'considering sustainable development in terms of its three primary aspects (economic, environmental and social), while meeting the requirements for technical and functional performance.' ISO/CD 21929-1 defines a framework for the improvement of buildings' sustainability indicators to assist the minimum functionality and performance of buildings with minimum environmental impact while improving economic and social aspects at the local and global levels (ISO 2011). Individual buildings are believed to impact seven core protection areas of sustainable development: cultural heritage, economic capital, economic prosperity, ecosystem, natural resources, health and well-being, and social equity. The purpose is to protect the areas of sustainability development and the scope is the building's life cycle. ISO/CD 21929-1 defines a list of core indicators to assess, diagnose, compare, and monitor sustainable performance-utilizing indicators. The indicators are presented in three levels: the location-specific level, the site-specific level, and the building-specific level (Figure 1.7).

1.5.2 REGULATORY APPROACH

A regulatory approach, in the form of the building standards and codes, provides a minimum amenity, safety, health, and sustainability standard in the design and construction phases of new

TABLE 1.2

Suite of Related International Standards for Sustainability in Buildings and Civil Engineering Works

	Environmental Aspects	Social Aspects	Economic Aspects
Methodological basics	ISO 15392: General principles ISO/TR 21932: Terminology		
	ISO 21929-1: Sustainability indicators-Part 1: Framework for the development of indicators and a core set of indictors for buildings		
Buildings	ISO 21931-1: Framework for methods of assessment of the environmental performance of construction works - Part I: Buildings		
Building products	ISO 21930: Environmental declaration of building products		

Source: https://www.iso.org/obp/ui/#iso:std:iso:21929:-1:ed-1:v1:en

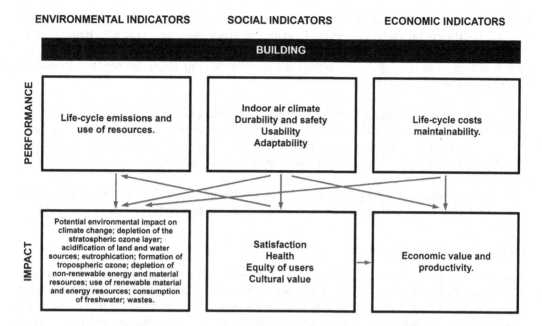

FIGURE 1.7   Building level core indicators ISO. Linework by Ar Shubham Satija

buildings. The regulatory approach involves federal, state, and local jurisdictions concerning mandatory building regulations and requirements. The purpose is to prescribe minimum acceptable performance standards in a limited number of criteria, and the scope is to design and construct the building. Federal involvement with the high-performance building can be highlighted by key milestones.

The Energy Policy Act of 2005 (US Govt. 2005) defines goals and standards for reducing energy use in existing and new federal buildings. The Act requires the application of sustainable design principles to new and replacement federal buildings. It sets an energy consumption target for new federal buildings of 30% below existing standards. The Act establishes an ENERGY STAR® labeling program. ENERGY STAR is a voluntary program of the U.S. Environmental Protection Agency (EPA) and the Department of Energy (DOE) to identify and promote energy-efficient products and buildings to reduce energy consumption, improve energy security, and reduce pollution through voluntary labeling of, or other forms of communication about, products and buildings that meet the highest energy conservation standards. Buildings can receive a 1-100 ENERGY STAR score; this score compares the building's energy performance to similar buildings nationwide. A score of 50 represents median energy performance, while a score of 75 means the building performs better than 75% of all similar buildings nationwide and may be eligible for ENERGY STAR certification. ENERGY STAR labeling and scoring are meant for both existing buildings and new buildings. On average, ENERGY STAR–certified buildings use 35% less energy and generate 35% fewer greenhouse gas emissions than their peers.

The Energy Independence and Security Act of 2007 (US Govt. 2007) defines attributes of high-performance buildings, which include reduction of energy, water, material, and fossil fuel use; improved indoor environmental quality for occupants; improved worker productivity; and lower life cycle costs when compared to baselines for building performance. EISA 2007 requires federal agencies to use a green building certification system for new construction and major renovations of buildings. It sets general water-conservation guidelines and stormwater runoff requirements for property development. The Act requires new buildings and major renovations to reach zero-net energy use by 2030. Building standards developed by non-profit organizations over the years have been adopted by state and local governments into their building codes to aid in the design and energy-efficient operation of high-performance or green buildings. Typically, building standards establish minimum requirements developed through consensus processes, for example, the American Society of Heating, Refrigerating, and Air-Conditioning Engineers (ASHRAE) and the International Code Council (ICC):

i) ASHRAE Energy Standard 90.1-2010 for buildings except low-rise residential
ii) ASHRAE Standard 189.1-2017 for the design of high-performance, green buildings except low-rise residential, now integrated with
iii) Other American National Standards Institute (ANSI)–accredited standards

Another development is to prefer performance-based code compliance options over prescriptive methods so that higher energy and green building standards can be achieved through innovative means.

1.5.3 Rating System Approach

A rating system approach is represented by the green building certification systems that differ from building standards in that they typically take a 'whole building' approach. Sustainable building rating systems are defined as methodologies that examine the performance or expected performance of a 'whole building' and translate that examination into an overall assessment that allows for comparison against other buildings (Fowler and Rauch 2006). Sustainable building rating systems are designed to foster and recognize different aspects of sustainable practices during the design, construction, and operation of a building and to incorporate the best practices in reducing the adverse impacts of the building on the environment. There are more than 600 different rating systems in use or being developed worldwide (BRE 2008). Table 1.3 summarizes some of these rating systems according to regions.

Cole (2013) distinguishes building environmental assessment methods from sustainability assessment systems. The field of building environmental assessment has developed remarkably since the introduction of BREEAM; the majority of extant building environmental assessment methods assess environmental performance improvement relative to business as usual practice (or regulatory minimum standards), either implicitly or explicitly, while sustainability assessment systems have been introduced that expand on the range of performance issues to explicitly include social and economic criteria and thus attempt to assess 'sustainability.'

A comparison of well-established building environmental assessment methods BREEAM, LEED, CASBEE, and GREEN STAR is presented in Table 1.4. A comparison of the salient features of four established sustainability assessment systems – LBC, SBTool, SPeAR, and DGNB – is presented in Table 1.5. Each of these rating systems differs in structure, terminologies, and performance assessment methods.

A building's energy performance has the maximum weightage in sustainability rating systems because of its high environmental impacts, but it is the least achieved one in sustainability assessments (Berardi 2012). The building sector represents a great potential to reduce energy consumption in both new and existing buildings by an estimated 30–50% (UNEP 2009).

1.6 BIOCENTRIC (ECOLOGICAL) DESIGN PARADIGM

In contrast to the anthropocentric view, from a biocentric perspective within an ecological worldview, 'value' is added to an ecological system, 'increasing its systemic capability to generate, sustain and evolve increasingly higher orders of vitality and viability for the life of a particular place' (Mang and Reed 2015). Sustainability may also be positioned as having as a prerequisite the maintenance of the functional integrity of the ecosphere so that it can remain resilient to human-induced stresses and remain biologically productive (Rees 1991). The ecological theories and life cycle assessment methods are discussed in this context.

TABLE 1.3
Building Environmental Assessment Methods and Sustainability Assessment Systems

Region	Country	Name	Owner/Management
	Global	Green buildings	World Green Building Council http://worldgbc.org/
	Global	Sustainable buildings and climate initiative	United Nations Environmental Programme https://energies2050.org/sustainable-buildings-and-climate-initiative-unep-sbci/?lang=en
Europe	Europe	CRISP	http://cic.vtt.fi/eco/crisp/
	Czech Republic	SBTool CZ	Technical and Test Institute for Construction Prague and Building Research Institute – certification company ltd. http://sbtool.cz/
	France	Haute Qualite Environnementale (HQE) Method	*Association pour la Haute Qualité Environnementale* http://hqegbc.org/accueil/
	France	Certivéa	www.certivea.fr/
	France	ESCALE	http://cstb.fr/
	Finland	PromisE	VTT (Technical Research Centre of Finland) http://vtt.fi/
	Germany	DGNB System	German Sustainable Building Council http://dgnb.de/en
	Italy	LEED	Green Building Council Italia www.gbcitalia.org/
	Italy	Protocollo ITACA	iiSBE Italy http://itaca.org/
	Norway	Envir. Programming of Urban Development	SINTEF (Skandinavias storste uavhengige forskningsorganisasjon) http://lidera.info/
	Portugal	LiderA (Leadership for the Environment in Sustainable Building)	Instituto Superior Técnico, Lisbon http://lidera.info
	Portugal	SBTool PT	iiSBE Portugal http://iisbeportugal.org/
	Poland	LEED	Polish Green Building Council http://plgbc.org
	Romania	LEED	Romania Green Building Council http://rogbc.org/en
	Spain	VERDE	Green Building Council españa http://gbce.es/
	Sweden	EcoEffect	Royal Institute of Technology http://ecoeffect.se
	Swiss	Minergie	Minergie Switzerland http://minergie.ch/
	The Netherlands	BREEAM-NL	Dutch Green Building Council http://dgbc.nl/
	UK	BREEAM (Building Research Establishment Environmental Assessment Method)	Building Research Establishment http://breeam.org
	Europe	LEnSE (Label for Environmental, Social & Economic building)	Belgian Building Research Institute and others https://cordis.europa.eu/project/rcn/78620/reporting/en
North America	United States	LEED® (Leadership in Energy and Environmental Design)	United States Green Building Council http://usgbc.org/
		Green Globes	Green Building Initiative http://greenglobes.com/
	Canada	LEED-Canada	Canada Green Building Council http://cagbc.org
		Green Globes	ECD Canada http://greenglobes.com/
	Mexico	SICES	Green Building Council Mexico https://sicesmexico.mx/
Asia	China	Chinese Green Building Evaluation Label (GBEL) three star	Ministry of Housing and Urban-Rural Construction http://cngb.org.cn/
	Hong Kong	BEAMPlus HK-BEAM	HK-BEAM Society http://beamsociety.org.hk/
		CEPAS (Comprehensive Environmental Performance Assessment Scheme)	HK Building Department

(Continued)

TABLE 1.3 (CONTINUED)
Building Environmental Assessment Methods and Sustainability Assessment Systems

Region	Country	Name	Owner/Management
	India	GRIHA (Green Rating for Integrated Habitat Assessment)	GRIHA Council http://grihaindia.org/
		IGBC Green rating	Indian Green Building Council https://igbc.in/igbc/
	Japan	CASBEE (Comprehensive Assessment System for Building Environmental Efficiency)	Japan Sustainable Building Consort. http://ibec.or.jp/CASBEE/
	Korea	G-SEED (Green Standard for Energy and Environmental Design)	Korea Research Institute of Eco-Environmental Architecture http://kriea.re.kr/home_eng/
	Singapore	Green Mark	Singapore Building & Construction Authority (BCA) http://bca.gov.sg/GreenMark/
	Taiwan	EEWH (Ecology, Energy, Waste and Healthy)	ABRI (Architecture & Building Research Institute) http://abri.gov.tw/en
	UAE	LEED Emirates	http://emiratesgbc.org/
		ESTIDAMA-Pearl	Abu Dhabi Urban Planning Council https://dpm.gov.ae/en/Urban-Planning/Pearl-Rating-System-Process
	Vietnam	LOTUS	Vietnam Green Building Council https://vgbc.vn/en/lotus-en/rating-systems/
Southern Hemisphere	Australia	Green Star	Australian Green Building Council http://new.gbca.org.au/green-star/
		NABERS (National Australian Building Environmental Rating Scheme)	https://nabers.gov.au/
	Argentina	LEED	http://argentinagbc.org.ar/
	Brazil	LEED-Brazil	GBC Brazil http://gbcbrasil.org.br/
		HQE	Fundação Vanzolini
	New Zealand	Green Star NZ	New Zealand Green Building Council http://nzgbc.org.nz/main/
	South Africa	Green Star SA	South African Green Building Council http://gbcsa.org.za/
		SBAT (Sustainable Building Assessment Tool)	CSIR (Council for Scientific and Industrial Research)https://csir.co.za/
Generic		GBTool/SBTool	iiSBE (International Initiative for a Sustainable Built Environment) https://iisbe.org/
		SPeAR (Sustainable Project Assessment Routine)	Ove Arup Ltd. https://arup.com/projects/spear
		Living Building Challenge	International Living Future Institute https://living-future.org › lbc

1.6.1 ECOLOGICAL THEORIES

Disparate disciplines have contributed to the discourse on ecological processes and the underlying principles of ecology. As bioregionalist Doug Aberley explains in *Futures by Design*, 'Ecology itself was formally created by Ernest Haeckel (1834–1919), for whom the belief that humans and nature were inextricably linked became the centrepiece of a unified philosophy, science, arts, theology, and politics.' The environmentalists and designers have scrutinized configurations and drawn prescriptive lessons from such scrutiny. Patrick Geddes, Henry David Thoreau, Ralph Waldo

TABLE 1.4

Salient Features of Building Environmental Assessment Methods: BREEAM, LEED, CASBEE, and GREEN STAR

Comparison items	BREEAM (2016)	LEED (2018)	CASBEE (2014)	Green Star (2020)
Location, year	UK, 1990	US, 1998	Japan, 2001	Australia (2003)
Developed by:	BRE (non-profit third party)	USGBC (non-profit third party)	Japan Sustainable Building Consortium (JSBC), Institute for Building Environment & Energy Conservation (IBEC)	GBCA (Green Building Council of Australia)
Sustainable categories	Management, health and wellbeing, energy, transport, materials, water, waste, land use, and ecology, pollution, and innovation	Sustainable site, indoor environmental quality, water efficiency, and resources, innovation, and regional priorities	Building environmental quality: indoor environment quality of service, outdoor environment on site; environmental load: energy, resources & materials, offsite environment	Management, indoor environment quality, energy, transport, water, materials, land use and ecology, emissions, innovation
Assessed building	Residence, retail, industrial unit, office, court, school, health care, prison, multifunction building, unusual building	Residence, school, retail, commercial building, multifunction building, healthcare	Residence (multi-unit), retail, industrial temporary construction, multifunction building	Design and as-built: schools, offices, universities, industrial facilities, public buildings, retail centers, and hospitals
Flexibility	Flexible in the UK and relatively overseas	Flexible in the USA and relatively overseas	Flexibility in Japan, and relative low flexibility overseas	Flexible in Australia and relatively overseas
Approach to scoring criteria	Additive pre-weighted credits approach	Additive simple approach (1 for 1)	Special	Additive pre-weighted credits approach
Ratings	Unclassified < 30 Pass ≥ 30 Good ≥ 45 Very good ≥ 55 Excellent ≥ 70 Outstanding ≥ 85	Certified 40–49 points Silver 50–59 points Gold 60–79 points Platinum 80+ points	BEE = 3.0 (excellent) BEE = ~1.5–3.0 (v. good) BEE = ~1.0–1.5 (good) BEE = ~0.5–1.0 (fairy poor) BEE = less than 0.5 (poor)	4 star = 45–59 5 star = 60–74 6 star = 75–100

Source: adapted from Parvesh Kumar (2018)

Emerson, Eugene and Howard Odum, James Lovelock, and many others down to the next-generation Frank Lloyd Wright, Rudolf Steiner, Buckminster Fuller, Malcolm Wells, Paolo Soleri, Ian McHarg, John Tillman Lyle, Wes Jackson, Amory and Hunter Lovins, John Todd and Nancy Jack Todd, Christopher Alexander, David Orr, Sim van der Ryn, and William McDonough have been the proponents of ecological design thinking (Guzowski 1999). Table 1.6 provides a summary and comparison of the development and design concepts and how to relate to each other.

John Tillman Lyle (1994), in his book *Regenerative Design for Sustainability*, discussed the dichotomy of design as degenerative and regenerative. The concept of regenerative design and development added a new dimension – a new intention to the broader theoretical context of sustainable design. A regenerative system provides for continuous replacement, through its functional processes, of the energy and materials used in its operation. Table 1.7 presents comparison of regenerative design support systems.

TABLE 1.5

Salient Features of Sustainability Assessment: LBC, SBTool, SPeAR, and DGNB

	SBTool Larsoon (2016)	Arup's Sustainable Project Assessment Routine (SPeAR)	Living Building Challenge LBC (2010)	DGNB Certificate Program
Location, year	Canada, 1998	UK, 2000	US, 2006	Germany, 2009
Developed by:	iiSBE (international non-profit collaboration)	ARUP	International Living Building Future (non-profit third party)	German Sustainable Building Council (DGNB)
Sustainable Categories	Site Selection, Project Planning, and Development, Energy and Resource, Environmental Loadings, Indoor Environmental Quality, Service Quality, Economic and Social aspects, Cultural and Perceptual Aspects	Environment & Natural Resources (60 indicators) Economic (26 indicators) Societal (34 indicators)	Place, Water, Energy, Health and Happiness, Materials, Equity, and Beauty	Environment Economic Socio-Cultural & Functional Technical Process Site
Assessed building	Almost any building	Almost any building	Renovation, landscape or infrastructure (non- conditioned development) almost any building	Almost any building
Flexibility	High flexibility around the world	High flexibility around the world	High flexibility around the world	High flexibility around the world
Approach to scoring criteria	Additive improved weighted scoring approach	Special (SPeAR diagram)	Actual recorded performance	Special (Performance Index)
Ratings	-1 = unsatisfied 0 = minimum acceptable performance 5 = best practice 1 to 4 = intermediate performance levels 2 = normal default	Score	Living certification (seven petals) Petal certification (three petals; one of which must be the Water, Energy, or Materials) Net-zero energy certification	Platinum (~80%) Gold (~65%) Silver (~50%) Bronze (~35% for existing building)

Source: Parvesh Kumar (2018)

1.6.2 Life Cycle Assessment

Life cycle assessment (LCA) is a technique for assessing the potential environmental loadings and impacts of buildings (Hobday 2001). LCA technique can be applied in the decision-making process concerning the design of a new building as well as the renovation of the old building stock. LCA systems measure the impact of the building on the environment by assessing the emission of one or more chemical substances related to the building construction and operation. LCA can have one or more evaluation parameters.

Many diverse methodologies have been developed for carrying out LCA of materials and products, which has resulted in difficulties in evaluating and comparing the environmental performance of materials and products. In response to these problems, the International Organization for Standardization (ISO) has developed consensus-based international standards for conducting and reporting life cycle analysis.

ISO 14040:2006 Environmental management-LCA – Principles and framework
ISO 14044:2006 Environmental management-LCA – Requirements and guidelines

TABLE 1.6

Comparison of Ecological Theories/Concepts

Ian McHarg (1969): Design with Nature	Malcolm Wells (1982): Wilderness-Based Checklist	Nancy Jack Todd and John Todd (1984): Living Machines	William McDonough (2002): Cradle-to-Cradle Design Philosophy	John Tillman Lyle: Regenerative Design Strategies	Brinda Vale & Robert Vale (1991): Principles of green architecture	Benyus (1997): Laws of Nature or Principles of Biomimicry	Sim Van der Ryn (1996): Ecological Principles
1. Negentropy	1. Creates pure air	1. Self-sustaining	1. Believes in repaying the earth in return for what it has given us	1. Letting nature do the work	1. Conserving energy	1. Nature runs on sunlight	10. Solutions grow from place
2. Apperception	2. Creates pure water	2. Based on the living relationship between our biotic and abiotic environment	2. Suggests to protect and enrich eco systems and nature's biological metabolism	2. Considering nature as both model and context	2. Working with climate	2. Nature uses only the energy it needs	11. Ecological accounting
3. Symbiosis	3. Stores rainwater	3. Based on ecosystem technologies	3. Categorizes all the material into 'technical' and 'biological'	3. Aggregating, not isolating	3. Minimizing new resources	3. Nature fits form to function	12. Design with nature.
4. Fitness and fitting	4. Produces its own food	4. Treat sewage and purify water with plants, animals, and microorganisms	4. Suggests the use of organic and technical nutrients	4. Seeking optimum levels for multiple functions	4. Respect for users	4. Nature recycles everything	13. Everyone is a designer
5. The presence of health or pathology	5. Creates rich soil	5. Maintain ecological balance in nature	5. Suggests removing dangerous technical materials from current life cycle	5. Matching technology to needs	5. Respect for site	5. Nature rewards cooperation	14. Making nature visible
	6. Uses solar energy			6. Using information to replace power	6. Holism	6. Nature banks on diversity	
	7. Stores solar energy			7. Providing multiple pathways		7. Nature demands local expertise	
	8. Creates silence			8. Seeking common solutions to disparate problems		8. Nature curbs excesses from within	
	9. Consumes its own wastes			9. Managing storage as a key to 'sustainability'		9. Nature taps the power of limits	
	10. Maintains itself			10. Shaping form to guide flow			
	11. Matches nature's pace			11. Shaping form to manifest process			
	12. Provides wildlife habitat			12. Prioritizing for sustainability			
	13. Provides human habitat						
	14. Moderates climate and weather						
	15. .... and is beautiful						

Source: Adapted from Guzowski (1999)

TABLE 1.7
Comparison of Regenerative Design Support Tools

	REGEN (Svec et al 2012)	Eco-Balance (Fisk 2009)	Perkins + Wills (2015)	LENSES (Plaut 2012)
Developer	BNIM	Plinky Fisk and Gail Vittori	Perkins + Wills	CLEAR
Types of Developer	Architectural firm	Non-profit organization	Architectural firm	Non-profit organization
Background	Practice + USGBC	The Center for Maximum Potential Building Systems	Practice + University of British Columbia	Institute for Built Environment at Colorado State University
What?	A data-rich, web-based tool	A design and planning tool	Issues and process-based frameworks	A process and a metrics tool
Audience	Professionals and community members	Professionals and businesses	Practitioners of Perkins+Wills	Professionals, business, government, students and non-profit teams
Mission	To guide dialogue and help professionals to engage with regenerative approaches	To provide principles for balancing life support systems across life cycle phases	To offer constructive direction to design teams and to generate dialogue for regenerative approaches	To cultivate, empower, and equip change makers to create a regenerative future
Goal	Transforming practice towards regenerative approaches	Supplying our needs in a regenerative manner	Expanding design for positive synergies	A thriving living environment
Structure	Linking specific strategies to the whole	Series of graphics	Challenging questions	Overlaid three lenses
Main categories	Robust and resilient natural systems, high performing constructed systems, prosperous economic systems and whole social systems	Air, water, food, energy and materials	Foundation, sandbox and toolbox	Foundation lens, flows lens and vitality lens

Source: Akturk (2016)

There are four phases in an LCA study:

i) the goal and scope definition phase
ii) the life cycle inventory analysis (LCI) phase
iii) the life cycle impact assessment (LCIA) phase
iv) the life cycle interpretation phase

The scope, including the system boundary and level of detail, of an LCA depends on the subject and the intended use of the study. The depth and the breadth of LCA can differ considerably depending on the goal of a particular LCA.

The life cycle inventory (LCI) analysis involves a collection of the input/output data necessary to meet the goals of the defined study.

The purpose of LCIA is to provide additional information to help assess a product system's LCI results to better understand its environmental significance.

In the life cycle interpretation the results of an LCI or an LCIA, or both, are summarized and discussed as a basis for conclusions, recommendations, and decision-making in accordance with the goal and scope definition.

For buildings, the LCA covers resource extraction and production of materials, through the construction and operation of the building to its disposal (Rashid 2015). The cradle-to-cradle assessment takes this further with recycling and reuse built into the assessment.

There are several examples when at the first instance an unfit existing building was planned for demolition and construction from scratch, but considerations of the environmental and social consequences changed the decision to renovate the existing building. A truly comprehensive, sustainable approach that includes economic, social, and cultural factors invariably has to give preferences to the preservation and adaptive reuse of existing structures. Renaissance Hall, a century-old building, was decided to be demolished but was renovated and reconstructed to the standard of LEED gold certification for adaptive reuse for the Department of Architecture and Landscape Architecture, North Dakota State University, Fargo (Figure 1.8).

On the other hand, new building stock should be designed and constructed considering the life cycle approach. The German engineer Prof. Werner Sobek suggests that future buildings should be characterized as follows:

- Zero energy: in total, they will not require energy for their annual operation.
- Zero emissions: they will not emit any harmful substances.
- Zero waste: all materials will be completely recyclable.

This 'triple-zero concept' is epitomized in his design for the Energy Plus experimental house in Berlin (Figure 1.9). According to Sobek, the ecologically conscious building can be extremely attractive and exciting, especially because it asks the designer to face new challenges.

FIGURE 1.8   Renaissance Hall, North Dakota State University, Fargo.© C. Kabre

FIGURE 1.9   Energy plus house, Berlin. © C. Kabre

1.6.3   SYSTEMS APPROACH

The well-known trajectory of an environmentally responsive design, developed over the last few decades, generally distinguishes technological (high-performance) and biocentric (ecological) design paradigms (Figure 1.10). While aiming for neutral or reduced environmental impacts in terms of energy, carbon, waste, or water are worthwhile targets, it is important that the built environment should aim for revitalization and regeneration to have net positive environmental benefits for

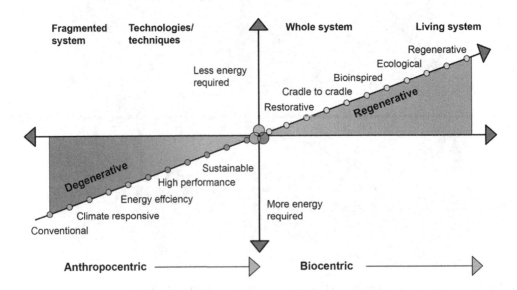

FIGURE 1.10   Environmentally responsive trajectory. Source: adapted from Reed (2006)

the living world. This implies that the built environment needs to produce more than it consumes, as well as remedy pollution and damage. It is a clear departure from the centuries-old outlook that the best-built environment can be 'neutral' in relation to the living world. It is increasingly apparent that society is entering a transition period between these technological (high performance) and biocentric (ecological) design paradigms. Thus a sustainability framework in synergy with nature or the ecosystem is necessary not only to preserve the environment but also to revitalize and regenerate the fragile and degraded environment. This implies a living or whole-systems approach to development which looks at the human and non-human ecology of the built environment. In taking a whole-systems approach, a more expansive notion of the built environment is required, one where dynamic relationships exist between a greater number of built and unbuilt elements and where a balanced, sustainable relationship between these elements is explored (Moffatt and Kohler 2008).

1.7 SYNERGISTIC DESIGN

Robinson (2004) suggests 'if sustainability is to mean anything, it must act as an integrating concept' and will require new concepts and tools 'that are integrative and synthetic, not disciplinary and analytic; and that actively creates synergy, not just summation.'

Synergistic design, therefore, can be visualized as an integrated systems approach to a sustainable built environment; conceptually it is illustrated in Figure 1.11. The social goal (*design goal*) of a built environment can be defined in terms of the design *objective* as comfort (thermal and luminous). This objective must be satisfied to achieve the design goal. The *performance variables*, such as temperature, humidity, solar radiation, and illumination levels, must acquire values within certain ranges that will satisfy the objective. These ranges may be stated in specific terms as constraints or general directional terms as a target; for example, achieving the illumination level of 300 lux can be a target. The *design variables*, such as length, width, height, and location of the window, materials properties of glazing, must be assigned some values to collectively describe a design (*system*). More generally, performance variables are related to the required functions and design variables to the form or structure of the design.

The crux of the design process lies in the correct mapping between the design and performance variables to achieve the objective or goal. A performance variable is often influenced by more than one design variable. The converse is also true: one design variable is likely to influence more than one performance variable; for instance, a window may influence not only the daylight admitted inside but also the solar heat gain and the light distribution. The performance and decision variables thus interact in complex ways, and the relationships between them are not always obvious.

The design process usually follows a basic generate-and-test approach (Mitchell 1990, pp. 179–81). In this process to find a solution requires applying decision rules to generate feasible solutions and then evaluating performance to determine whether feasible solutions are acceptable solutions. The basic structure of this model is illustrated in Figure 1.12.

In a completely manual design process, all the generation and testing are performed by the architect, or the tasks are divided among members of the design team. In a partially automated design process, the computer can either generate or test. For example, a computer can be used to generate alternative shading devices by mechanically applying the rules of solar geometry, with a human critic inspecting and testing the machine's proposals or a human designer generating alternatives which are then tested by computer programs. In a fully automated design process, the computer both generates and tests.

Given a problem, the efficiency with which a solution can be found will depend on the formulation of effective generation and test mechanisms. Generation requires decision rules to assure that only feasible alternatives are produced and to determine what alternative to try next. In addition to testing, optimization is necessary to evaluate and sort out the acceptable solutions by drawing inferences, to tradeoff conflicting performances, and to prescribe the best solution. A sustainable built environment is likely to be assessed by the way in which various systems (decisions) fulfill multiple

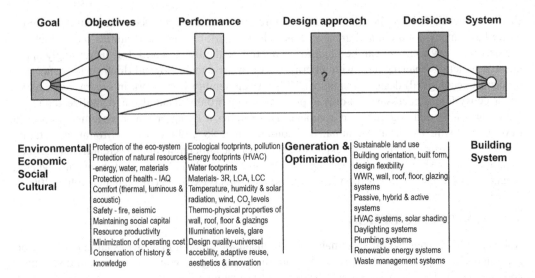

Goal	Objectives	Performance	Design approach	Decisions	System
Environmental Economic Social Cultural	Protection of the eco-system Protection of natural resources -energy, water, materials Protection of health - IAQ Comfort (thermal, luminous & acoustic) Safety - fire, seismic Maintaining social capital Resource productivity Minimization of operating cost Conservation of history & knowledge	Ecological footprints, pollution Energy footprints (HVAC) Water footprints Materials- 3R, LCA, LCC Temperature, humidity & solar radiation, wind, CO_2 levels Thermo-physical properties of wall, roof, floor & glazings Illumination levels, glare Design quality-universal accebility, adaptive reuse, aesthetics & innovation	Generation & Optimization	Sustainable land use Building orientation, built form, design flexibility WWR, wall, roof, floor, glazing systems Passive, hybrid & active systems HVAC systems, solar shading Daylighting systems Plumbing systems Renewable energy systems Waste management systems	Building System

FIGURE 1.11   A conceptual framework of synergistic design of the sustainable built environment. Source: adapted from Coyne et al. (1990)

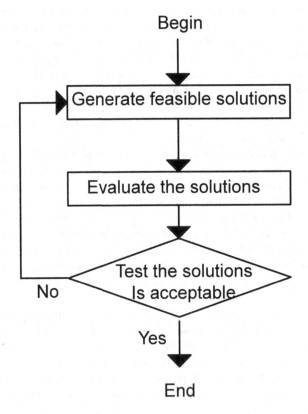

FIGURE 1.12   A basic generate-and-test process of design.

performances, and, indeed, it is typically only possible to achieve high environmental performance within demanding cost and time constraints through a creative synergy of systems (Figure 1.13).

While the design of the thermal and luminous environment is at the heart of a much larger sustainable built environment picture, it represents an intriguing point of intersection between external and internal environmental forces that are ultimately shaped by and given meaning through

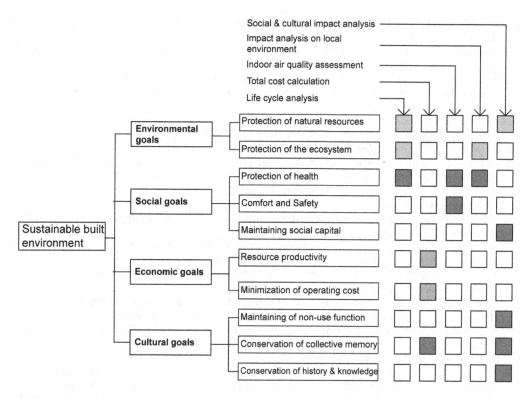

FIGURE 1.13   Goals for action toward sustainable building: test of sustainability can be achieved with a combination of various assessment methods. Source: arch plus, issue 184, October, Niklaus Kohler

architectural form. It should be the designer's aim to ensure the required indoor conditions with little or no use of energy, other than from ambient or renewable sources. As a result, a synergistic approach to a thermal and luminous environment embodies not only environmental but also rich aesthetic and innovation opportunities. Therefore the designer's task is to

1. Analyze the given climate conditions
2. Establish the limits of desirable or acceptable thermal comfort
3. Generate alternative design solutions for the thermal environment from the palette of passive, hybrid (low-energy), and active design strategies
4. Generate alternative design solutions for the luminous environment from the palette of passive and advanced daylighting strategies
5. Optimize energy requirements for the thermal and luminous environment
6. Consider the integration of renewable energy with the built environment

It is hoped that the synergy of passive, hybrid (low-energy), and active design strategies with renewable energy will lead to a greater realization of a comfortable thermal and luminous environment and will perhaps even make the built environments sustainable (regenerative in future).

REFERENCES

AIA (2007) AIA Committee on the environment. American Institute of Architects. http://www.aia.org/practi cing/groups/kc/AIAS077347. Accessed 15 August 2016.

Akturk A (2016) *Regenerative Design and Development for a Sustainable Future: Definitions and Tool Evaluation*. MS Thesis, University of Minnesota, Minneapolis.

Ali ZF, Yannas S (1999) Masters in the tropics: Environmental features of the buildings of Le Corbusier and Louis I. Kahn in Bangladesh and India. In: Szokolay SV (ed) *Proceedings of the Sixteenth International PLEA (Passive and Low Energy Architecture) Conference*, Brisbane, Australia, 22–24 September 1999, vol 1, pp 41–46.

ASHRAE (2006) *Green Guide: the Design, Construction, and Operation of Sustainable Buildings*. American Society of Heating, Refrigerating & Air-conditioning Engineers Inc., Butterworth-Heinemann, Amsterdam.

BEA (2019) Gross domestic product by industry accounts. U.S. Bureau of Economic Analysis. https://www.bea.gov/system/files/2019-07/gdpind119.xlsx. Accessed 15 October 2019.

Benyus JM (1997) *Biomimicry: Innovation Inspired by Nature*. Harper Collins, New York.

Berardi U (2012) Sustainability assessment in the construction sector: Rating systems and related buildings. *Sustainable Development*, 20(6), pp 411–424.

BRE (2008) *A Discussion Document Comparing International Environmental Assessment Methods for Buildings*. BRE, Glasgow.

Cole RJ (2005) Building environmental assessment methods: redefining intentions and roles. *Building Research and Information*, 33: 455–467.

Cole RJ (2013) Rating Systems for Sustainability, in Loftness, V and Haase, D (eds), *Sustainable Built Environments*, Springer Science + Business Media, New York, pp. 464–477.

Coyne RD, Rosenman MA, Radford AD, Balachandran M, Gero JS (1990) *Knowledge-Based Design Systems*. Addison Wesley, Reading, MA.

Dean AO (2002) Rural studio: Samuel Mockbee and an architecture of decency. Princeton Architectural Press, p 2. https://issuu.com/papress/docs/9781568982922/7. Accessed 11 January 2020.

Dodds F, Strauss M, Strong MF (2012) *Only One Earth: The Long Road via Rio to Sustainable Development*. Routledge, London.

Farmer J (1999) *Green Shift*. Architectural Press, Oxford.

Fisk P (2009) *The Eco-Balance Approach to Transect-based Planning: Efforts Taken at Verano*, a New Community and University in San Antonio, Texas. Center for Maximum Potential Building Systems, Austin, Texas. [Online] URL: http://www.cmpbs.org/sites/default/files/mp12_ecobalance_transect.pdf

Fowler KM, Rauch EM (2006) *Sustainable Building Rating Systems-Summary*. Pacific Northwest National Laboratory, US Department of Energy, USA, Richland, WA, report no. 15858. Available at https://s3.amazonaws.com/legacy.usgbc.org/usgbc/docs/Archive/General/Docs1915.pdf, accessed on 20 July 2020.

Frampton K (1985) *Modern Architecture: A Critical History*. Thames and Hudson Ltd., London.

GBCI (2016) Certification programmes. http://www.gbci.org/certification. Accessed 12 August 2016.

Gordon A (1972) Designing for survival: The President introduces his long life/loose fit/low energy study. *Royal Institute of British Architects Journal*, 79(9), pp 374–376.

Green KE (1979) Green climate and architecture. In: *Environmental Data and Information Service*, US Dept of Commerce, National Oceanic and Atmospheric Administration, vol. 10 (5), pp 6–10. Available at https://play.google.com/books/reader?id=Nm7zAAAAMAAJ&hl=en&pg=GBS.RA7-PA6, accessed on 5 August 2020.

Gropius W (1955) *Scope of Total Architecture*. Harper and Brothers, New York.

Guzowski M (1999) *Daylighting for Sustainable Design*. McGraw-Hill Professional, New York.

Hare B, Schaeffer M, Lindberg M, Höhne N, Fekete H, Jeffery L, Gütschow J, Sferra F, Rocha M (2014) Below 2°C or 1.5°C depends on rapid action from both Annex I and Non-Annex I countries. Climate Action Tracker Policy Brief. climateactiontracker.org/assets/publications/briefing_papers/CAT_Bonn_policy_update_jun2014-final_revised.pdf. Accessed 14 August 2016.

Hastings R, Wall M (2007) *Sustainable Solar Housing*, Vol. 1 *Strategies and Solutions*. Earthscan, London.

Hobday R (ed.) (2001) *Energy-Related Environmental Impact of Buildings*. Technical Synthesis Report Annex 31, International Energy Agency. Available at https://www.iea-ebc.org/Data/publications/EBC_Annex_31_tsr.pdf. Accessed on 14 October 2019.

Houghton JT, Jenkins GJ, Ephraums JJ (eds) (1990) Climate change. The IPCC scientific assessment. https://www.ipcc.ch/ipccreports/far/wg_I/ipcc_far_wg_I_full_report.pdf. Accessed 14 August 2016.

IEA (2018) *World energy statistics and balances*. International Energy Agency, Paris. https://webstore.iea.org/download/direct/2263?fileName=World_Energy_Balances_2018_Overview.pdf. Accessed 23 September 2019

IISD (2012) Sustainable development timeline. International Institute of Sustainable Development. https://www.iisd.org/pdf/2012/sd_timeline_2012.pdf. Accessed 12 August 2016.

ISO (2007) *Sustainability in Building Construction – Environmental Declaration of Building Products.* International Organization for Standardization, *ISO*/CD 21930, Geneva.

ISO (2008) *Sustainability in Building Construction – General Principles.* International Organization for Standardization, ISO Standard 15392, Geneva.

ISO (2010) *Sustainability in Building Construction – Framework for Methods of Assessment for Environmental Performance of Construction Works – Part 1: Buildings.* International Organization for Standardization, *ISO/CD* 21931–1, Geneva.

ISO (2011) *Sustainability in Building Construction: Sustainability Indicators—Part 1: Framework for the Development of Indicators and a Core Set of Indicators for Buildings.* International Organization for Standardization, *ISO/CD 21929–1*, Geneva.

Lyle JT (1994) *Regenerative Design for Sustainability.* John Wiley & Sons, Inc., New York.

Mang P, Reed W (2000) Designing from place: A regenerative framework and methodology. *Journal of Building Research and Information*, 40(1), pp 23–38.

Mang P, Reed W (2015) The nature of positive. *Journal of Building Research and Information*, 43(1), pp 7–10.

McDonough W, Braungart M (2002) *Cradle to Cradle: Remaking the Way We Make Things.* North Point Press, New York.

McHarg I (1969) *Design with Nature.* Natural History Press/Falcon Press, Philadelphia, PA.

MGI (2012) Urban world: Cities and the rise of the consuming class. McKinsey Global Institute. www.mckinsey.com/~/media/McKinsey/Global%20Themes/Urbanization/Urban%20world%20Cities%20and%20the%20rise%20of%20the%20consuming%20class/MGI_Urban_world_Rise_of_the_consuming_class_Full_report.ashx. Accessed 13 August 2016.

Mitchell WJ (1990) *The Logic of Architecture: Design, Computation, and Cognition.* The MIT Press, Cambridge, MA.

Moffatt S, Kohler N (2008) Conceptualizing the built environment as a social-ecological system. *Journal of Building Research and Information*, 36(3), pp 248–268.

Office of the Federal Environmental Executive (2008) *The Federal Commitment to Green Building: Experiences and Expectations, United States.* Office of the Federal Environmental Executive. Available at https://archive.epa.gov/greenbuilding/web/pdf/fedcomm_greenbuild.pdf. Accessed on 20 July 2020

Olgyay V (1953) *Bioclimatic Approach to Architecture.* In the BRAB Conference Report No 5. National Research Council, Washington, DC.

Olgyay V (1963) *Design with Climate, Bioclimatic Approach to Architectural Regionalism.* Princeton University Press: Princeton, NJ.

Oliver P (1987) *Dwellings: The House across the World.* University of Texas Press, Austin, TX.

Parvesh Kumar (2018) *Biocenosis: Rethinking Future of Sustainable Built Environment.* M. Tech. dissertation, DCR University of Science and Technology, Murthal (Sonepat), India.

Perkins +Will (2015) *Issues & Process-Based Frameworks for Regenerative Design*, cross reference in Akturk A (2016).

Plaut JM, Dunbar B, Wackerman A, Hodgin S (2012) Regenerative design: the LENSES framework for buildings and communities. *Journal of Building Research and Information*, 40 (1):112–122.

Rapoport A (1969) *House Form and Culture.* Prentice Hall, New Jersey, NJ.

Rashid AFA, Yusoff S (2015) A review of life cycle assessment method for building industry. *Renewable and Sustainable Energy Reviews*, 45, pp 244–248.

Reed B (2006) The trajectory of environmental design. http://www.integrativedesign.net/images/Trajectory_EnvironmentallyResponsibleDesign.pdf. Accessed 31 October 2012.

Rees WE (1991) Conserving natural capital: the key to sustainable landscapes. *International Journal of Canadian Studies*, 4 (Fall): 7–27.

Roaf S, Fuentes M, Thomas S (2001) *Ecohouse: A Design Guide.* Architectural Press, Oxford.

Robinson J (2004) Squaring a circle? Some thoughts on the idea of sustainable development. *Ecological Economics*, 48(4), pp 369–384.

Rudofsky B (1964) *Architecture Without Architects.* Academy Editions, London.

Sassi P (2016) Built environment sustainability and quality of life (BESQoL) assessment methodology. In: Filho WL, Brandli L (eds) *Engaging Stakeholders in Education for Sustainable Development at University Level.* Springer International Publishing, Switzerland, pp 21–32.

Svec P, Berkebile B, Todd JA (2012) REGEN: Toward a tool for regenerative thinking. *Journal of Building Research and Information*, 40(1): 81–94.

Todd N, Todd J (1984) *Bioshelters, Ocean Arks, City Farming: Ecology as the Basis of Design.* Sierra Club Books, San Francisco, CA.

UN Environment and International Energy Agency (2017) Towards a zero-emission, efficient, and resilient buildings and construction sector. Global Status Report (2017). Available at https://www.worldgbc.org/sites/default/files/UNEP%20188_GABC_en%20%28web%29.pdf. Accessed 27 September 2019.

UNEP (2009) Buildings and climate change: Summary for decision makers. Sustainable Buildings & Climate Initiative. United Nations Environment Programme. http://www.unep.org/sbci/pdfs/SBCI-BCCSummary.pdf. Accessed 26 January 2016.

UNSD (1992) Agenda 21, United Nations for Sustainable Development, United Nations Conference on Environmental & Development, Rio de Janeiro, Brazil, 3 to 14 June. https://sustainabledevelopment.un.org/content/documents/Agenda21.pdf. Accessed 24 August 2016.

US Government (2005) *Energy Policy Act of 2005*. Washington. Available at https://www.govinfo.gov/content/pkg/BILLS-109hr6enr/pdf/BILLS-109hr6enr.pdf. Accessed on 20 July 2020.

US Government (2007) *Energy Independence and Security Act of 2007*. Washington. Available at https://www.govinfo.gov/content/pkg/BILLS-110hr6enr/pdf/BILLS-110hr6enr.pdf. Accessed on 20 July 2020.

Vale B, Vale R (1991) *Green Architecture, Design for a Sustainable Future*. Thames and Hudson, London.

Van der Ryn S, Cowan S (1996) *Ecological Design*. Island Press, Washington, DC.

Vitruvius MP (1960) *The Ten Books on Architecture*, (Morgan MH, translator). Dover Publications, Inc., New York, p 170.

Ward B, Dubos R (1972) *Only One Earth: The Care and Maintenance of a Small Planet*. Penguin, Harmondsworth.

WCED (1987) Our common future. World Commission on Environment and Development, Oxford GB. Oxford University Press. conspect.nl/pdf/Our_Common_Future-Brundtland_Report_1987.pdf. Accessed 30 August 2012.

Wells M (1982) *A Regeneration-Based Checklist for Design and Construction, Gentle Architecture*. McGraw-Hill, New York.

2 Climate and Thermal Comfort

2.1 INTRODUCTION

Traditionally built environments evolved in response to the climate of each region. New England's saltbox cottage with small windows, low ceiling, and long north-facing roofs that deflected winter winds and bore snow loads well was an ideal solution to a harsh climate. New Mexico's adobe ranch house provided the time lag necessary to ease the southwest region's wide and uncomfortable diurnal temperature swings. The airy piazzas of Charleston ameliorated hot-humid conditions, while Nebraska's early sodhouses provided insulation against the arctic winter winds. However, since the Industrial Revolution, most houses and offices have been poorly designed, sited, and constructed in response to climate and energy conservation. To quote New York architect Richard Stein, too many buildings are 'glass skinned heat percolators.' They admit and trap the heat of the summer sun and transmit manmade heat outdoors in winter. Consequently, such a climatically poor design has been compensated by brute-force air conditioning and heating, leading to overconsumption of energy. A few years ago, as much as one-third of all the energy used in the United States went to heat, cool, and operate homes and other buildings. Another 10% went into their construction. Estimates are that the potential strategy against this gross energy consumption and climate change is to design, site, and construct built environments by applying climatic data to minimize adverse environmental impacts and to maximize the impact of beneficial environmental elements. The climate data needed to do this are available from different sources, including the Energy Plus (energyplus.net), Department of Energy, US Government. The most crucial strategy for energy-conscious and sustainable design of the built environments is to design and build in response to the climate. Hence the design of built environment entails an analysis of the climate of the given place and desirable thermal comfort conditions:

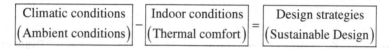

$$\begin{bmatrix} \text{Climatic conditions} \\ (\text{Ambient conditions}) \end{bmatrix} - \begin{bmatrix} \text{Indoor conditions} \\ (\text{Thermal comfort}) \end{bmatrix} = \begin{bmatrix} \text{Design strategies} \\ (\text{Sustainable Design}) \end{bmatrix}$$

This chapter intends to present an overview of the earth and its atmosphere, classification of climate, elements of climate, solar geometry, and parameters of thermal comfort. The next section discusses the earth and its atmosphere. The third section discusses the globally prevalent Köppen classification of climate and 17 climate zones in the United States. The fourth section on elements of climate presents sources and the derivation of climate data given in Chapter 7. The fifth section on solar geometry deals with the apparent movement of the sun and the method to graphically present the sun-path. The last section on thermal comfort defines parameters of physiological thermal comfort and the method to delineate the comfort zones for a given city.

2.2 EARTH AND ITS ATMOSPHERE

Earth and its vital statistics are given in Figure 2.1. Earth is enveloped by an atmosphere of 80–85 km depth. There are three different ways of classifying the atmosphere into layers based on composition, temperature, and radiation (Figure 2.2). The first classification distinguishes the homosphere and heterosphere based on the composition of the atmosphere. *Homosphere* is below about 80 km; the composition of the atmosphere here is an almost uniform mixture of nitrogen (~78%), oxygen (~21%), and argon (~1%), with traces of other gases and a variable amount of water vapor and particulate matter. Of the other gases, carbon dioxide, ozone, and methane are of particular significance

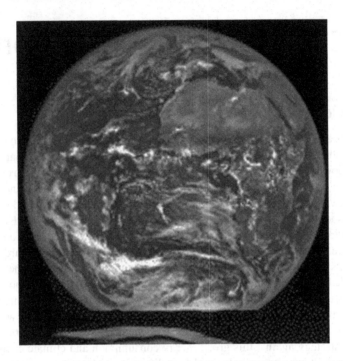

FIGURE 2.1 Earth and its vital statistics. Source: http://nssdc.gsfc.nasa.gov/planetary/factsheet/earthfact. html (Public domain)

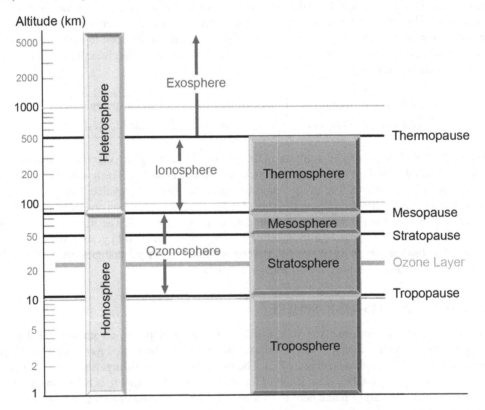

FIGURE 2.2 Three ways of delineating the Earth's atmosphere. Source: After Diston (2009), linework by Ar Sakshi Singhal

in the discussion of climate change. *Heterosphere* is above about 80 km; the composition changes with altitude as the residual air tends to stratify according to the molecular mass of the component gases. Nitrogen is virtually absent above 200 km, above which the dominant components tend to be atomic oxygen (up to 1000 km), helium (up to 2000 km), and hydrogen (above 2000 km).

The second classification based on temperature delineates four atmospheric layers: the troposphere, the stratosphere, the mesosphere, and the thermosphere. The upper boundaries of each layer are defined by the tropopause, stratopause, mesopause, and thermopause, respectively. Note that the mesopause marks the top of the homosphere.

The troposphere extends upward from ground level to an altitude of about more than 20 km at the equator and less than 10 km at the poles. It contains about 90% of the air mass and is warmest near ground level and cools gradually the higher up in it one goes, the temperature is down to about −50°C at the tropopause (at mid-latitudes). Most of the climatic phenomena take place within this thin layer, and most clouds are to be found in this layer. Convection currents keep the gases in the troposphere well mixed. It houses practically all biosphere.

The stratosphere extends upward from the top of the troposphere to an altitude of about 50 km. Some high-altitude clouds can be found in this layer. The stratosphere is strongly influenced by ozone. Atmospheric flow pattern causes a high concentration of ozone at around 25 km, which is known as the *ozone layer*. Because ozone absorbs ultraviolet radiation from the sun, the temperatures increase with altitude in the upper stratosphere (above ozone layer). In the lower stratosphere, the temperature is virtually constant, and circulation is limited, making this layer very stable. Jet airliners fly in this layer, for it is far less turbulent than the underlying troposphere.

The top of the mesosphere lies about 80–85 km above the earth's surface. The mesosphere is characterized by rapid temperature decline with increasing altitude (as was the case in the troposphere), falling as low as −100°C (−146°F) in the upper mesosphere. This is mainly due to the role of carbon dioxide absorbing infrared radiation from the surface of the earth (which is thought to be a critical factor in global warming). It is also part of the process of ozone creation that operates in the stratosphere. This layer is relatively poorly studied, for it is above the reach of most aircraft but below the altitude where satellites orbit. Most meteors burn up in the mesosphere. The pressure is practically zero at the *mesopause* (80 km), and it increases downward to about 101 kPa at the ground level.

The thermosphere extends to somewhere between 100 and 500 km above the earth's surface. It is a region of rapid temperature increase with altitude, leading to a kinetic temperature of about 1500 K at 300 km. This layer is so thin that many satellites orbit within it. Energy cannot be re-radiated at this level and so must follow a slow conduction path to lower altitudes (below 100 km altitude) from where it cannot be radiated at night. Note that the atmosphere at these extreme altitudes is so rarefied that, while the kinetic temperature is high, the specific heat capacity is practically zero. In other words, the external environment is cold. Many of the atoms and molecules in the thermosphere (and above) have lost electrons, thus becoming electrically charged ions, so the motions of particles in the upper atmosphere are partially influenced by electrical currents and the earth's magnetic field.

Third classification based on radiation designated atmosphere into three layers: the *ionosphere*, the *ozonosphere*, and the *exosphere*. In the ionosphere, cosmic and solar radiation ionizes gas molecules and also causes the dissociation of molecular oxygen into atomic oxygen. This layer has high electrical conductivity (which affects radio communication, a feature that people used to send messages beyond the line-of-sight range of the horizon before the advent of satellites) and can be energized by the solar wind (which creates aurorae in the skies around polar latitudes). The ozonosphere is a layer that corresponds with the stratosphere and the mesosphere, in which ultraviolet radiation dissociates any stray water vapor from lower layers into hydrogen and oxygen atoms. The process of ozone creation occurs by the reaction between atomic and molecular oxygen. The exosphere is essentially everything outside, and it extends indefinitely. This contains the magnetosphere, which marks the extent of the geomagnetic field. Figure 2.3 shows the sectional structure of the atmosphere and temperature profile.

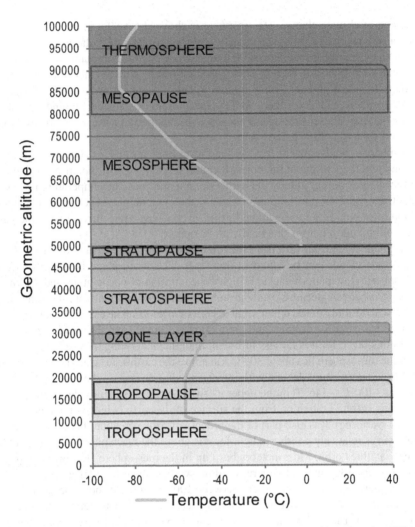

FIGURE 2.3 The Earth's atmosphere and temperature profile. Source: Plotted based on data NOAA (NASA) and USAF (1976)

2.3 SOLAR RADIATION

The climate of the earth, although complex, can be understood in a general way through several relatively simple relationships. First, energy is supplied to the surface of the earth from only two sources: the sun and interior of the earth. The sun generates energy through the fusion of hydrogen atoms to form helium, releasing large amounts of energy in the process. The interior of the earth generates energy through the radioactive decay of unstable isotopes. Of these two, the sun is by far the more significant source of energy for the surface of the earth. The climate of the earth, therefore, is driven by the energy input from the sun.

Incoming ultraviolet, visible, and a limited portion of infrared energy (together sometimes called 'shortwave radiation') from the sun drive the earth's climate system. Its quantity can be measured in two ways:

1. Irradiance, in W/m², i.e., the instantaneous flux or energy flow density, or 'power density.'
2. Irradiation, in J/m² or Wh/m², i.e., an energy quantity integrated over a specified period of time (hour, day, month, or year)

The surface of the sun is at a temperature of some 5500°C; thus, the peak of its radiant emission spectrum is around 550 nm wavelength.

$$E = \sigma T^4$$

E = the amount of energy released in W/m²
T = the temperature of the sun in Kelvins
σ = the Stefan–Boltzmann constant of 5.67×10^{-8} W/m²/K⁴
(E = 51,884,043.75 W/m²)

The spectrum of solar radiation extends from 20 to 2300 nm (nanometer = 10^{-9} m) (Figure 2.4). Human means of perception can distinguish:

A. Ultraviolet (UV) radiation, 290–380 nm (most of the UV below 200 nm is absorbed by the atmosphere; thus, some sources give 200 nm as the lower limit), which produces photochemical effects, bleaching, sunburn, etc.
B. Light, or visible radiation, from 380 (violet) to 780 nm (red)
C. Short infrared radiation, 780–2300 nm, or thermal radiation, with some photochemical effects

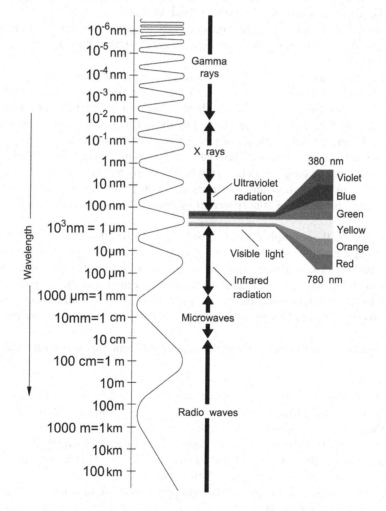

FIGURE 2.4 The radiant energy (electromagnetic) spectrum of the sun. Linework by Ar. Shiva Bagga

The 'solar constant' I_{sc} is defined as the intensity of solar radiation on a surface normal to the sun's rays, just beyond the earth's atmosphere, at the average earth–sun distance. One frequently used value is that proposed by the World Meteorological Organization in 1981, $I_{sc} = 1367$ W/m^2 (Iqbal 1983). Because the earth's orbit is slightly elliptical, the extraterrestrial radiant flux I_o varies throughout the year, reaching a maximum of 1412 W/m^2 near the beginning of January, when the earth is closest to the sun (aphelion) and a minimum of 1322 W/m^2 near the beginning of July, when the earth is farthest from the sun (perihelion). Extraterrestrial solar irradiance incident on a surface normal to the sun's ray can be approximated with equation 2.1:

$$I_o = I_{sc}\left\{1 + 0.033\cos\left[360°\frac{(n-3)}{365}\right]\right\} \qquad (2.1)$$

where n is the day of the year (1 for January 1, 32 for February 1, etc.), and the argument inside the cosine is in degrees. Table 2.1 tabulates the values of I_o for the 21st day of each month.

As the earth's radius is 6371 km (6.371×10^6 m), its circular projected area is $(6.371 \times 10^6)^2 \times 3.14 \approx 127 \times 10^{12}$ m^2; it continuously receives a radiant energy input of $1.367 \times 127 \times 10^{12} \approx 173.609 \times 10^{12}$ kW. The maximum irradiance at the earth's surface is around 1000 W/m^2, and the annual total horizontal irradiation varies from about 400 kWh/m^2 y near the poles to a value in excess of 2500 kWh/m^2 y in the Sahara desert or north-western inland Australia.

Since the atmosphere is, however, almost opaque to the longwave radiation emitted by the earth, this is mostly absorbed in the atmosphere and continuously re-radiated to outer space – this being a condition of equilibrium.

As Figure 2.5 shows, some 30% of solar radiation arriving at the earth is reflected; the remaining 70% enters the terrestrial system. Some are absorbed in the atmosphere, and about 51% reaches the earth's surface. Ultimately all of it is re-radiated, this being a condition of equilibrium. The energy received, reflected, absorbed, and emitted by the earth system are the components of the earth's radiation budget. Based on the physics principle of the conservation of energy, this radiation budget represents the accounting of the balance between incoming radiation, which is almost entirely solar radiation, and outgoing radiation, which is partly reflected solar radiation and partly radiation emitted from the earth system, including the atmosphere. A budget that's out of balance can cause the temperature of the atmosphere to increase or decrease and eventually affect our climate.

Greenhouse gases in the atmosphere (such as water vapor and carbon dioxide) absorb most of the earth's emitted longwave infrared radiation, which heats the lower atmosphere. In turn, the warmed atmosphere emits longwave radiation, some of which radiates toward the earth's surface, keeping our planet warm and generally comfortable. Increasing concentrations of greenhouse gases such as carbon dioxide and methane increase the temperature of the lower atmosphere by restricting the outward passage of emitted radiation, resulting in 'global warming,' or, more broadly, global climate change.

There are significant variations in irradiation among different locations on the Earth for three reasons:

1. The angle of incidence: according to the cosine law (Figure 2.6), the irradiance received by a surface is the normal irradiance times the cosine of the angle of incidence (INC).
2. Atmospheric depletion, a factor varying between 0.2 and 0.7, mainly because at lower altitude angles the radiation has to travel along a much longer path through the atmosphere (especially through the lower, denser, and most polluted layer) but also because of variations in cloud cover and atmospheric pollution (Figure 2.7).
3. Sunshine duration i.e., the length of daylight period (sunrise to sunset) and, to a lesser extent, also on local topography.

TABLE 2.1

Approximate Astronomical Data for the 21st Day of Each Month

Month	January	February	March	April	May	June	July	August	September	October	November	December
Day of year	21	52	80	111	141	172	202	233	264	294	325	355
I_o, W/m^2	1410	1397	1378	1354	1334	1323	1324	1336	1357	1380	1400	1411
Equation of time (ET), min	−10.6	−14.0	−7.9	1.2	3.7	−1.3	−6.4	−3.6	6.9	15.5	13.8	2.2
Declination δ, degrees	−20.1	−11.2	−0.4	11.6	20.1	23.4	20.4	11.8	−0.2	−11.8	−20.4	−23.4

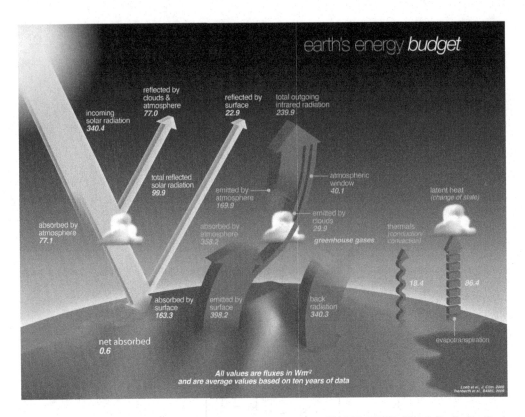

FIGURE 2.5 The earth's energy budget. Credit: Image courtesy NASA's ERBE (Earth Radiation Budget Experiment) program. Source: https://nasa.gov/feature/langley/what-is-earth-s-energy-budget-five-questions-with-a-guy-who-knows

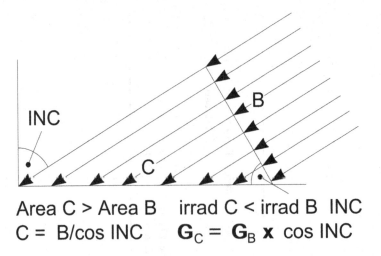

FIGURE 2.6 Angle of incidence (INC). Source: after Szokolay (2008), linework by Ar. Shiva Bagga

2.4   GLOBAL CLIMATE

At the global level, climates are formed by the differential solar heat input and the almost uniform heat emission over the earth's surface. Equatorial regions receive a much higher energy input than areas nearer to the poles. Up to about 30° N and S latitudes, the radiation balance is positive (i.e., the solar 'income' is greater than the radiant loss), but at higher latitudes, the heat loss far exceeds

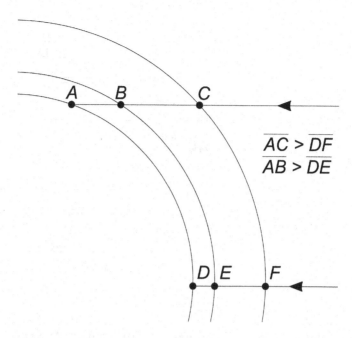

$$\overline{AC} > \overline{DF}$$
$$\overline{AB} > \overline{DE}$$

FIGURE 2.7 Radiation path-lengths through the atmosphere. Source: after Szokolay (2008) linework by Ar. Shiva Bagga

the solar input. Differential heating causes pressure differences, and these differences are the main driving force of atmospheric phenomena (winds, cloud formations, and movements), which provide a heat transfer mechanism from the equator toward the poles. In the absence of such heat transfer, the mean temperature at the north pole would be −40°C, rather than the present −17°C, and at the equator, it would be about 33°C and not 27°C as at present.

At points of intense heating, the air rises, and at a (relatively) cold location, it sinks. The movement of air masses and moisture-bearing clouds is driven by temperature differentials but strongly influenced by the Coriolis force (Figure 2.8).

The tropical front or ITCZ (inter-tropical convergence zone) moves seasonally north and south (with a delay of about 1 month behind the solar input, thus extremely north in July and south in January), as shown in Figure 2.9. Note that the movement is much more significant over continents than over the oceans. The ITCZ also reflects precipitation; this zone experiences two rainy seasons each year and thus is beneficial for agriculture and human habitation; about half the world population lives here. The other half lives in the temperate zone where the annual evaporation equals rainfall, thus retaining ground moisture for agriculture. In between these two areas extends an arid zone where few people live. To the north 60°N and south 50°S, it is too cold for plant productivity; therefore few people live over there.

The atmosphere is a very unstable three-dimensional system; thus, small differences in local heating (which may be due to topography and ground cover) can have significant effects on air movements and influence the swirling patterns of low and high pressure (cyclonic and anticyclonic) zones.

A 'stationary' air mass at the equator moves with the earth's rotation, and it has a certain circumferential velocity (some 1600 km/h or 463 m/s); hence, it has a moment of inertia. As it moves toward the poles, the circumference of the earth (the latitude circle) reduces; therefore, it will overtake the surface.

An air mass at a higher latitude has a lesser velocity and inertia, and when moving toward the equator (a larger circumference), it lags behind the earth's rotation. This mechanism causes the N/E and S/E trade winds.

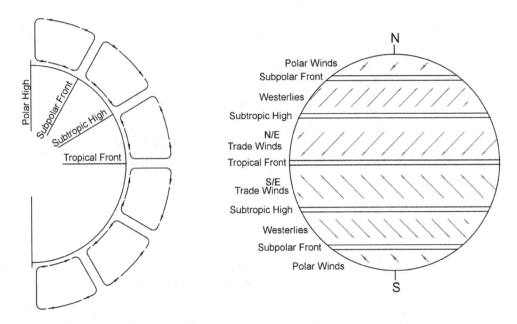

FIGURE 2.8   Global wind pattern. Source: after Szokolay (2008) linework by Ar. Shiva Bagga

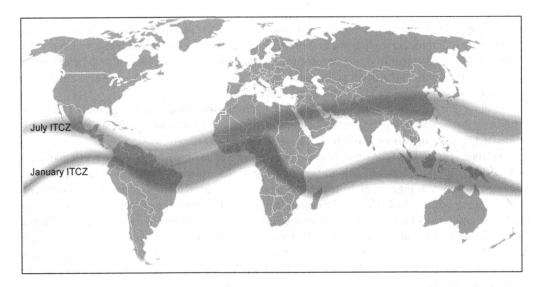

FIGURE 2.9   North-south shift of the ITCZ. Source: By Mats Halldin – Own work. Based on Image: ZICT en janvier.jpg, Image: ZICT en juillet.jpg, and Image: BlankMap-World.png, Public Domain: https://commons. wikimedia.org/w/index.php?curid=1456454

2.5   CLIMATE AND ITS CLASSIFICATION

The essential characteristics of any planet are controlled by its climate. The word 'climate' comes from the Greek *klima*, which means the slope of the earth with respect to the sun. Climate is defined as 'region with certain conditions of temperature, dryness, wind, light, etc. of a region' (*Oxford Dictionary*). Climate is also defined as 'an integration in time of the weather conditions, characteristics of a certain geographical location.' The weather is the set of atmospheric conditions, including temperature, rainfall, wind, humidity, and sky conditions prevailing at a given place and time. The climate, on the other hand, is the general weather conditions over a long period of time. In totality, the climate is the sum of all the statistical weather information that helps describe a place or region.

Different regions of the earth have diverse characteristic climates. Both natural and manmade factors determine the climate of a place or region. The natural elements include the atmosphere, geosphere, hydrosphere, and biosphere; while the human factors include land use and consumption of other natural resources. Changes in any of these factors can cause local, regional, and even global changes in the climate.

Since Aristotle's time, attempts at climate classification have been made chiefly by biologists who realized that the natural vegetation represents an excellent indication of the climate of a place. The well-known vegetation-based classification of climates by Wladimir Köppen, a German biologist trained in St. Petersburg, was first published in 1900; this system has been improved over time and is still prevalent. The Köppen system is based on monthly mean temperature, monthly mean precipitation, and mean annual temperature. The Köppen scheme of classification distinguishes five primary climate zones based on vegetation: the equatorial zone (A), the arid zone (B), the warm temperate zone (C), the snow zone (D), and the polar zone (E). A second letter in the classification considers the precipitation (e.g., Df for snow and fully humid), and a third letter the air temperature (e.g., Dfc for snow, fully humid with cool summer). The annual mean near-surface (2 m) temperature is denoted by T_{ann}, and the monthly mean temperatures of the warmest and coldest months by T_{max} and T_{min}. P_{ann} is the accumulated annual precipitation, and P_{min} is the precipitation of the driest month. Additionally, P_{smin}, P_{smax}, P_{wmin}, and P_{wmax} are defined as the lowest and highest monthly precipitation values for the summer and winter half-years, respectively, on the hemisphere considered. All temperatures are given in °C, monthly precipitations in mm/month, and P_{ann} in mm/year. In addition to these temperature and precipitation values, a dryness threshold P_{th} in mm is introduced for the arid climates (B), which depends on $\{T_{ann}\}$ the absolute measure of the annual mean temperature in °C and on the annual cycle of precipitation:

$$\begin{cases} \{T_{ann}\} & \text{if at least 2/3 of the annual precipitation occurs in winter} \\ 2\{T_{ann}\}+28 & \text{if at least 2/3 of the annual precipitation occurs in summer} \\ 2\{T_{ann}\}+14 & \text{Otherwise} \end{cases}$$

Köppen-Geiger classification enlists some 25 climate types (Figure 2.10; Table 2.2).

The United States is a significant geographical unit, which has almost 23° of latitudinal extent (between 25.84°N and 49.38°N) and 58° of longitudinal extent (between 66.95°W and 124.67°W). With its vast size of about 9.8 million km^2, the United States has sharp contrasts in its climatic conditions. The US Department of Energy Building America Program distinguishes 17 climatic zones for the design of buildings. The US Building America's climate regions described here are based on the climate designations used by the International Energy Conservation Code (IECC) and the American Society for Heating, Ventilating, and Air-Conditioning Engineers (ASHRAE). The climatic map included in the ASHRAE Standard 169 (2013) for the design of buildings is given in Figure 2.11 and Table 2.3. Column 1 contains the alphanumeric designation for each zone. The numeric part of the designation relates to the thermal properties of zones 0–8. The letter part indicates the major climatic group to which the zone belongs; A indicates humid, B indicates dry, and C indicates marine. Zones 1B and 5C have been defined but are not used for the United States. Column 2 contains a descriptive name for each climate zone and the major climate type. Column 3 contains definitions for the zone divisions based on the degree of day cooling and/or heating criteria. The humid/dry/marine divisions must be determined first before these criteria are applied. The definitions in Tables 2.3 and 2.4 contain the logic capable of assigning a zone designation to any location with the necessary climate data anywhere in the world, but this classification is focused on the 50 United States.

The Köppen climate classification system inspired the climate zone classifications used in the United States (Table 2.5).

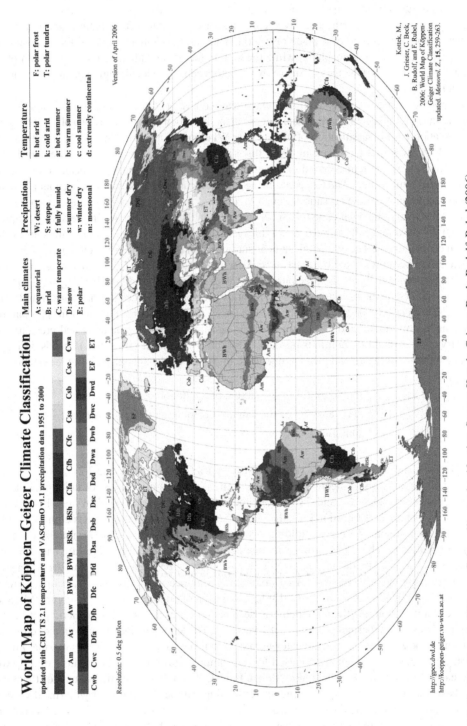

FIGURE 2.10 World map of Köppen–Geiger climate classification. Source: Kottek, Grieser, Beck, Rudolf, Rubel (2006)

TABLE 2.2

Köppen's Major Climates (Kottek, Grieser, Beck, Rudolf, Rubel 2006)

Type	Main Group Climates	Sub-Group: Precipitation	Second Sub-Group: Temperature
A	Equatorial climates ($T_{min} \geq +18°C$)		
Af		Equatorial rainforest, fully humid ($P_{min} \geq 60$ mm rainy, all seasons)	
Am		Equatorial monsoon ($P_{ann} \geq 25\ (100 - P_{min})$)	
Aw		Equatorial savannah with dry winter ($P_{min} < 60$ mm in winter)	
As		Equatorial savannah with dry summer ($P_{min} < 60$ mm in summer)	
B	Arid climates ($P_{ann} < 10P_{th}$)		
Bs		Semi-arid steppe climate ($P_{ann} > 5P_{th}$)	
Bsh			Hot steppe/desert ($T_{ann} \geq +18°C$)
Bsk			Cold steppe/desert ($T_{ann} \leq +18°C$)
Bw		Desert climate ($P_{ann} \leq 5P_{th}$)	
Bwh			Hot steppe/desert ($T_{ann} \geq +18°C$)
Bwk			Cold steppe/desert ($T_{ann} \leq +18°C$)
C	Warm temperate climates ($-3°C < T_{min} < +18°C$)		
Cw		Warm temperate climate with dry winter ($P_{wmin} < P_{smin}$ and $P_{smax} > 10P_{wmin}$)	
Cwa			Hot summer ($T_{max} \geq +22°C$)
Cwb			Warm summer (not 'a' and at least 4 $T_{mon} \geq +10°C$)
Cs		Warm temperate climate with dry summer ($P_{smin} < P_{wmin}$, $P_{wmax} > 3P_{smin}$, and $P_{smin} < 40$ mm)	
Csa			Hot summer ($T_{max} \geq +22°C$)
Csb			Warm summer (not 'a' and at least $4T_{mon} \geq +10°C$)
Cf		Neither Cs nor Cw (moist all seasons)	
Cfa			Hot summer ($T_{max} \geq +22°C$)
Cfb			Warm summer (not 'a' and at least $4T_{mon} \geq +10°C$)
Cfc			Cool summer and cold winter (not 'b' and $T_{min} > -38°C$)
D	Snow climates ($T_{min} \leq -3°C$)		
Ds		Snow climate with dry summer ($P_{smin} < P_{wmin}$, $P_{wmax} > 3P_{smin}$, and $P_{smin} < 40$ mm)	
Df		Snow climate, fully humid (neither Ds nor Dw)	
Dfa			Hot summer ($T_{max} \geq +22°C$)
Dfb			Warm summer (not 'a' and at least $4T_{mon} \geq +10°C$)

(Continued)

TABLE 2.2 (CONTINUED)
Köppen's Major Climates (Kottek, Grieser, Beck, Rudolf, Rubel 2006)

Type	Main Group Climates	Sub-Group: Precipitation	Second Sub-Group: Temperature
Dfc			Cool summer and cold winter (not 'b' and $T_{min} > -38°C$)
Dfd			Extremely continental (like 'c' but $T_{min} \leq -38°C$)
Dw		Snow climate with dry winter ($P_{wmin} < P_{smin}$ and $P_{smax} > 10\,P_{wmin}$)	
Dwa			Hot summer ($T_{max} \geq +22°C$)
Dwb			Warm summer (not 'a' and at least $4T_{mon} \geq +10°C$)
Dwc			Cool summer and cold winter (not 'b' and $T_{min} > -38°C$)
Dwd			Extremely continental (like 'c' but $T_{min} \leq -38°C$)
E	Polar climate ($T_{max} < +10°C$)		
ET		Tundra climate short summer allows tundra vegetation ($0°C \leq T_{max} < +10°C$)	
EF		Frost climate (perpetual ice and snow) ($T_{max} < 0°C$)	

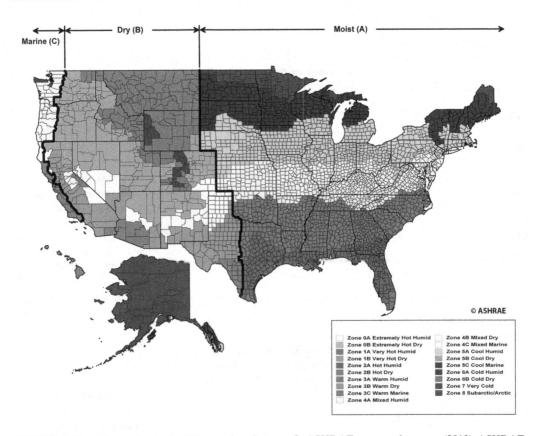

FIGURE 2.11 Climate zones for US counties. Source: © ASHRAE, www.ashrae.org (2013) ASHRAE Standard 169-2013, figure B-1

TABLE 2.3

Thermal Criteria for Climate Zone Definitions

Thermal Zone	Name	SI Units	I-P Units
0	Extremely hot	6000 < CDD 10°C	10,800 < CDD 50°F
1A and 1B	Very hot-humid (1A), dry (1B)	5000 < CDD 10°C ≤ 6000	9000 < CDD 50°F ≤ 10,800
2	Hot-humid (2A), dry (2B)	3500 < CDD 10°C ≤ 5000	6300 < CDD 50°F ≤ 9000
3A and 3B	Warm-humid (3A), dry (3B)	2500 < CDD 10°C < 3500 and HDD 18°C ≤ 2000	4500 < CDD 50°F ≤ 6300 and HDD 65°F ≤ 3600
3C	Warm-marine (3C)	CDD 10°C ≤ 2500 and HDD 18°C ≤ 2000	CDD 50°F ≤ 4500 and HDD 65°F ≤ 3600
4A and 4 B	Mixed-humid (4A), dry (4B)	1500 < CDD 10°C < 3500 and 2000 < HDD18°C ≤ 3000	2700 < CDD 50°F ≤ 6300 and 3600 < HDD 65°F ≤ 5400
4C	Mixed-marine	CDD 10°C ≤ 1500 and 2000 < HDD18°C ≤ 3000	CDD 50°F ≤ 2700 and 3600 < HDD 65°F ≤ 5400
5A and 5B	Cool-humid (5A) and cool-dry (5B)	1000 < CDD 10°C ≤ 3500 and 3000 < HDD 18°C ≤ 4000	1800 < CDD 50°F ≤ 6300 and 5400 < HDD 65°F ≤ 7200
5C	Cool-marine	CDD 10°C ≤ 1000 and 3000 < HDD 18°C ≤ 4000	CDD 50°F ≤ 1800 and 5400 < HDD 65°F ≤ 7200
6A and 6B	Cold-humid (6A), dry (6B)	4000 < HDD 18°C ≤ 5000	7200 < HDD 65°F ≤ 9000
7	Very cold	5000 < HDD 18°C ≤ 7000	9000 < HDD 65°F ≤ 12,600
8	Subarctic/arctic	7000 < HDD 18°C	12,600 < HDD 65°F

Source: ASHRAE Standard 169-2013, Table A-3: Thermal Climate Zone Definitions

TABLE 2.4

Major Climate Type Definition

Marine (C) zone definition – locations meeting all four criteria:

a. Mean temperature of the coldest month between −3°C (27°F) and 18°C (65°F)

b. Warmest month mean < 22°C (72°F)

c. At least four months with mean temperatures over 10°C (50°F)

d. The dry season in summer. The month with the heaviest precipitation in the cold season has at least three times as much precipitation as the month with the least precipitation in the rest of the year. The cold season is October through March in the Northern Hemisphere and April through September in the southern hemisphere.

Dry (B) definition – locations meeting the following criteria:

a. Not marine (C)

b. If 70% or more of the precipitation, P, occurs during the high sun period, then the dry/humid threshold is $P < 20.0 \times (T + 14)$ [SI units] or $P < 0.44 \times (T - 7)$ [I-P units]

c. If 30% and 70% of the precipitation, P, occurs during the high sun period, then the dry/humid threshold is $P < 20 \times (T + 7)$ [SI units] or $P < 0.44 \times (T - 19.5)$ [I-P units]

d. If 30% or less of the precipitation, P, occurs during the high sun period, then the dry/humid threshold is $P < 20 \times T$ [SI units] or $P < 0.44 \times (T - 32)$ [I-P units]

where:

P = annual precipitation, inches (mm)

T = annual mean temperature, °C (°F)

Summer or April through September in the northern high sun period hemisphere and October through March in the southern hemisphere

(Continued)

TABLE 2.4 (CONTINUED)
Major Climate Type Definition

Winter or October through March in the northern cold season hemisphere and April through September in the southern hemisphere

Moist (A) definition – locations that are not marine (C) and not dry (B)

Warm-humid definition – moist (A) locations where either of the following wet-bulb temperature conditions shall occur during the warmest six consecutive months of the year:

- 19.4°C (67°F) or higher for 3,000 or more hours or
- 22.8°C (73°F) or higher for 1,500 or more hours

Source: ASHRAE Standard 169-2013

TABLE 2.5
Thermal Climate Zone Definitions and Köppen Classification

Thermal Zone	Climate Zone Type	Köppen Classification
1A	Very hot-humid	Aw
1B	Very hot-dry	BWh
2A	Hot-humid	Cfa
2B	Hot-dry	BWh
3A	Warm-humid	Cfa
3B	Warm-dry	BSk/BWh/H
3C	Warm-marine	Cs
4A	Mixed-humid	Cfa/Dfa
4B	Mixed-dry	BSk/BWh/H
4C	Mixed-marine	Cb
5A	Cool-humid	Dfa
5B	Cool-dry	BSk/H
5C	Cool-marine	Cfb
6A	Cold-humid	Dfa/Dfb
6B	Cold-dry	BSk/H
7	Very cold	Dfb
8	Subarctic	Dfc

Source: Briggs RS, Lucas RG, Taylor ZT (2003).

2.6   ELEMENTS OF CLIMATES

The main elements of the climate are regularly measured by the meteorological organization and published in summary form. Further, data are also obtained from satellites. Comprehensive weather data over several decades are statiscally analyzed to build the hourly weather data files to be used for energy modelling. Common formats for weather files include Typical Meteorological Year (TMY), Test Reference Year (TRY) and Weather Year for Energy Calculations (WYEC). The Department of Energy's EnergyPlus website provides data for all locations in the United States derived from the 1991 to 2005 period of record in the TMY3 data set for download in EnergyPlus Weather (EPW) format. In addition to the EPW file, each station has a summary data STAT file. The TMY3s are data sets of hourly values of solar radiation and meteorological elements for a 1-year period. The location's latitude, longitude, and altitude above mean sea level are taken from these data. In the absence of observed data, some of the climate data can also be estimated based on standard algorithms or extrapolated from measured data of a nearby location; for example sun shine hour. The following sections discuss the elements of climate needed for building design.

2.6.1 TEMPERATURE AND HUMIDITY

Air temperature is expressed by the dry-bulb temperature (DBT), measured with a standard thermometer in the shade, usually in a ventilated box, the Stevenson screen, 1.2–1.8 m above ground level. The bulb of thermometer should not get wet – if it were wet, the evaporation of moisture from its surface would affect the reading and give something closer to the wet-bulb temperature. Air temperature can also be measured by the thermocouple or resistance temperature devices. It is usually given in degrees Celsius (°C) or Fahrenheit (°F); however, its true SI unit is Kelvin (K). On the Kelvin scale, 0°K equals –273°C. In conformity with the accepted usage of the SI system, the symbol C is used to denote a specified point on the temperature scale, but K (degree Kelvin) is used for a range, a span, or a difference of temperature, i.e., a length of the scale, without specifying its position.

Air temperature at a site depends on both the incoming air flows driven by large-scale weather systems and local climatic energy inputs. The latter has a significant effect on the daily air temperature swing close to the ground. The lowest temperature prevails just before sunrise when the cooling of the earth's surface is maximal; the highest temperature prevails in the afternoon.

Atmospheric humidity is the moisture content of the atmosphere, and it can be expressed by six psychrometric parameters: (i) wet-bulb temperature (WBT, °C/°F); (ii) dew-point temperature (DPT, °C/°F), (iii) absolute humidity (AH, g/kg, or lb/lb), (iv) humidity ratio (HR) (v) vapor pressure (p, kPa/psi), and (vi) relative humidity (RH, %).

Atmospheric humidity is usually measured by the wet-and-dry bulb (whirling) psychrometer or an aspirated psychrometer. These contain two thermometers; one has its bulb wrapped in gauze, which is kept moist from the small water container. When whirled around (or the fan is operated) until their readings become steady to obtain the maximum possible evaporation, this produces a cooling effect, showing the WBT. The other thermometer measures the air- or dry-bulb temperature. The difference between DBT and WBT is referred to as the wet-bulb depression, and it is indicative of the humidity. Evaporation is inversely proportional to humidity. In saturated air, there is no evaporation and no cooling; thus WBT = DBT. In the dry air, the moisture rapidly evaporates to produce a significant depression, which indicates low humidity. In moisture-laden air, evaporation is less, and a small wet-bulb depression occurs, which indicates high humidity.

For any particular dry-bulb temperature, there is only a certain amount of moisture vapor that can be absorbed in the air before it becomes saturated and precipitation occurs. The density of water vapor is the actual amount of moisture in the sample air and is termed as the AH and is measured in g/kg (or lb/lb). The dew-point temperature (°C/°F) refers to the maximum amount of moisture that the air can hold at a given temperature. The RH is the ratio of the actual density of water vapor in the air to the maximum density of water vapor that such air could contain, at the same temperature, if it were 100% saturated. Relative humidity may be measured directly or derived from DBT and WBT. At 100% relative humidity, DBT and WBT are equal. In general, the relative humidity varies inversely with temperature during the day, tending to be lower in the early afternoon and higher at night.

The maximum and minimum temperatures and relative humidities for all the cities have been taken from the TMY3 data

Both the graph and the table give the mean minimum and mean maximum temperatures. The monthly mean temperatures shown on the graph as well as in the table are calculated as in Equation 2.2:

$$\overline{T} = \frac{T_{\text{meanmax}} + T_{\text{meanmin}}}{2} \qquad (2.2)$$

Annual averages of all three temperature values are given in the table, found as in Equation 2.3.

$$\sum_{1}^{12} \frac{\overline{T}}{12} \tag{2.3}$$

The hourly values of temperature for a typical winter and summer day for all the cities are also taken from the TMY3 data.

The average diurnal range of temperature is the difference between the monthly mean maximum and the monthly mean minimum. The annual mean range of temperature is the difference between the highest monthly mean maximum and the lowest monthly mean minimum.

The recommended outdoor 'design conditions' (summer DBT and WBT and winter DBT) have been adopted from the ASHRAE (2009).

The dry-bulb temperature of the ambient air is the primary basis for developing cooling and heating strategies depending on the conditions at a location. Similarly, the relative humidity and absolute humidity of the ambient air may be used for developing humidifying and de-humidifying methods or adiabatic cooling strategies depending on the conditions of the location. The analysis of annual mean temperatures at a location is useful to establish the near-surface soil temperature and strategies to harness renewable energy using shallow geothermal energy applications, thermal labyrinths, earth tubes, slinky collectors, and water-source heat pumps.

2.6.2 CLOUD AND SUNSHINE

Cloud cover based on visual observation is expressed as a fraction of the sky hemisphere ('octas' eighths or more recently tenths) covered by clouds. The cloud cover data (in percentage) for all the cities have been adopted from the TMY3 data.

Sunshine duration, i.e., the period of clear sunshine (when a sharp shadow is a cast), is measured by a sunshine recorder, in which a lens burns a trace on a paper strip, shown as hours per day or month. The sunshine data for some cities have been collated from the World Meteorological Organization (WMO), and for the remaining cities, it was estimated. The sunshine duration can be estimated using the standard algorithm (Muneer 2004, pp. 36–39; Equation 2.4):

$$n = \frac{N}{b}\left(\frac{\overline{G}}{\overline{E}} - a\right) \tag{2.4}$$

where $\overline{G}$ and $\overline{E}$ are the monthly-averaged daily terrestrial and extraterrestrial irradiation on a horizontal surface (W/m²), a and b are empirical coefficients (Table 2.6), n is the average daily hours of bright sunshine duration (hours), and N is the day length (hours).

TABLE 2.6
Coefficients for Use in Equation 2.4

Location	a	b
Charleston, South Carolina	0.480	0.090
Atlanta, Georgia	0.380	0.260
Miami, Florida	0.420	0.220
Madison, Wisconsin	0.300	0.340
El Paso, Texas	0.540	0.200
Albuquerque, New Mexico	0.410	0.370

Source: Lof et al. (1966), cross reference in Muneer (2004).

The length of the day can be obtained by Equation 2.5:

$$\omega_s = \cos^{-1}\left(-\tan LAT \times \tan DEC\right) \tag{2.5}$$

$$N = \left(\frac{2\omega_s}{15}\right) \tag{2.6}$$

where

ω_s = sunset hour angle degrees

LAT = latitude degrees (southern hemisphere –ve)

DEC = solar declination degrees (varies from a maximum value of +23.45 on June 22 to a minimum value of –23.45 on December 22. It is zero on the two equinox days of March 21 and September 22).

Cooper (1969) has given the following simple equation for calculating declination on any day of the year (Equation 2.7):

$$DEC = 23.45 \times \sin\left[\frac{360}{365}(284 + DN)\right] \tag{2.7}$$

where

DN = Julian date counted January 1 DN = 1 to December 31 as DN = 365

Another more accurate expression (2.8) is given by (Aydinli 1981)

$$DEC = 23.45 + \sum_{i=1}^{3} a_i \times \cos(i\frac{2\pi}{365}DN + b_i) \tag{2.8}$$

a_i and b_i are as follows:

i	a_i	b_i radians
1	–23.2559	0.1582
2	–0.3915	0.0934
3	–0.1764	0.4539

Another algorithm (Dogniaux 1975) (Equation 2.9):

$$DEC = 0.33281 + \sum_{i=1}^{3} a_i \times \cos(i\frac{2\pi}{366}DN) + b_i \times \sin\left(i\frac{2\pi}{366}DN\right) \tag{2.9}$$

where a_i and b_i are as follows:

i	a_i	b_i radians
1	–22.984	3.7872
2	–0.34990	0.03205
3	–0.13980	0.07187

The extraterrestrial irradiation, E, may be calculated by:

$$E = \frac{0.024}{\pi} I_{sc} \left[1 + 0.033 \times \cos\left(360 \times \frac{DN}{365} \right) \right]$$

$$\times \left[\cos LAT \times \cos DEC \times \sin \omega_s + \left(\frac{2\pi\omega_s}{360} \right) \times \sin LAT \times \sin DEC \right] \tag{2.10}$$

In the above equation I_{sc} is the solar constant (= 1367 W/m^2).

Page et al. (1984) identified the particular day in each month for which the extraterrestrial radiation is nearly equal to the monthly mean value; Table 2.7 gives the solar declinations for representative dates. These dates can also be taken for computing the monthly average values of instantaneous hourly radiation.

- Cloud over (table only) given in %
- Monthly mean sunshine hours (both table and graph).

The sunshine duration and cloud cover data are useful in assessing the potentials of active solar system applications, such as solar thermal systems and/or photovoltaic.

2.6.3 IRRADIATION

Solar radiation is measured by a pyranometer (solarimeter), on an unobstructed horizontal surface and recorded either as the continuously varying irradiance in W/m^2 or through an electronic integrator as irradiation over the hour or day in Wh/m^2. If the hourly value of irradiation is given in Wh/m^2, it will be numerically the same as the average irradiance (W/m^2) for that hour. As an energy unit, the Wh (Watt-hour) is used for solar radiation, although it is only a 'tolerated' unit in the SI.

Monthly irradiation data, as well as the hourly values of direct and diffuse irradiation for a typical winter and summer day (in Wh/m^2) for all the cities, have been adopted from the TMY3 data. The hourly values of irradiation for a typical winter and summer day for all the cities are adopted from the TMY3 data.

The annual solar radiation of a particular location as a result of the apparent movement of the sun (azimuth in the horizon and altitude in the sky) is the basis for an analysis of the potentials of

---

TABLE 2.7

Solar Declination for Representative Dates and Associated Day Number (DN)

Date	Day (DN)	Solar Declination (DEC)	Date	Day (DN)	Solar Declination (DEC)
January 17	17	−20.71	July 17	198	21.16
February 15	46	−12.81	August 16	228	13.65
March 16	75	−1.80	September 16	259	2.89
April 15	105	9.77	October 16	289	−8.72
May 15	135	18.83	November 15	319	−18.37
June 11	162	23.07	December 11	345	−22.99

Note: The values of monthly mean solar declination are the average of the individual daily values calculated using the algorithm given by Aydinli (1981). For use in the southern hemisphere the sign should be reversed (assuming that the latitude is given a positive value) (Page et al. 1984).

---

the passive solar system, the design for summer solar shading, and active solar system applications, such as solar thermal systems and/or photovoltaic.

2.6.4 Wind

Air movement, i.e., wind, is generally measured at 10 m above ground in open country but higher in built-up areas to avoid obstructions; both velocity and direction are recorded. Wind velocity and direction are measured by a cup-type or propeller anemometer.

For all the cities, wind data have been taken from the TMY3 data. For building design, wind data are best represented graphically in the form of a wind rose. Annual wind roses for 50 locations are drawn. These roses have eight sides, corresponding to the four cardinal and four semi-cardinal points of the compass, giving directions from which the wind comes. Each side has 12 lines, corresponding to the 12 months, from January to December in a clockwise direction, where the length of a line is proportionate to the frequency (% of observation) of wind from that direction in that month. Mean wind speed in m/s is shown in tables.

The speed of the wind and prevailing wind directions may be used not only to determine the possibilities for natural ventilation of indoor spaces but also for an assessment of the degree to which wind can be harnessed for power generation.

2.6.5 Precipitation

Precipitation, i.e., the total amount of rain, hail, snow, or dew, is measured in the tipping bucket rain gauges, i.e., calibrated receptacles, and expressed in mm per unit time (day, month, or year). Values indicating the total precipitation for each month of the year (and as many years' average) would show the pattern of dry and wet seasons. The mean monthly precipitation data have been obtained from the National Oceanic and Atmospheric Administration (NOAA 2017). The mean monthly rain data have been obtained from the World Meteorological Organization (2020), which is based on monthly averages for the 30-year period, 1981–2010.

The amount of annual rainfall and its distribution over the year forms the basis for the analysis of the potential of evaporative cooling strategies as well as greywater usage.

2.7 SOLAR GEOMETRY

The earth is almost spherical, some 6371 km in radius, and it revolves around the sun in a slightly elliptical (almost circular) orbit. The earth–sun distance is approximately 150×10^6 km, varying between:

152.10×10^6 km (at *aphelion*, on July 1)
147.09×10^6 km (at *perihelion*, on January 1)

The full revolution takes 365.24 days (365 days 5 h 48′ 46″ to be precise), and as the calendar year is 365 days, an adjustment is necessary: one extra day every four years (the 'leap year'). This would mean 0.25 days per year, which is too much. The excess 0.01 day a year is compensated by a one-day adjustment per century.

The plane of the earth's revolution is referred to as the *ecliptic*. The earth's axis is not normal to the plane of its orbit but tilted by 23.45°. Consequently, the angle between the earth's equatorial plane and the earth–sun line (or the ecliptic, the plane of the earth's orbit) varies during the year (Figure 2.12). This angle is known as the declination (DEC) and varies as

- + 23.45° on June 22 (northern solstice)
- 0° on March 21 and September 22 (equinox dates)
- −23.45° on December 22 (southern solstice)

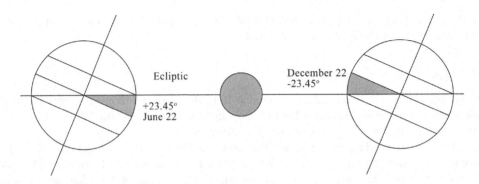

FIGURE 2.12 Elevation section of the earth's orbit and solar declination (DEC).

Geographic latitude (LAT) of a point on the earth's surface is the angle subtended between the plane of the equator and the line connecting the center with the surface point considered. The latitude of the equator is LAT = 0°; the north pole is +90°, and the south pole is −90°. The convention is to use the negative sign for southern hemisphere latitudes. The extreme latitudes where the sun reaches the zenith at mid-summer are the 'tropics' (Figure 2.13).

LAT = +23.45 is the Tropic of Cancer
LAT = −23.45 is the Tropic of Capricorn

The *heliocentric* view is necessary for explaining the sun and earth relationship; in building design, the *lococentric* view is necessary. In this view, the observer's location is at the center of the sky hemisphere on which the sun's position can be determined by two angles (Figure 2.14):

Solar altitude (ALT): measured upward from the horizon, 90 being the zenith
Solar azimuth (AZM): measured in the horizontal plane from the north (0°), through the east
    (90°), south (180°), and west (270°) to the north (360°)

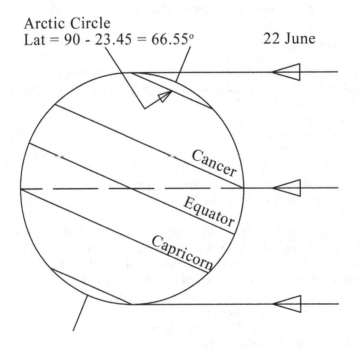

FIGURE 2.13 Definitions of tropics.

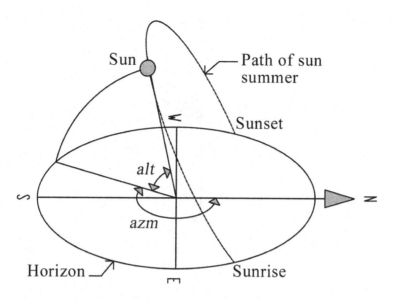

FIGURE 2.14 Altitude and azimuth angles.

Sun-path diagrams or solar charts are the simplest practical tools for depicting the sun's apparent movement. The sky hemisphere is represented by a circle (the horizon). Solar azimuth angles (i.e., the direction of the sun) are given along the perimeter, and solar altitude angles (from the horizon up) are shown by a series of concentric circles, 90° (the zenith) being the center. Several methods are in use for the construction of these charts: orthographic, equidistant, wall diagram, and stereographic projections. The stereographic projection (developed by Phillips 1948) is widely used. These are constructed by a radial projection method (Figure 2.15), in which the center of projection is vertically below the observer's point, at a distance equal to the radius of the horizon circle (the nadir point). The sun-path lines are plotted on this chart for a given latitude for the solstice days, the equinoxes, and any intermediate dates. The date-lines (sun-path lines) are intersected by hour lines. The vertical line at the center is noon. The solar time (apparent local time) is used on solar charts, which coincide with clock time only at the reference longitude of each time zone.

The local apparent time can be obtained from the standard time observed on a clock by applying two corrections. The first correction arises because of the difference between the longitude of a location and the meridian on which the standard time is based. The correction has a magnitude of 4 minutes for every degree difference in longitude. The second correction is called the equation of time correction because the earth's orbit and the rate of rotation are subject to small perturbations (Table 2.8). Thus,

$$\text{Local apparent time} = \text{Standard time}$$
$$\pm 4 \left(\text{standard time longitude} - \text{longitude of location} \right) \qquad (2.11)$$
$$+ \text{Equation of time correction}$$

The negative sign in the first correction is applicable for the eastern hemisphere, while the positive sign is applicable for the western hemisphere. Standard time longitude is given in Table 2.9 for US locations.

The path of the sun across the sky on any day is a circle on the stereographic projection whose radius (R_s) and position of its center (D_s) depend on the latitude of the place for which the diagram

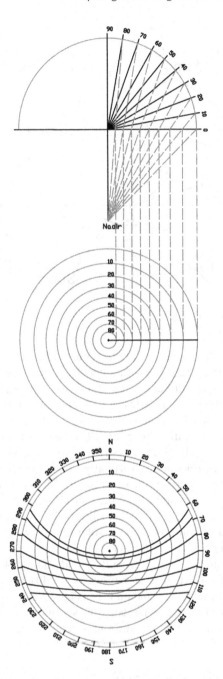

FIGURE 2.15 The stereographic projection method for the sun-path diagram.

is drawn and the declination of the day (Figure 2.16). These and the radius and the distance of hour circles (R_h and D_h^1, D_h^2) can be computed from the equations given below (Lim et al. 1979):

$$R_s = \frac{R \times \cos DEC}{\sin LAT + \sin DEC} \qquad (2.12)$$

$$D_s = \frac{R \times \cos LAT}{\sin LAT + \sin DEC} \qquad (2.13)$$

TABLE 2.8
Equation of Time Correction

Day	1–5	6–10	11–15	16–20	21–25	26–30	31
January	−3″14‴	−5′33″	−7′41″	−9′34″	−11′10″	−12′28″	−13′26″
February	−13′35″	−14′09″	−14′22″	−14′16″	−13′51″	−13′11″	
March	−12′39″	−11′37″	−10′24″	−9′02″	−7′34″	−6′03″	−4′32″
April	−4′14″	−2′45″	−1′21″	−0′03″	+1′06″	+2′04″	
May	+2′50″	+3′22″	+3′41″	+3′46″	+3′37″	+3′14″	+2′38″
June	+2′29″	+1′41″	+0′45″	−0′17″	−1′21″	−2′26″	
July	−3′28″	−4′24″	+5′11″	−5′47″	−6′11″	−6′22″	−6′18″
August	−6′15″	−5′53″	−5′15″	−4′23″	−3′18″	−2′01″	−0′35″
September	−0′17″	+1′20″	+3′03″	+4′49″	+6′35″	+8′19″	
October	+9′59″	+11′33″	+12′58″	+14′11″	+15′10″	+15′52″	+16′16″
November	+16′19″	+16′20″	+16′01″	+15′21″	+14′19″	+12′56″	
December	+11′16″	+9′19″	+7′08″	+4′47″	+2′20″	−0′10″	−2′38″

Source: Smithsonian Meteorological Tables (List 2000, pp. 445–446).

TABLE 2.9
Standard Time Longitude

Zone	Position	Standard Meridian (Longitude)
+4	Atlantic	60° W
+5	Eastern	75° W
+6	Central	90° W
+7	Mountain	105° W
+8	Pacific	120° W

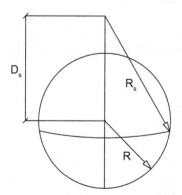
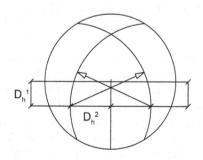

FIGURE 2.16 Drawing of the stereographic sun-path diagram.

$$R_h = R \times \sec LAT \times \csc \omega \qquad (2.14)$$

$$D_h^{\,1} = R \times \tan LAT \qquad (2.15)$$

$$D_h^{\,2} = \pm R \times \sec LAT \times \cot \omega \qquad (2.16)$$

where

DEC = declination

LAT = latitude $D_h^{\,2} = \pm R \times \sec LAT \times \cot \omega$

ω = hour angle (noon = 0)

R = radius of the stereographic projection of the horizon circle

Chapter 7 presents sun-path diagrams for 50 cities in the United States drawn using the software Winshade (Kabre 1999).

2.8   THERMAL COMFORT

A principal purpose of sustainable building design is to provide conditions for human thermal comfort, 'condition of mind that expresses satisfaction with the thermal environment' (ASHRAE 2017). This definition emphasizes that judgment of comfort is a cognitive process involving many parameters influenced by physical, physiological, psychological, and other processes (ASHRAE 2009).

2.8.1   THERMAL BALANCE OF HUMAN BODY

The human body continuously produces heat through metabolic processes that must be dissipated and regulated to maintain normal body temperatures. The heat output of a resting adult is about 100 W, but it can range from about 70 W (in sleep) to about 700 W in heavy activity (playing tennis). The deep-body temperature is about 36.8°C at rest in comfort and rises with activity to about 37.4°C when walking and 37.9°C when jogging, while the skin temperatures associated with comfort at sedentary activities are 33–34°C and decrease with increasing activity (Fanger 1967). The metabolic heat production can be of two kinds: *basal* metabolism due to biological processes (assimilation and utilization of food), which are continuous and non-conscious and *muscular* metabolism while carrying out work, which is consciously controllable (except in shivering). The heat is dissipated to the environment by conduction, convection, radiation, or evaporation (Figure 2.17).

The human body interacts with its thermal environment through sensible heat loss or gains by conduction, convection, radiation, and latent heat loss by evaporation. Thermal comfort is achieved when there is a balance between metabolic heat production and heat dissipation. The thermal balance of the human body, i.e., thermal interaction with its environment, can be expressed as equation 2.17.

$$M \quad W \pm R \pm C \pm K - E - S \qquad (2.17)$$

where

M = rate of metabolic heat production, W/m²

W = rate of mechanical work accomplished, W/m²

S = any surplus or deficit heat stored, W/m²

C = sensible heat flow (loss or gain) by convection (including respiration), W/m²

R = sensible heat flow (loss or gain) by radiation, W/m²

K = sensible heat flow (loss or gain) by conduction, W/m²

E = latent heat loss by evaporation (including respiration), W/m²

Thermal balance or comfort exists when external heat gains and heat produced by the body are entirely dissipated to the environment and a condition of equilibrium prevails; i.e., ΔS is zero.

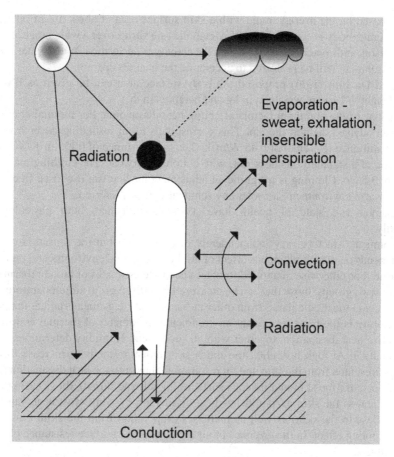

FIGURE 2.17 Thermal balance of the human body; thermal interaction with its environment.

2.8.2 Parameters of Thermal Comfort

It is well established empirically that air temperature, relative humidity, radiant temperature, and airspeed all affect human thermal comfort. Several non-environmental factors like the level of activity, clothing, and acclimatization are also crucial for the determination of an optimum thermal environment. The parameters of thermal comfort are classified into three categories: personal, environmental, and other, in Table 2.10.

Metabolic activity is defined in terms of the rate of heat produced, expressed in W/m² of the body surface. A unit used to express the metabolic rate per unit Du Bois area is the *met*, defined as the metabolic rate of a sedentary person (seated quite), 1 met = 58.2 W/m² (ASHRAE 2009). The

TABLE 2.10

Parameters of Thermal Comfort

Personal	Environmental	Other
Metabolic rate (level of activity)	Air temperature	Food and drink
		Living habits
Clothing insulation	Humidity	Body shape
State of health	Airspeed	Subcutaneous fat
Acclimatization	Radiant temperature	Age and gender

calculation is based on the average male with a skin surface area of about 1.8 m^2. Thus, the heat output of an average body is about 104.76 W. Metabolic rate varies over a wide range, depending on the activity, person, and conditions under which the activity is performed. With higher levels of met, the cooler environment will be preferred to accelerate the heat dissipation.

Du Bois and Du Bois (1916) proposed the body surface area can be given as BSA= 0.202 * weight$^{0.425}$ * height$^{0.725}$, where weight is in kg and height is in m.

Clothing is one of the dominant factors affecting heat dissipation. For thermal comfort studies, a unit has been advised, named the clo. This corresponds to an insulating cover over the whole body of a transmittance (U-value) of 6.45 W/m^{2}K (i.e., the resistance of 0.155 m^{2}K/W). 1 clo is the insulating value of a standard business suit, with cotton underwear. The clothing may range from 0 to more than 3.5 clo. Clothing is a significant adjustment mechanism if chosen freely, but if it is restricted in a warm environment, a cooler environment will be preferred.

Acclimatization and state of health have a strong influence, both physiologically and psychologically.

The environmental factors vary independently of each other, but the sensation of comfort or discomfort depends on the simultaneous effect of all these. Thermal environmental parameters that must be measured or otherwise quantified to obtain accurate estimates of human thermal response are divided into two groups: those that can be measured directly, e.g., (a) air temperature, (b) humidity, (c) airspeed, and those calculated from other measurements, e.g., mean radiant temperature.

Air temperature is the most significant environmental parameter of thermal comfort; it determines convective heat dissipation, together with air movement. Humidity determines heat dissipation by evaporation. At high humidity, too much skin moisture tends to increase discomfort as evaporation is prevented from the skin and, in respiration, thus curb the heat dissipation mechanism. Low humidity can dry the skin and mucous surfaces (mouth and throat), thus causing discomfort. Air movement across the skin accomplishes heat dissipation by convective heat transfer (removing warm air close to the skin) as well as increases evaporation from the skin, thus producing a physiological cooling effect. In the presence of air movement, the surface resistance of the body (or clothing) is reduced.

Precise relationships between increased airspeed and improved comfort have not been established. Under hot conditions, 1 m/s is pleasant, and indoor air velocities up to 1.5 m/s are acceptable. Above this, light objects may be blown about; thus, indirect nuisance effects may be created. Under cold conditions, in a heated room, 0.25 m/s velocity should not be exceeded, but even in a heated room, stagnant air (velocities < 0.1 m/s) would be judged as 'stuffy.' A draft is an undesirable local cooling of the human body by air movement, and it is a serious comfort problem.

An additional effect is that with no air movement, a practically saturated air layer is formed at the body surface, which prevents (reduces) further evaporation. The air movement would remove this saturated air envelope. The skin is surrounded by a thin, still air layer, which is close to the skin temperature and insulates the body from its surroundings. Air movement decreases the thickness of this insulating layer and thus provides a cooling effect, provided that the vapor pressure of air is lower than the skin vapor pressure, even if the dry-bulb temperature is higher than the skin temperature within a specific limit. Increased air movement reduces the amount of moisture-laden air close to the skin, thereby increasing evaporation. The effect of air movement is, therefore, two-fold: the convection heat loss coefficient of the body (or clothing-) surface (h_c) is a function of air velocity, but evaporation from the skin, thus the evaporation heat loss coefficient (h_e), is also increased by moving air.

Radiation exchange depends on the mean radiant temperature (T_{mrt}), the average temperature of the surrounding surfaces, each weighted by the solid angle it subtends at the measurement point. If the temperature of the surrounding surfaces is lower than the skin temperature, then the body will radiate heat. Surroundings that are hotter than the skin temperature will radiate heat, causing the skin temperature to increase (Vernon 1932). This effect is accentuated when lighter clothing is worn (for example, in summer). Radiation exchange with the surroundings can have a significant effect on human comfort. Measurements of globe temperature (T_g), air temperature (T_a), and air velocity (v) can be combined to estimate the mean radiant temperature. The globe thermometer is a mat black copper sphere, usually of 150 mm diameter, with a thermometer located at its center. Positioned in a room, after equilibrium is reached (in 10–15 minutes), the globe will respond to the net radiation to or from the surrounding surfaces. If radiation is received, then $T_g > T_a$; $T_g < T_a$ indicates that if the surrounding surfaces are cooler than the air, radiation is emitted. In still air $T_{mrt} = T_g$, but a correction for air movement of v velocity (in m/s) is possible:

$$T_{mrt} = T_g \times \left(1 + 2.35\sqrt{v}\right) - 2.35 \times T_a \sqrt{v} \qquad (2.18)$$

In warm climates (with light clothing) the mean radiant temperature is twice as important as indoor dry-bulb temperature (T_{ai}), which is accounted in environmental temperature, T_{env} (CIBSE 1999):

$$T_{env} = \frac{2}{3}T_{mrt} + \frac{1}{3}T_{ai} \qquad (2.19)$$

However, in cooler climates (with heavier clothing) the mean radiant temperature has the same influence as the dry-bulb temperature, which is expressed as the dry resultant temperature (T_{drt}):

$$T_{drt} = 0.5 \times T_{ai} + 0.5 T_{mrt} \qquad (2.20)$$

In addition to independent personal and environmental parameters influencing thermal comfort, other factors may have some effect. Food and drink consumed may influence metabolism, thus affecting heat production and dissipation. The body shape is significant in that heat production is proportional to body mass, but heat dissipation depends on the body surface area. Age and gender may have a modicum influence in the preferred temperature.

2.8.3 THERMOREGULATION

The human thermoregulatory system attempts to maintain a constant deep-body temperature of 36.8°C. The *hypothalamus*, located in the brain, controls various physiological processes to regulate body temperature. Its control behavior is primarily proportional to deviations from deep-body temperatures with some integral and derivative response aspects. The most important and often-used physiological process is regulating blood flow to the skin: *vasodilation* (in extreme heat when internal temperatures rise above a set point) – more blood is directed to the skin to transport internal heat to elevate the skin temperature and increase heat dissipation to the environment; *vasoconstriction* (in extreme cold when body temperatures fall below the deep-body temperature) – skin blood flow is reduced to conserve heat. The effect of the maximum vasoconstriction is equivalent to the insulating effect of a heavy sweater. Vasodilation and vasoconstriction are known as vasomotor adjustments.

At temperatures less than the set point, muscle tension increases to generate additional heat; where muscle groups are opposed, this may increase to visible shivering, which can increase resting heat production to 4.5 met. At elevated internal temperatures, sweating occurs. This defense mechanism is a powerful way to cool the skin and increase heat loss from the core.

Insufficient heat loss leads to overheating (hyperthermia), and excessive heat loss results in body cooling (hypothermia).

2.8.4 THERMAL NEUTRALITY

Adaptive models predict the almost constant conditions under which people are likely to be comfortable in buildings. In general, people naturally adapt and may also make various adjustments to themselves and their surroundings to reduce discomfort and physiological strain. Auliciems (1981) proposed the psycho-physiological model of thermal perception, which is the basis of the adaptive models. It has been empirically established that, through adaptive actions, an acceptable degree of comfort in residences and offices is possible over a range of air temperatures from about 17°C to 31°C (Humphreys and Nicol 1998). An ASHRAE sponsored study (de Dear et al. 1997) compiled an extensive database from past field studies to study, develop, and test adaptive models (ASHRAE 2009).

Adaptive adjustments are typically conscious actions such as altering dress codes, posture, flexible activity schedules or levels, rate of work, diet, ventilation, air movement, and local temperature. They may also include long-term unconscious changes to physiological set points and gains for the control of shivering, skin blood flow, and sweating, as well as adjustments to body fluid levels and salt loss after a few days of exposure up to about six months. In a hot climate, this may consist of increased blood volume, which improves the effectiveness of vasodilation, enhanced performance of the sweat mechanism, as well as the readjustment of set point. In a cold climate, the vasoconstriction may become permanent, with reduced blood volume, while the body metabolic rate may increase (Szokolay 2008). The adjustment of seasonal preferences can be quite significant, even over a period of a month.

The term 'thermal neutrality' refers to a specific value of the indoor thermal environmental index (e.g., operative temperature) corresponding to a mean thermal sensation vote of zero on the seven-point scale (i. e. 'neutral'). Values of thermal neutrality were calculated by using an empirical correlation function developed by deDear and Brager (2002) as an improved version of the function earlier proposed by Auliciems (1981) and Humphreys (1978). Thermal neutrality (T_c) is expressed as a function of mean monthly temperature ($\bar{T}_o$):

$$T_c = 17.8 + 0.31 \times \bar{T}_o \tag{2.21}$$

If the mean monthly outdoor temperature is less than 10°C or greater than 33.5°C, this option may not be used (ASHRAE 2010). Then the comfort zone can be taken up as given by ASHRAE (Figure 2.18).

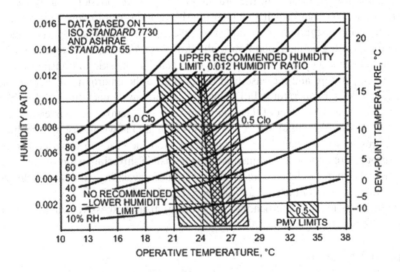

FIGURE 2.18   ASHRAE summer and winter comfort zones. © ASHRAE, www.ashrae.org (2009) *ASHRAE Handbook of Fundamentals.*

2.9 ENVIRONMENTAL INDICES AND COMFORT ZONE

The comfort zone is defined in terms of a range of thermally acceptable conditions within which the average person would feel comfortable. The environmental conditions required for comfort are not the same for everyone. However, extensive laboratory and field data have been collected that provide the necessary statistical data to define conditions that a specified percentage of occupants will find thermally comfortable. The comfort parameters are usually derived to satisfy about 80–90% of people in a space.

As four environmental parameters influence thermal comfort, attempts have been made to create a single index. An environmental index combines two or more parameters (e.g., air temperature, mean radiant temperature, humidity, air velocity) into a single variable to express the thermal response. Indices simplify the description of the thermal environment and the stress it imposes. Since the early 1900s a large number of thermal indices have been developed in various countries throughout the world. Environmental indices may be classified according to how they are developed. Empirical indices are based on field measurements with the subject under defined environmental conditions or simplified relationships that do not necessarily follow theory. Rational or analytical indices are based on the theoretical concepts of the thermal exchanges with the environment, i.e., heat flow paths from metabolic heat production to the environment and resistances to such flows.

The earliest empirical index, effective temperature, was developed at ASHVE Pittsburgh research laboratories (Houghten and Yaglaglou 1923a, 1923b). It is defined as the temperature of a still, saturated atmosphere, which would, in the absence of radiation, produce the same effect as the atmosphere in question. It is represented by a set of equal *comfort lines* drawn on the psychrometric chart. There are about 30 different such indices developed over the years by various research workers, all based on different studies, all with different derivations and names (Auliciems and Szokolay 2007).

Effective temperature (***ET****) is an analytical index, and it has the broadest range of application in practice. It is defined as the temperature (DBT) of a uniform enclosure at 50% relative humidity, which would produce the same net heat exchange by radiation, convection, and evaporation as the environment in question. It combines temperature and humidity into one single index, so two environments with the same ET* should evoke the same thermal response even though they have different temperatures and humidities, as long as they have the same air velocities. Because ***ET**** depends on clothing and activity, it is not possible to generate a universal ***ET**** chart.

A standard set of conditions representative of typical indoor applications is used to define a standard effective temperature *SET*, defined as the equivalent air temperature of an isothermal environment at 50% relative humidity in which a subject, wearing clothing standardized for the activity concerned, has the same heat stress and thermoregulatory strain as in the actual environment. It is interpreted as a sub-set of *ET** under standardized conditions: clothing standardized for given activities. At the sea level, under the above standard environmental conditions *SET = ET**. The *SET*, thus, defined combines the effect of temperature and humidity, the two most important determinants. The slope of the *SET* lines indicates that at higher humidities, the temperature tolerance is reduced, while at lower humidities, higher temperatures are acceptable.

ASHRAE used the psychrometric chart for the definition of the comfort zone since 1966. Current and past studies periodically reviewed to update the ***ASHRAE Handbook of Fundamentals***, which specifies conditions or comfort zones where 80% of sedentary or slightly active persons find the environment thermally acceptable. The 1966 version gave the temperature limits by DBT (vertical) lines and the humidity limits by two RH curves. In 1974 the side boundaries were changed to ***ET**** lines, and the humidity boundaries were defined in terms of the vapor pressure (or the corresponding AH or RH (horizontal) lines).

The *ASHRAE Handbook of Fundamentals* (2009) specifies summer and winter comfort zones (Figure 2.20), appropriate for clothing insulation levels of 0.5 [0.078 (m^2 K/W)], and

1 clo [0.155 (m² K/W)], respectively. It is assumed that a winter business suit has about 1 clo of insulation, and a short-sleeved shirt and trousers have about 0.5 clo. This is justified by needing an 'objective' reason rather than a nebulous notion of 'acclimatization.' The warmer and cooler temperature borders of the comfort zones are affected by humidity and coincide with the lines of constant ET*. In the middle of a zone, a typical person wearing the prescribed clothing would have a thermal sensation at or very near neutral. Near the boundary of the warmer zone, a person would feel about +0.5 warmer on the ASHRAE thermal sensation scale; near the boundary of the cooler zone, that person may have a thermal sensation of −0.5. In general, comfort temperature for other clothing levels can be approximated by decreasing the temperature border of the zone by 0.6 K for each 0.1 clo increase in clothing insulation and vice versa. Similarly, a zone's temperatures can be decreased by 1.4 K per met increase in activity above 1.2 met.

ASHRAE (2009) specifies the upper humidity limit of 0.012 $kg_w/kg_{dry\ air}$ or 12 g/kg because it restricts the evaporation and thus the cooling effect. There is no lower limit specified, but the accepted lower humidity limit is 4 g/kg for non-thermal comfort factors such as skin drying, irritation of mucus membranes, dryness of the eyes, and static electricity generation (Liviana et al. 1988).

The comfort zone can be plotted on a psychrometric chart that will vary with the climate and be different for each month. The procedure may be as follows:

The thermal neutrality temperature (T_c) as equation 2.21: $T_c = 17.8 + 0.31 \times T_o$ is used as a threshold to articulate comfort zone for both the summer and winter months. The temperature limits of such a comfort zone are taken as $(T_c − 2.5)$°C to $(T_c + 2.5)$°C for 90% acceptability. The SET coincides with DBT at the 50% RH curve; these points are marked on the 50% RH curve. These will define the 'side' boundaries of the comfort zone as the corresponding SET lines. The humidity limits (top and bottom) will be 12 and 4 g/kg, respectively (1.9 and 0.6 kPa vapor pressure).

Up to 14°C, the SET lines coincide with the DBT. Above that the slope of these isotherm lines is progressively increasing, with the slope coefficient taken as DBT/AH × 0.023 × (DBT − 14) which gives the deviation from the corresponding vertical DBT line for each g/kg AH, positive below the 50% and negative above it (Szokolay 2008).

Figures 2.19 and 2.20 shows the summer and winter comfort zones for Los Angeles and Austin, respectively. It is noteworthy that Los Angeles has a minimal seasonal variation (a warm-dry climate), while in Austin, there is a significant difference between winter and summer.

2.10   COOLING AND HEATING DEGREE-DAYS

Degree-days (DD or Kd, Kelvin-days) are relatively simple forms of climatic data, useful as an index of climatic severity as it affects energy use for space cooling or heating. Degree-days are calculated as the difference between the prevailing external, dry-bulb temperature, and a 'base temperature.' Traditionally used base temperatures to calculate HDD and CDD are 18.3°C in the United States (ASHRAE 2009). This is the external temperature at which, in theory, no artificial cooling (or heating) is required to maintain an acceptable internal temperature. If the mean temperature of a day is $\bar{T}_o$, then for the day, we have $T_b − \bar{T}_o$ degree-days. (When $\bar{T}_o = T_b$ the degree-day number is zero.) This number can then be summed for any given period, e.g., a month or a year. The number multiplied by 24 gives the degree-hours number. Degree-days are used in energy estimating methods.

Two types of degree-days are used in building design. The cooling degree-days (K-day) or cooling degree-hours (K-h) indicate the warmth of the summer and hence cooling requirements. The heating degree-days (K-day) indicate the severity of the heating season and therefore heating energy requirements.

Cooling and heating degree-days (base 18.0°C) are calculated as the sum of the differences between the daily average temperature and the base temperature. The number of cooling degree-days

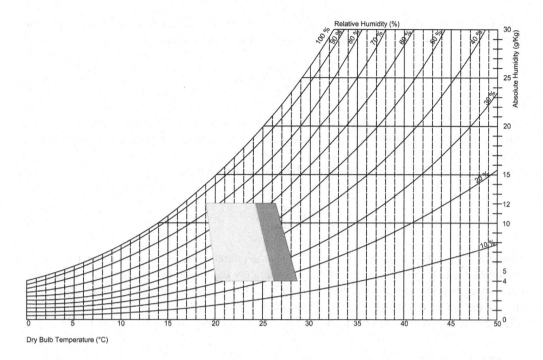

FIGURE 2.19 Summer (dark) and winter (light) comfort zones for Los Angeles, California.

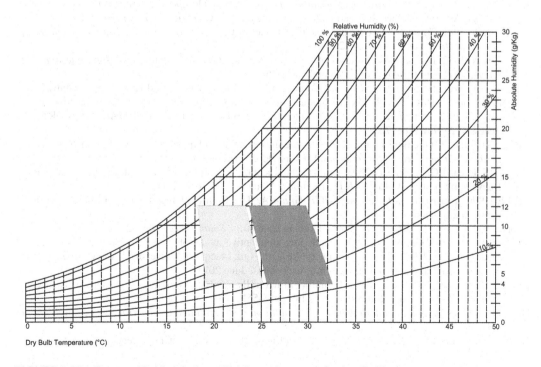

FIGURE 2.20 Summer (dark) and winter (light) comfort zones for Austin, Texas.

(CDD) is defined as 'cumulative temperature excess' above an agreed reference level or base temperature. For example, the number of CDD in the month is calculated as:

$$CDD = \sum_{i=1}^{N} \left(\overline{T_o} - T_b \right)^+ \tag{2.22}$$

The concept of heating degree-days is somewhat similar to the cooling degree-days, but here the definition would be 'cumulative temperature deficit' below an agreed reference level or base temperature (T_b).

Similarly, monthly heating degree-days (HDD) are calculated as:

$$HDD = \sum_{i=1}^{N} (T_b - \overline{T_o})^+ \tag{2.23}$$

where N is the number of days in the month, T_b is the reference temperature to which the degree-days are calculated, and T_o is the mean daily temperature calculated by adding the maximum and minimum temperatures for the day, then dividing by 2. The + superscript indicates that only positive values of the bracketed quantity are taken into account in the sum.

The primary source of cooling and heating degree-days (base 18.0°C) data is the TMY3.

REFERENCES

ASHRAE (2009) *Handbook of Fundamentals.* American Society of Heating Refrigerating and Air Conditioning Engineers, Atlanta, Chapter 9.

ASHRAE (2013) *Climatic Data for Building Design Standards Standard 169–2013.* American Society of Heating, Refrigerating and Air-Conditioning Engineers, Atlanta.

ASHRAE (2017) *Thermal Environmental Conditions for Human Occupancy.* ANSI/ASHRAE Standard 55-2017, American Society of Heating, Refrigerating and Air-Conditioning Engineers, Atlanta.

Auliciems A (1981) Towards a psycho-physiological model of thermal perception. *International Journal of Biometeorology*, vol. 25(2): 109–22.

Auliciems A (1982) Psycho-physiological criteria for global thermal zones of building design. *International Journal of Biometeorology*, vol. 26(Supplement): 69–86.

Auliciems A, Szokolay SV (2007) *Thermal Comfort.* PLEA Note 3. Passive and Low Energy Architecture International, Design Tools and Techniques in association with University of Queensland Dept of Architecture. http://plea-arch.org/wp-content/uploads/PLEA-NOTE-3-THERMAL-COMFORT.pdf. Accessed 26 November 2012.

Aydinli S (1981) Uber die Berechnung der zur Verfugung stehenden Solarenergie und des Tageslichtes. *Verein Deutscher Ingenieure.* (VDI-Verlag GmbH Dusseldorf), vol. 6(79): 10.

Briggs RS, Lucas RG, Taylor ZT (2003) Climate classification for building energy codes and standards: Part 1-development process. *ASHRAE Transactions*, Atlanta, vol. 1:4610–4611.

CIBSE (1999) *Environmental Design: CIBSE Guide A.* The Chartered Institution of Building Services Engineers, London.

Cooper PI (1969) The absorption of solar radiation in solar stills. *Solar Energy*, vol. 12: 3.

de Dear R, Brager G, Cooper D (1997) Developing an adaptive model of thermal comfort and preference. *Final Report ASHRAE RP-884*, Macquarie Research Ltd. Macquarie University, Sydney. http://aws.mq.edu.au/rp-884/ashrae_rp884_home.html. Accessed 02 June 2015.

de Dear RJ, Brager GS (2002) Thermal comfort in naturally ventilated buildings: Revisions to ASHRAE Standard 55. *Energy and Buildings*, vol. 34(6): 549–561.

Diston DJ (2009) *Computational Modeling and Simulation of Aircraft and the Environment.* John Wiley & Sons Ltd, Chichester, England.

DOE (n.d.) *EnergyPlus Weather Data.* Department of Energy, US Government, https://energyplus.net/weather.

Dogniaux R (1975) Variations geographiques et climatiques des expositions energetiques solaries sur des surfaces receptrices horizontales et verticals. Institut Royal Meteorologique de Belgique, *Misc. Ser. B*, vol. 38: 7.

Fanger PO (1967) Calculation of thermal comfort: Introduction of a basic comfort equation. *ASHRAE Transactions*, vol. 73(2): III.4.1.

Hawas M, Muneer T (1983) Correlation between global radiation and sunshine data for India. *Solar Energy*, vol. 30(3): 289.

Houghten FC, Yaglaglou CP (1923a) ASHVE Research Report No. 673 Determination of the comfort zone. *ASHVE Transactions*, vol. 29: 361–79.

Houghten FC, Yaglaglou CP (1923b) Determining the lines of equal comfort. *Journal of ASHVE*, vol. 29: 165.

Humphreys M (1978) Outdoor temperatures and comfort indoors. *Building Research and Practice*, 6(2): 92–105

Humphreys M, Nicol JF (1998) Understanding the adaptive approach to thermal comfort. *ASHRAE Technical Data Bulletin*, vol. 14(1): 1–14.

IESNA (2013) *The IESNA Lighting Handbook: Reference and Application. Illuminating Engineering Society of North America, National Bureau of Standards (1991) The International System of Units (SI)*, 6th edition, NBS Special Publication 330, National Bureau of Standards, Gaithersburg, MD.

Iqbal M (1983) *An Introduction to Solar Radiation*. Academic Press, Toronto.

Kabre C (1999) WINSHADE: A computer design tool for solar control. *Building and Environment*, vol. 34(3): 263–274.

Kottek M, Grieser J, Beck C, Rudolf B, Rubel F (2006) World Map of the Köppen-Geiger climate classification updated. *Meteorologische Zeitschrift*, vol. 15(3): 259–263.

Lim BP, Rao KR, Tharmaratnam K, Mattar AM (1979) *Environmental Factors in the Design of Building Fenestration*. Applied Science Publishers Ltd., London.

List RJ (2000) *Smithsonian Meteorological Tables*, 6th revised edition. Smithsonian Miscellaneous Collections, vol. 114, Smithsonian Institution Press, Washington, DC. Available at https://repository.si.edu/bitstream/handle/10088/23746/1951%20%20smc%20%20vol%20114.pdf?sequence=1&isAllowed=y

Liviana JE, Rohles FH, Bullock OD (1988) Humidity, comfort, and contact lenses. *ASHRAE Transactions*, vol. 94(1): 3–11.

Lof GOG, Duffie JA, Smith CO (1966) World distribution of solar radiation. *Engineering Experiment Station Report 21*, University of Wisconsin, Madison, WI.

Muneer T (2004) *Solar Radiation and Daylight Models*. Elsevier Butterworth Heinemann, Amsterdam.

NOAA (2017) *Climate Normal*. National Oceanic and Atmospheric Administration, US Government. Available at https://ncdc.noaa.gov/data-access/land-based-station-data/land-based-datasets/climate-normals.

NOAA, NASA, and USAF (1976) *U.S. Standard Atmosphere. National Oceanic & Atmospheric Administration (NOAA)*. National Aeronautics & Space Administration (NASA) and United States Air Force (USAF). Available at https://ntrs.nasa.gov/search.jsp?R=19770009539.

Olgyay V (1963) *Design with Climate, Bioclimatic Approach to Architectural Regionalism*. Princeton University Press, Princeton, NJ.

Page JK, Thompson JL, Simmie J (1984) *Algorithms for Building Climatology Applications. A Meteorological Data Base System for Architectural and Building Engineering Designers*, Handbook. Department of Building Science, University of Sheffield, Sheffield, England.

Phillips RO (1948) *Sunshine and Shade in Australia*. TS 23, also Bulletin 8, 1963.

Szokolay SV (1992) *Architecture and Climate Change*. RAIA, Red Hill, Canberra

Szokolay SV (2008) *Introduction to Architectural Science: The Basis of Sustainable Design*. Architectural Press/Elsevier Science, Oxford.

Vernon HM (1932) The measurement of radiant heat in relation to human comfort. *Journal of Industrial Hygiene*, vol. 14: 95–111.

Warner CD (1897) Editorial. Hartford courant of Connecticut. 27 Aug. http://quoteinvestigator.com/2010/04/23/everybody-talks-about-the-weather. Accessed 27 November 2016.

WMO (2010) *World Meteorological Organization Standard Normal*. United Nations Statistics Division. Available at http://data.un.org/Explorer.aspx.

WMO (2020) *World Weather Information Service*. World Meteorological Organization. Available at http://worldweather.wmo.int/en/home.html.

3 Thermal Environment Design Strategies

3.1 INTRODUCTION

Victor Olgyay (1963) said,

> We do not expect to solve the problems of uncomfortable conditions by natural means only. The environmental elements aiding us have their limits. But it is expected that the architect should build the shelter in such a way as to bring out the best of the natural possibilities.

The quintessential step toward the design of a sustainable built environment is space heating and cooling by natural means, i.e., 'passive,' the term that was first introduced in the United States. Passive space heating may be defined as harnessing the thermal energy of the sun to heat the indoor environment. Passive space cooling may be defined as the removal of excess heat from an indoor environment by utilizing the natural processes of disposing thermal energy to a natural heat sink, namely, the ambient air, the water, the ground, and the sky, when they are at a lower temperature, by convection, conduction, radiation, and/or evaporation. Hybrid or low energy cooling utilizes those same heat sinks, with the assistance of pumps or fans to circulate a heat transfer fluid, such as air or water between the cooled space and the heat sink. These heat sinks are also the thermal dump for all active and mechanical cooling systems (Cook 1989).

The success of a design can be measured in terms of the cooling or heating energy requirements, by hand calculation using a steady-state method or by computer modeling using a non-steady-state or dynamic heat flow method. If energy is excessive, the design can be modified and tested again; the process is repeated until the thermal environment is optimized in terms of thermal comfort and energy. This chapter aims to present the qualitative and quantitative analysis of climate, thermal comfort, and thermal behavior of building to develop passive, hybrid, and active thermal environment design strategies for a sustainable built environment.

The first step is the assessment of climate and comfort zone using bioclimatic analysis, which is followed by the extension of comfort to develop a design hypothesis, first using passive design strategies (presented in the second section), secondly using hybrid design strategies (dealt in the third section), and thirdly using energy-efficient active design strategies for the residual task of space conditioning (discussed in the fourth section). The last section is one of the important considerations of energy conservation that pertains to 'solar control design.' The climate data and sun-path diagrams given in Chapter 7 are used as an aid for the design and decision-making process.

3.2 PASSIVE DESIGN STRATEGIES

The conceptual design stage involves the following design parameters that have the most significant influence on the thermal performance of the building:

i) **Site and orientation**: the longer axis of the building should be in an east-west direction with openings on the north and south sides; also, position the building on-site to facilitate breeze access. The term 'aspect ratio' denotes the ratio of the longer side of a rectangular plan to the shorter side. In most cases, the ratio may be taken between 1.3 and 2.0 depending on the climate (Olgyay 1963).

ii) **Building form**: the surface-to-volume ratio determines the heat loss or gain through the building envelope. It is advisable to design the built form with the least surface area for a given volume; the hemisphere is the most efficient form, but a compact plan is always better than a spread-out plan.

iii) **Building envelope**: the thermo-physical properties of the materials determine the heat gain or loss through the building components. The ASHRAE 90.1 (2010) has given prescriptive building envelope requirements (roofs, walls, floors, fenestration, doors, etc.) in terms of the maximum U-value of the overall assembly and the minimum R-value of the insulation alone for the eight climate zones in the United States. The standard also prescribes the maximum shading heat gain coefficient (SHGC) and U-value for vertical glazings and skylights. There are specific minimum requirements for visible light transmittance (VLT) of vertical fenestration as a function of the window to wall ratio (WWR).

The subsequent sub-sections describe the process of design decision making.

3.2.1   BIOCLIMATIC ANALYSIS

The next step in passive design is to delineate the summer and winter comfort zones and overlay the monthly temperatures and humidities to diagnose the nature of the climatic problem. Comfort zones and climatic conditions can be represented graphically by either the 'bioclimatic chart' or the 'psychrometric chart.'

Olgyay (1963) devised his bioclimatic chart to show the effects of four environmental variables on human comfort (Figure 3.1). Dry-bulb temperature is on the vertical axis, relative humidity on the horizontal, and a comfort zone is delineated in terms of these two variables. Lines above this indicate an extension of the upper comfort limit by air movement, and lines below it show a downward extension of the comfort zone by solar radiation. Arens et al. (1980) revised the bioclimatic chart based on the J. B. Pierce laboratory human thermal model.

Givoni (1969) used the psychrometric chart to delineate his bioclimatic chart. On this, he delineates the comfort zone and its extensions by various active and passive techniques. Milne and Givoni (1979) further developed this chart, and it is used in the Climate Consultant software developed by the Energy Design Tools Group, the University of California (Figure 3.2). Watson and Labs (1983) further developed Givoni's system. Szokolay's (1986, 2008) control potential zone (CPZ) method defines the comfort zone on the psychrometric chart using the Auliciems (1982) expression for thermal neutrality. The zone is then taken as 5 K wide (±2.5°C to thermal neutrality) plotted on the 50% RH (relative humidity) line, and the side boundaries are taken as the corresponding sloping ET^* (or standard effective temperature) lines.

The psychrometric chart is chosen as an analytical tool in this book, as it offers a more pedantic way and allows the representation of many other parameters, such as wet-bulb temperature, relative humidity, enthalpy, and density. This analysis tool is recommended for 'envelope dominated buildings,' e.g., residences and small office buildings. The method comprises three steps:

1. Delineate the 'comfort zone' for summer and winter (as in Section 2.9, Figures 2.19 and 2.20). Note that the words 'summer' and 'winter' are used to refer to the overheated (warmest) and underheated (coldest) periods of the year and not necessarily to the calendar months.
2. Plot the ambient climatic conditions and mark on the chart two points for each of the 12 months: one taking mean maximum temperature with minimum RH and one taking mean minimum temperature with maximum RH. Connect two points by a line. The 12 lines would indicate the median zone of climatic conditions. A comparison of comfort zones with these lines would ascertain the nature of the climatic problem (Figures 3.3, 3.4, 3.5, and 3.6).

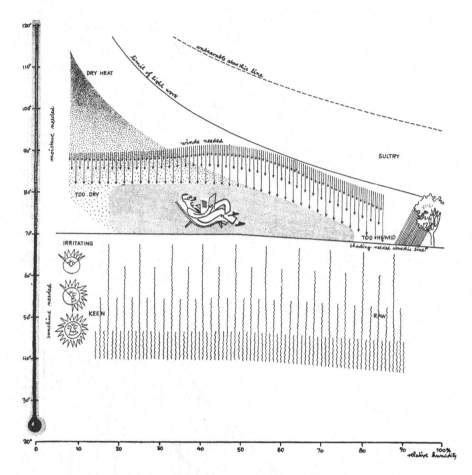

FIGURE 3.1 Olgyay's bioclimatic chart. (Olgyay 1963)

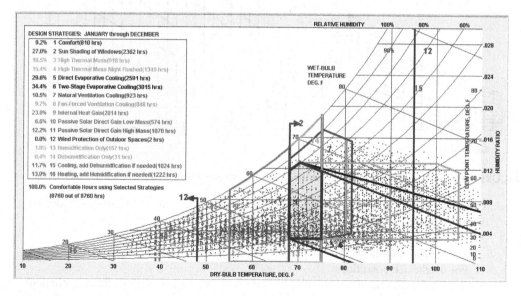

FIGURE 3.2 Psychrometric chart showing design strategies. © 2016 The Regents of the University of California

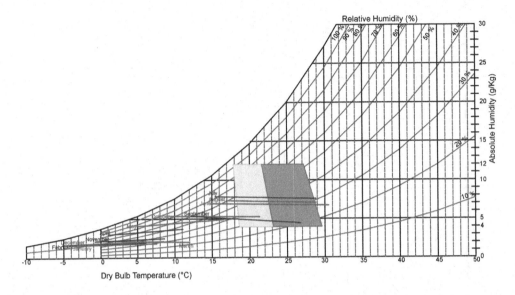

FIGURE 3.3   Denver, Colorado: Cool dry comfortable summer, cool in winter but not severe.

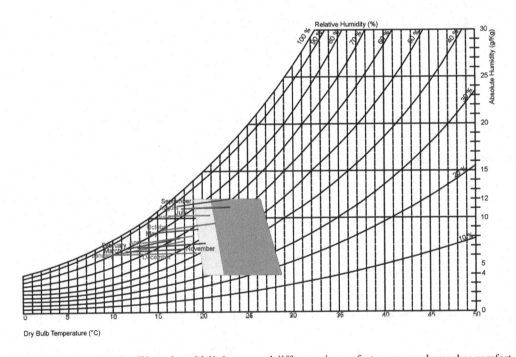

FIGURE 3.4   Los Angeles: Warm dry with little seasonal difference in comfort zones, rarely reaches comfort.

3. To identify the design strategies, i.e., the range of outdoor conditions within which passive design strategies, such as can achieve indoor comfort
   - For underheated conditions:
     – Passive solar heating (direct gain, Trombe–Michel wall, greenhouse)
   - For overheated conditions:
     – Passive thermal mass (summer and winter, for summer with night flush ventilation)
     – Comfort ventilation (physiological cooling)
     – Evaporative cooling (direct and indirect)

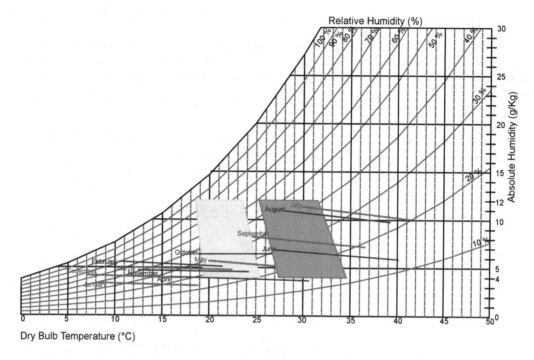

FIGURE 3.5 Phoenix: Hot and dry with the large seasonal difference in comfort zones, large diurnal ranges, hot summer and cold winter nights.

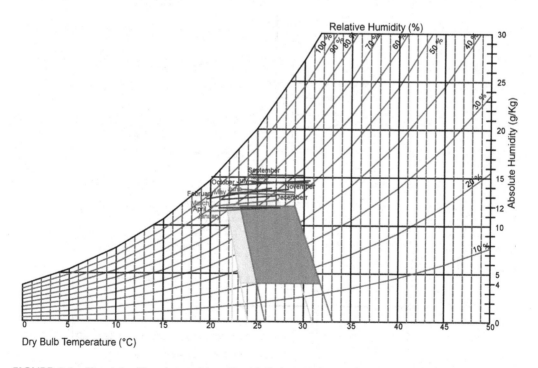

FIGURE 3.6 Honolulu: Very hot and humid with little variation in summer and winter comfort zones, mostly humid, rarely reaches a comfort.

3.2.2 PASSIVE SOLAR HEATING

The passive solar heating strategy is applicable in a cold climate. Every passive solar heating system has at least two elements: a collector consisting of equator-facing glazing and an energy storage element that consists of passive thermal mass. Three basic types of passive solar heating system are as given below and shown in Figure 3.7:

  i) Direct solar gain
 ii) Trombe–Michel wall
iii) Sunspaces (also known as solar greenhouses or conservatories)

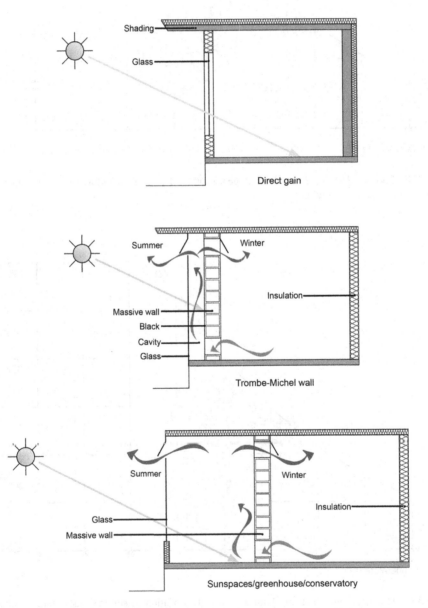

FIGURE 3.7 Passive solar heating: direct gain, Trombe–Michel wall, and greenhouse, line work by Ar Vaibhav Ahuja.

A window facing the equator creates a direct solar gain system. It allows the short-wave infra-red radiation but intercepts the heat from escaping and thus causing the greenhouse effect. The passive thermal mass (massive walls and floors) inside the building then stores this heat, both to reduce daytime overheating and to release at night time. A horizontal shading device may be provided to intercept the summer sun and to allow entry of the beneficial solar radiation in the winter.

The Trombe–Michel wall was named after professor Felix Trombe and architect Jacques Michel, who developed this system in France in 1966. In this passive system, a massive wall is placed behind the equator-facing glazing, leaving an air gap of 100–150 mm. The wall is painted black for higher heat absorption. As the greenhouse effect traps the solar radiation, the wall surface behind the glass is heated and it will heat the air in the gap. The wall incorporates vent openings near the floor and the ceiling. As the heated air rises, it would enter the room through the top vent, drawing in cooler air from the room near the floor level, forming a thermosiphon circulation. In summer, the vents are closed, and the vent in the glass is opened to exhaust hot air. This system can be used when the sun's heat and not its light is desired.

A sunspace (greenhouse or conservatory) is an attached space of about 2 m or more; the thermal function is similar to the Trombe–Michel wall, but it serves dual purposes of heating the room behind it and providing a secondary living space for daytime use. At night time sunspace can lose heat, so it is essential to close off the room it serves.

The passive solar heating can be estimated as a function of the mean daily total irradiation on an equator-facing vertical plane (H_v in Wh/m^2) for the coldest month. The climatic data tables in Chapter 7 include the daily total global irradiation (H_h) and diffuse irradiation on a horizontal surface (H_d); the beam component (H_b) can be found ($H_b = H_h - H_d$) and then equation 3.1 can calculate the value of H_v:

$$H_v = R \times H_b + H_d/2 + r * H_h/2 \tag{3.1}$$

r = reflectance of the foreground, normally taken as 0.2.

Mathematically, R is expressed in equation 3.2 (Muneer 2004)

$$R = \frac{\cos(LAT - 90) \times \cos DEC \times \sin \omega s' + \omega s' \times \sin(LAT - 90) \times \sin DEC}{\cos LAT \times \cos DEC \times \sin \omega s + \omega s \times \sin LAT \times \sin DEC} \tag{3.2}$$

where
sunset hour angle (ω_s):

$$\omega_s = \cos^{-1}(-\tan LAT \times \tan DEC) \tag{3.3}$$

And the sunset hour angle (ω_s') on a vertical plane:

$$\omega_s' = \min\left[\omega_s, \cos^{-1}\left(-\tan(LAT - 90) \times \tan DEC\right)\right] \tag{3.4}$$

LAT = latitude degrees (southern hemisphere –ve)

DEC = solar declination degrees (varies from a maximum value of +23.45 on June 22 to a minimum value of −23.45 on December 22. It is 0 on the two equinox days of March 21 and September 22)

The limiting condition will be the lowest temperature at which the solar gain can match the heat losses under steady-state conditions, which are given by Equation 3.5.

$$H_v \times A \times \eta = q \times (T_i - T_o) \times 24 \tag{3.5}$$

where

H_v = vertical irradiation (Wh/m²day)
A = area of solar aperture
η = efficiency (utilizability), taken as 0.5
$q = q_c + q_v$, building conductance (W/K)
T_i = indoor temperature limit, taken as $T_c - 2.5$
T_o = the limiting temperature to be found

Assume a simple office of 100 m² floor area and 20% (= 20 m²) solar window and a building conductance of 115 W/K. Substituting,

$$H_v \times 20 \times 0.5 = 115 \times (T_i - T_o) \times 24$$

and rearranging for T_o

$$T_i - T_o = \frac{H_v \times 20 \times 0.5}{115 \times 24}$$

$$T_o = T_i - \frac{H_v \times 20 \times 0.5}{115 \times 24}$$

In Denver in February H_v = 3897.53 Wh/m²day on a south-facing vertical surface, $\bar{T}_o$ = 0.5°C, RH = 55.5 % and taking T_i = 18.4°C (for T_c = 20.9°C ~ T_o = 10°C the lowest limit of thermal neutrality) with η = 0.5, the lowest T_o that the passive solar heating can provide comfort for

$$T_o = 18.4 - 0.0036 \times 3897.53 = 4.3°C$$

Figure 3.8 delineates the comfort zone for passive solar heating.

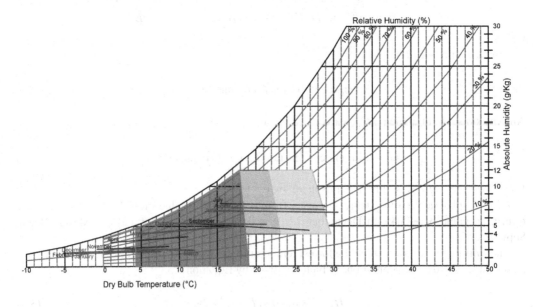

FIGURE 3.8   Comfort zone for passive solar heating, Denver, Colorado.

3.2.3 PASSIVE THERMAL MASS

The passive thermal mass is the appropriate passive solar strategy for the hot-dry climate, characterized by a large diurnal temperature, and the mean outdoor temperature of the day is within the comfort zone. This technique involves the use of high thermal mass materials, e.g., brick, stone, and concrete, within the building, both in the external envelope and internally. This has a capacitive insulation effect, which tends to attenuate and delay heat transmission through a wall or roof and even out internal both diurnal and seasonal internal temperature fluctuations. In other words, it would ensure the indoor temperature is practically constant and at about the level of mean outdoor temperature of the 24 hours.

The passive thermal mass can be coupled with night flush ventilation where the mean outdoor temperature of the day is higher than the comfort limit, and night external air temperatures are relatively cold. Figure 3.9 illustrates the principle of night flush ventilation, and a detailed discussion is in Section 3.3.3.

Effectiveness of 0.5 may be assumed for this strategy; hence, the mean range of temperature for the hottest month is determined ($T_{max} - T_{min}$), and 0.5 times the mean range will be added to

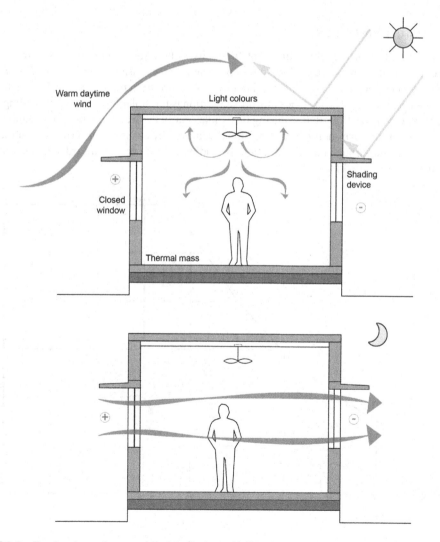

FIGURE 3.9 Passive thermal mass with night flush ventilation. Line work by Ar Vaibhav Ahuja.

the upper comfort limit. The upper humidity is the 14 g/kg line, truncated by the relative humidity curve corresponding to the lower comfort temperature.

In a cold climate characterized by air temperatures below comfort, the mass effect can be supplemented with passive solar heating to improve the indoor conditions.

In Phoenix in the hottest month, July,

$T_{o.max} = 41.4°C$, $T_{c.min} = 29.9°C$, $\overline{T_o} = 35.7°C$,
$T_c = 28.2°C$, hence upper comfort limit = 30.7°C
Amplitude = $(41.4–29.9) \times 0.5 = 5.75$ K
Limit of the extended comfort zone = $30.7 + 5.75 = 36.45°C$
Similarly in the coldest month, December,

$T_{o.max} = 17.2°C$, $T_{c.min} = 7.1°C$, $\overline{T_o} = 12.2°C$,
$T_c = 21.6°C$, hence lower comfort limit = 19.1°C
Amplitude = $(17.2 – 7.1) \times 0.5 = 5.05$ K
Limit of the extended comfort zone = $19.1 – 5.05 = 14.05°C$

All these temperatures are taken at the 50% RH curve, and the corresponding SET lines are the boundaries of the comfort zone for mass effect (Figure 3.10).

3.2.4 COMFORT VENTILATION

The term 'ventilation' serves three different functions in building: supply of fresh air (to fulfill oxygen requirement), convective cooling to remove heat from the inside of a building ($T_o < T_i$ using stack effect) and physiological cooling by evaporating moisture from the surface of the skin to dissipate heat (using cross ventilation). The comfort ventilation is physiological cooling, providing direct human comfort by natural ventilation during the whole year in the warm-humid climate and certain periods of the year in warm-humid seasons. This is achieved by ensuring air movement at the body level of the occupant through cross ventilation (providing inlet on the windward side and outlet on the leeward side) (Figure 3.11).

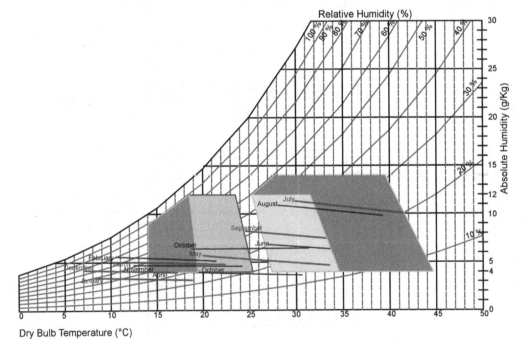

FIGURE 3.10   Comfort zone for passive thermal mass, Phoenix.

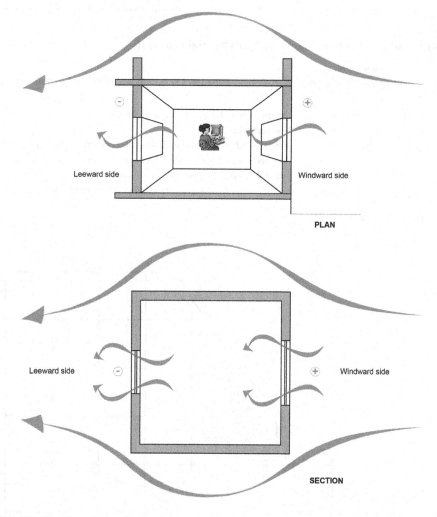

FIGURE 3.11 Cross ventilation for comfort, line work by Ar Vaibhav Ahuja.

Climatic data tables provide wind rose diagrams to find the prevailing wind direction, and openings may be provided in this direction to ensure natural air movement. However, comfort ventilation can rarely rely on passive means because, in most climates, winds are not sufficient to create the necessary indoor air velocities. Mechanical means electric fans are usually required to supplement the wind.

Air velocities and equivalent temperature reduction can be taken from Table 3.1 to define the comfort zone for comfort ventilation. Szokolay (2008) recommends that air velocity of 1 and 1.5 m/s will offset the increase in temperature by up to 3.8 and 5.1 K, respectively. However, ASHRAE (2009, p. 9.12) suggests the high airspeed may be used to offset an increase in temperature by up to 3 K above the warm-temperature boundary. The upper boundary is 90% RH.

The effect of comfort ventilation can be defined by adding the equivalent temperature reduction values to the upper comfort limit along the 50% RH curve. Above that, the boundary will be the corresponding SET line, but below 50% there is a cooling effect even without air movement, as the air is dry, so the additional effect of the air movement is taken as only half of the above: the boundary line will be nearer to the vertical.

In Honolulu, the hottest month is August, with $\overline{T}_o = 31.1°C$; hence $T_c = 26.3°C$ and upper comfort limit = 28.8°C.

The limit of the comfort ventilation will hence be = 28.8 + 3 = 31.8°C, as illustrated in Figure 3.12.

TABLE 3.1

Air Velocities and Subjective Reactions and Thermal Comfort

Air Velocity	Equivalent Temperature Reduction	Subjective Reactions and Effect on Comfort
m/s	K	
0.25	1.3	Design velocity for outlets that are near occupants
0.4	1.9	Noticeable and pleasant
0.8	2.8	Very noticeable but acceptable in certain high activity areas if the air is warm
1	3.3	Awareness, the upper limit for air-conditioned spaces, good air velocity for natural ventilation in hot and dry climates
2.0	3.9	Draughty, Good air velocity for ventilation in hot and humid climates

Source: Lechner (2009, table 10.8, p.281)

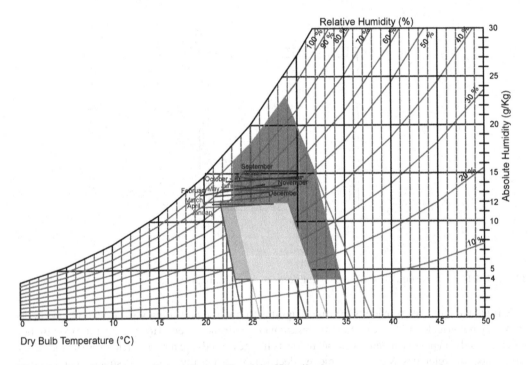

FIGURE 3.12   Comfort zone for ventilation, Honolulu, Hawaii.

3.2.5 EVAPORATIVE COOLING

The evaporative cooling is much less energy-intensive than conventional cooling and, therefore, can be considered as part of a passive system in the hot, dry climate. When water evaporation occurs, it draws a large amount of sensible heat from its surroundings and converts this sensible heat into latent heat in the form of water vapor at a constant wet-bulb temperature. As a result, the dry-bulb temperature can be reduced by about 70–80% of the wet-bulb depression (Givoni 1991), which is defined as the difference between dry-bulb temperature (DBT) and the wet-bulb temperature (WBT). This principle is used for cooling buildings in two very different ways (Figure 3.13):

  i) Direct evaporative cooling
  ii) Indirect evaporative cooling

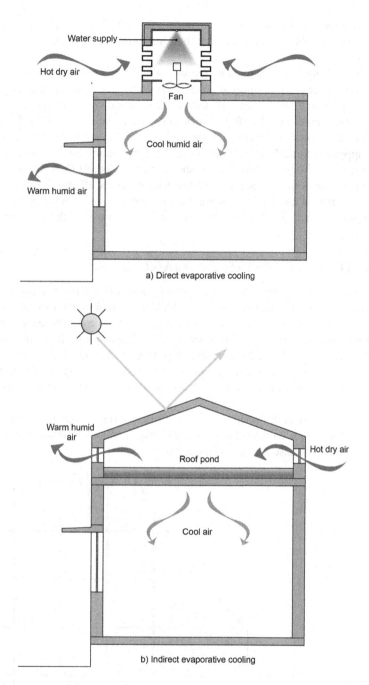

FIGURE 3.13 Evaporative cooling, line work by Ar Vaibhav Ahuja.

In direct evaporative cooling, the water evaporates from the ventilation air within an enclosed space, and, consequently, the dry-bulb temperature of the air may decrease, but the humidity will increase. The total heat content of the system does not change; i.e., it is said to be adiabatic. Conventional desert cooler, misting, and other systems work on this principle.

In indirect evaporative cooling, the building is cooled by evaporation without adding moisture, or the moist air is removed by ventilation. In indirect evaporative cooling, secondary air removes heat from primary air using a heat exchanger. Indirect evaporative coolers are now commercially available as package units. A simple example to explain the principle of evaporative cooling can be a roof pool or a spray over the roof or some other building surface, which then becomes a heat sink to cool the interior.

The evaporative cooling can be defined by the wet-bulb temperature line tangential to the upper right, and lower-left corner of the comfort zone on the psychrometric chart and a vertical line at the $T_c + 0.8 \times (T_c - T_w)$, where T_w is the wet-bulb temperature. In indirect evaporative cooling, a slight increase in humidity tolerance (to 14 g/kg) can be accepted if the air is cooled; hence, the upper boundary of the comfort zone is a horizontal line at this level, but not beyond the RH curve corresponding to the upper-left corner of the comfort zone. The temperature limit would be at $T_c + 15$, and the high humidity, the high-temperature corner, should be rounded off (Wooldridge Pescod 1976). The evaporation potential of the humid air primarily determines the effectiveness of this strategy.

In Phoenix, in the hottest month, July, the wet-bulb temperature at the lower-left corner is 27.4°C and the upper-right corner is 31.5°C, and the vertical line is drawn at the dry-bulb temperature 39.0°C (Figure 3.14).

3.3   HYBRID (LOW ENERGY) DESIGN STRATEGIES

The term 'passive cooling' generally denotes the dissipation of heat from buildings by the natural process of radiation, convection, and evaporation, which do not require the expenditure of any non-renewable energy. In many cases, evaporation and convection can be significantly enhanced by the use of motor-driven fans or pumps, which consume a small amount of electrical energy, and the word 'hybrid' has been adopted to characterize such processes (Yellott 1982).

Considering the viability of low energy systems to provide comfort should be the next step toward achieving sustainable design. Once this is assessed, any residual cooling loads could be met using active strategies.

There are many non-refrigerant-based alternative comfort systems such as radiant cooling, evaporative cooling, and structure cooling that find application in various climate zones as illustrated with contemporary building case studies in Chapter 6.

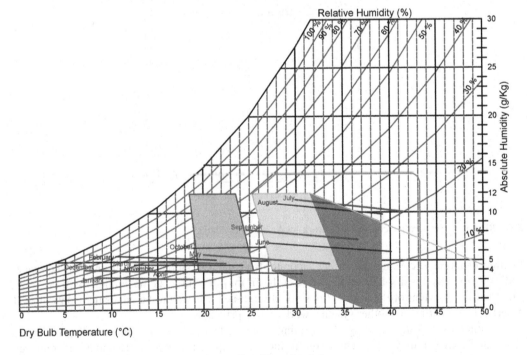

FIGURE 3.14   Comfort zone of evaporative cooling, Phoenix, Arizona.

3.3.1 Earth-Sheltered Design

The moderate and stable temperatures which always prevail in subsurface earth have been used for millennia by the people who live in caves and other subterranean spaces. The high thermal capacity of earth dampens the oscillation of ambient temperature and results in a relatively constant temperature of the earth at a certain depth while being affected moderately by seasonal changes. Relatively stable temperatures of the earth are a useful heat source, sink, thermal storage medium, and wind protection.

The earth-sheltered design employs the earth as thermal energy storage media to reduce building envelope heat loss in winter and heat gain in summer, thus to dampen daily and seasonal peak energy demands of buildings. The earth-sheltered design applications can be classified into two primary systems:

i) The direct system, which conditions the indoor environment by conduction
ii) The indirect system, which conditions the indoor environment by convection

In the direct system, the building envelope is integrated with the earth, and conduction through the building components (primarily walls and floor) provides direct passive cooling in hot-dry regions with mild winters. The direct earth-sheltered design can be accomplished in three ways (Watson 1997):

i) Recess structure below grade or raise existing grade for earth-berming.
ii) Use slab-on-grade construction for ground temperature heat exchange.
iii) Use earth-covered or sod roofs.

High temperatures and low ground moisture in the summer (moist soil conditions are preferable for best conductivity) must be addressed for optimal benefits from earth sheltering. The soil can be shaded from the scorching sun with a layer of 100 mm (4 inches) thick mulch, while enabling cooling by water evaporation from the soil surface. The cooling performance of a roof with moist soil shaded by 10 cm (4 inches) of pebbles in a full-scale room in an existing building in Riyadh, Saudi Arabia, was 4–5 K cooler than a 'control' room without any earth cooling (Givoni 1998).

The Cesar Chavez Regional Library, Laveen, Phoenix, Arizona, exemplifies the principles of earth-sheltered design by earth-berming (Zone 2B Hot Dry, COTE 2008) (Figure 3.15).

FIGURE 3.15 Built up earth berms provide thermal mass along the curved masonry wall, view from the southeast, Cesar Chavez Regional Library, Laveen, Phoenix.Photo Credit: Bill Timmerman

The indirect system conditions indoor spaces of an insulated building by ventilating air through an array of prefabricated metal, PVC, or concrete pipes with a diameter of 100 mm (4 inches) embedded in the earth. The indirect system can be used for winter pre-heating the air when the outdoor air temperature is lower than that of the earth, but auxiliary heating systems may be required. Conversely, for summer pre-cooling the air when the outdoor air temperature is higher than that of the earth, but auxiliary cooling systems may be required, thus resulting in significant energy savings in space conditioning. Circulating the indoor air through air tubes embedded in the cool soil can keep the indoor temperature about 10 K below the outdoor average maximum air temperature (Givoni 1994).

The indirect system is also found applicable in a wide range of climates with substantial temperature differences between summers and winters, as well as between days and nights. The indirect system is a low energy active system since it requires electrical fans to provide the airflow driving forces. Figure 3.16 gives a schematic overview of an indirect system. High humidity levels may create mood problems in direct and indirect earth-sheltered systems. Proper ventilation and dehumidification can reduce these problems in indirect systems.

3.3.2 SOLAR CHIMNEY

The solar chimney is a kind of organized natural ventilation technology because it can effectively guide natural ventilation for improving the indoor thermal environment. The method of using solar radiation to enhance natural ventilation appeared in the 16th century in Italy. It is known as Scirocco rooms (Cristofalo et al. 1989).

The solar chimney consists of a glass plate, a dark-colored heat-absorbing plate, a layer of thermal-protective material, metal support, a blind metal flange, air inlets, and air outlets (Figure 3.17). The ventilating duct is between the glass plate and the heat-absorbing plate. These chimneys are generally placed on a wall facing equator, i.e., the south wall in the northern hemisphere and north wall in the southern hemisphere.

When the solar radiation falls on the side of the chimney, the column of air inside the chimney is heated. If the top exterior vents of the chimney are closed, the heated air is forced back into the living space. This provides a type of convective air heating. As the air cools in the room, it is pulled back into the solar chimney, heating once again. Conversely, if the vent at the top of the chimney is kept open, the heated air is pulled up and out of the chimney, pulling new air in from the outside and creating a sort of 'draft' that provides cool, fresh air into the building. Thus a solar chimney can operate as a passive solar heating and cooling system.

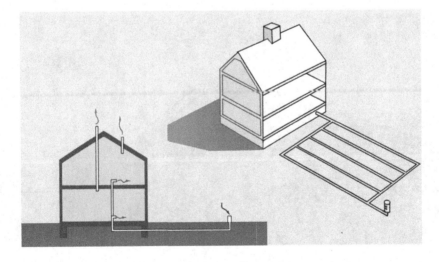

FIGURE 3.16   Indirect earth-sheltered system, line work by Ar Vaibhav Ahuja.

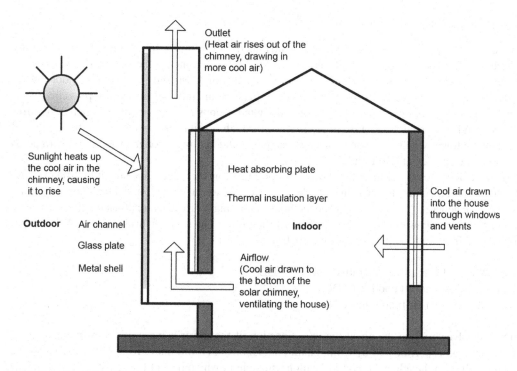

Outlet
(Heat air rises out of the
chimney, drawing in
more cool air)

Sunlight heats up
the cool air in the
chimney, causing
it to rise

Heat absorbing plate

Thermal insulation layer

Cool air drawn
into the house
through windows
and vents

Outdoor   Air channel   **Indoor**

Glass plate

Metal shell   Airflow
                                           (Cool air drawn to
                                           the bottom of the
                                           solar chimney,
                                           ventilating the house)

FIGURE 3.17   Solar chimney, line work by Ar Sakshi Singhal.

3.3.3 NIGHT FLUSH COOLING

The night flush cooling is a two-stage cooling strategy for an insulated high-mass building; applicable mainly in regions with a diurnal temperature swing of more than 15 K (27°F); mainly arid regions where the daytime temperatures are between 32°C and 36°C (89.6–96.8°F), and the night temperatures are about or below 20°C (68°F) (Givoni 1998). In stage one, the building is opened up and ventilated at nighttime with high volume to airflow; thus the heat stored in the structural mass will be dissipated, and the indoor temperature is reduced down to near the outdoor minimum; bypassing the thermal resistance of the envelope. It may be achieved by natural means through windows and other openings or by mechanical means through an exhaust fan operated when $T_o <$ T_i. If the building is occupied at night, like residences, the ventilation should not be so cold as to be uncomfortable for occupants.

In stage two, the building should be closed (unventilated) during the daytime to prevent hotter outdoor air from heating the interior. Thus, the cooled mass serves as a heat sink during the daytime, provided the mass is of sufficient amount and surface area and is adequately insulated from the outdoors. It absorbs the heat transmitted and generated inside the building, by radiation and natural convection, and thus reduces the rate of indoor temperature rise. As a result, the maximum indoor temperature in such buildings can be appreciably lower than either the outdoor maximum or the maximum indoor temperature of a similar building not ventilated at night. From the climatic aspect, night flush ventilative cooling would be preferred in regions where the daytime temperature in summer is above the upper limit of the comfort zone – with an indoor airspeed of about 1.5 m/s (300 fpm). Figure 3.9 illustrates the principle of night flush cooling, and Figure 3.10 delineates the comfort zone extension for this strategy.

The Research Support Facility, National Renewable Energy Laboratory, Golden, Colorado employs night flushing to cool thermal mass of the building; see Section 6.5.

3.3.4   PASSIVE DOWNDRAFT COOLING

The concept of passive downdraft cooling is attributed to Middle Eastern traditional architecture, which featured wind towers or wind catchers or wind scoops to induce natural interior ventilation; these towers are also known as Baud–Geer in Iraq and Iran or Malqaf in Egypt (Bahadori 1985). These towers rise above the roofs of buildings to pull the air into the interior by pressure differences that result from the wind blowing over the wind scoops and the building as per Bernoulli's theorem. While partial cooling of air is achieved through sensible heat exchange with the thermal mass of the tower wall (bricks or stones) and supplemental cooling through evaporation with pools and fountains placed at tower outlets.

Passive downdraft cooling is the process of cooling air at high levels, causing it to drop by gravity due to its becoming heavier, into the living spaces at lower levels. Passive downdraft cooling systems are generally categorized by the methods of cooling air to generate downdraft: wetted pads, misting nozzles, showerheads, porous media, chilled water pipes, and hybrid systems. A taxonomy of downdraft cooling in literature is given below:

   i)   Passive downdraft evaporative cooling (PDEC)
  ii)   Active downdraft cooling (ADC)
 iii)   Hybrid downdraft cooling (HDC)

PDEC involves cooling the air at the upper portion of the tower using a direct evaporative cooling system, as experimented by Cunnigham and Thomson (1986) and Givoni (1994). Evaporative cooling relies on the adiabatic humidification principle in which part of the sensible heat of the air stream is transferred to latent heat. This is due to the transfer of energy (required to induce evaporation) in the form of heat from the air to water. As a result, air supplied is not only cooler but is also more humid. Figure 3.18 illustrates the principle of PDEC. According to various studies, the PDEC system is such that a temperature reduction of up to 80% of the difference between ambient dry-bulb temperature and wet-bulb temperature (Equation 3.6).

$$TT = DBT - 0.8\left(DBT - WBT\right) \qquad\qquad (3.6)$$

where
   TT = tower supply air temperature
   DBT = ambient dry-bulb temperature
   WBT = ambient wet-bulb temperature

Psychrometric charts can be used to test the climatic applicability of PDEC to achieve summer thermal comfort in different climate regions of the United States. PDEC in various forms has been applied in a wide variety of contemporary building types (including offices, laboratories, educational buildings, factories, and rail stations) (Ford 2012). The PDEC can be integrated with dehumidification for a hot-humid climate. The NOAA Daniel K. Inouye Regional Centre features an innovative passive downdraft cooling system in the very hot, humid climate of Honolulu; Section 6.2.

When there is no wind, and downdraft airflow relies solely on buoyancy forces, in this case, there is a need to use fans to enhance the air distribution.

Active downdraft cooling (ADC) is achieved by using chilled water coils or panels and driving air over evaporative cooling pads directly into the building.

HDC system cools the air at the upper portion of the enclosure using at least two stages of cooling: a sensible cooling stage using chilled water coils or pipes (ADC) and a direct evaporative cooling stage (PDEC). ADC and HDC are alternative to conventional air conditioning because energy is consumed via the mechanical system needed for air cooling or the fans required for the air circulation.

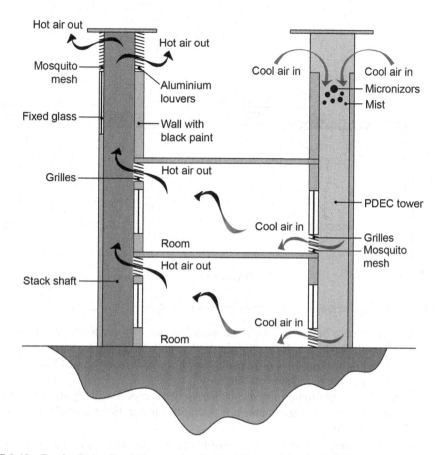

FIGURE 3.18 Passive Down Draft Evaporative Cooling, line work by Ar Vaibhav Ahuja.

3.3.5 Passive Radiant Cooling

The passive radiant cooling system is a specialized design to utilize the cooling of the exterior surfaces of the building envelope by the emission of long-wave radiation toward the sky. There are two variants of passive radiant cooling:

- Direct passive radiant cooling
- Indirect passive radiant cooling

In direct passive radiant cooling, the external surfaces of a highly conductive building envelope (e.g., dense concrete roof, brick walls) are exposed to the sky during the night but kept well insulated externally during the daytime (using retractable insulation). Such building envelope dissipates heat at night, both by long-wave radiation emissions to the visible sky and by convection to the outdoor air. The cooled mass of the building envelope then serves as a heat sink and absorb; through the internal surfaces (e.g., ceiling and walls) the heat penetrating and generated inside the building's interior during the daytime hours. During the daytime, the (installed) external insulation minimizes the heat gain from solar radiation and the hotter ambient air.

Harold Hay (1978) designed and built the 'Skytherm,' radiant cooling system combined with passive solar heating. In this system, sealed transparent plastic bags filled with water are placed above the (horizontal) metal deck roof with a movable insulation panel (Figure 3.19). In summer, the water bags are exposed and cooled during the night and are insulated during the daytime. The cooled bags are in direct thermal contact with the metal deck and thus the ceiling serves as a cooling

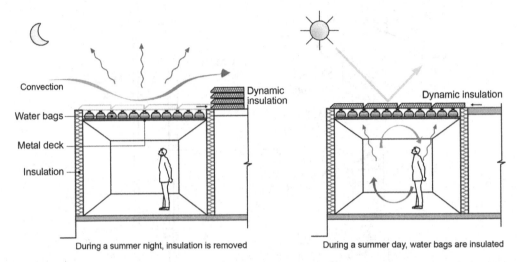

FIGURE 3.19   Direct passive radiant cooling, 'Skytherm', line work by Ar Vaibhav Ahuja.

element over the whole space below. In winter, the water bags are exposed to the sun during the day and covered by the insulation panels during the nights, thus providing passive solar heating. Several buildings utilizing various variations of this system have been built in the United States. The main challenge in this system is the retractable insulation.

Recent advances in nanophotonics and metamaterials have experimentally demonstrated daytime radiant sky cooling achieving sub-ambient temperatures under direct sunlight by the concept of 'cool roof' (Zhao et al. 2019).

The indirect passive radiant cooling design consists of a metallic layer placed over the ordinary insulated, horizontal or slightly inclined roof, with an air space of about 50–100 mm (2–4 inches) underneath it (Figure 3.20). The metallic layer attains sub-ambient temperature even before the sunset. The term 'stagnation temperature' denotes the temperature attained by the roof when the heat loss by radiation to the sky just equals the convective heat gain from the air. The radiant cooling

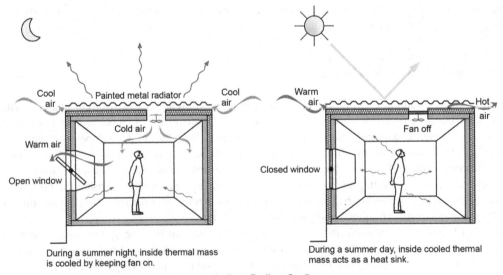

FIGURE 3.20   Indirect passive radiant cooling, line work by Ar Vaibhav Ahuja.

system functions when the radiator's stagnation temperature is lower than the ambient temperature by some minimum temperature drop – e.g., by at least 5 K (9°F) (Givoni 1998). The cool air is blown with a fan through the interior space to cool the mass of the building. The cooled mass then serves during the following day as a sink for heat penetrating and generated inside the sealed building.

3.4 THERMAL BEHAVIOR OF THE BUILT ENVIRONMENT

Design of an energy-efficient (sustainable) built environment primarily depends on:

  i) Climatic conditions outside
 ii) Comfort conditions required inside
iii) Thermal behavior of the built environment
 iv) Occupancy pattern

The thermal performance of a building is governed by a series of heat inputs and outputs (Figure 3.21). The thermal balance of building as determined by mechanisms of energy transfer, radiation, conduction, and convection can be expressed by equation 3.7.

$$\pm Q_c + Q_s \pm Q_v + Q_i - Q_e = \Delta Q \tag{3.7}$$

where

Q_c = conduction heat gain or loss
Q_s = solar heat gain
Q_v = convection heat gain or loss
Q_i = internal heat gain
Q_e = evaporation heat loss

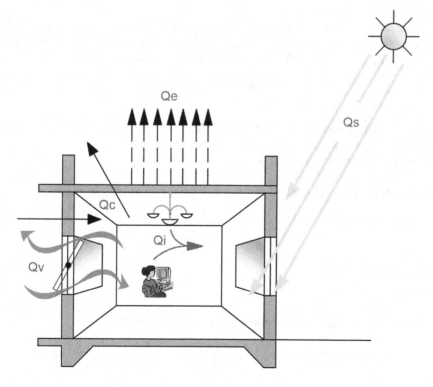

FIGURE 3.21 Heat input and output of a building.

ΔQ indicates the energy balance of the building. If ΔQ is zero, thermal balance is achieved, no cooling or heating is required. If ΔQ is negative, there is heat loss, and the inside temperature is falling below the comfort level, the building requires heating. If ΔQ is positive, there is heat gain, the inside temperature is rising above the comfort level, and the building requires cooling. A building that is effectively designed to limit heat loss and gain may cost less to build, equip, and maintain and can reduce initial and life-cycle costs.

Thermal performance of a building can be mathematically analyzed following two approaches:

i) Steady-state heat flow, the inside, and the outside conditions are assumed to be steady, non-changing.
ii) Dynamic heat flow, the inside, and the outside conditions are assumed to be unsteady, changing.

The above methods provide useful quantitative information, which helps the designers to understand the relationship between design decisions and the energy performance of a building. The first section discusses thermo-physical properties, U-value, and SHGC of building elements. The sun is the primary source of energy, and solar radiation is the most significant energy input into a building. The next section presents the concept of sol–air temperature to account for solar energy input in a building. This will be followed by an analysis of the heating and cooling requirements of a building adopting the steady-state method. Finally, dynamic methods are discussed for energy performance modeling of a building.

3.4.1 THERMO-PHYSICAL PROPERTIES

Conduction heat flow depends on the thermo-physical properties of materials. Conductivity (k) is measured as the heat flow density (W/m^2) in a 1 m thick material with unit temperature difference, in-unit of W/m K. Typical thermal properties of the conventional building and insulating materials design values for a mean temperature of 24°C is published in *Handbook of Fundamentals* by ASHRAE (2009, ch 26, Table 4).

The total thermal resistance to one-dimensional heat flow through building construction, e.g., roof or wall, is the numerical sum of the resistance (R_T) of all parts of the construction in series (Equation 3.8). In buildings, the air film and cavity resistances must be added.

$$R_T = R_{si} + R_m + R_c + R_{so} \tag{3.8}$$

where
R_{si} = the inside film or surface resistance (Table 3.2)
R_{so} = the outside film or surface resistance Table 3.2)
R_c = the air space resistance (Table 3.3)

Mathematically R_m, the total material resistance, is expressed as in Equation 3.9.

$$R_m = \frac{l_1}{k_1} + \frac{l_2}{k_2} + \frac{l_3}{k_3} \ldots \tag{3.9}$$

where
l_1 = thickness of layer (m)
k_1 = conductivity of the material (W/m K)

The U-value (air-to-air thermal transmittance) is the reciprocal of R_T,

$$U = \frac{1}{R_T} \tag{3.10}$$

Surface resistance to heat flow is a function of the combined radiant and convective components of heat transfer (Equation 3.11). The convection heat transfer coefficient is dependent on airspeed and direction of heat flow. The radiation heat transfer coefficient is dependent on the view factor and the emittance of the radiation and absorbing surfaces. In the analysis of heat transfer from air into a body such as a wall, roof, or vice versa, it is convenient to use the published standard values (Table 3.2).

$$R_{so} \text{ or } R_{si} = \frac{1}{f_r} + \frac{1}{f_c} \tag{3.11}$$

where

f_r = radiation coefficient (W/m²K)
f_c = convection coefficient (W/m²K)

For pitched roofs containing a horizontal ceiling, the R-value is measured with respect to the plane of the ceiling and has to be corrected for the roof pitch, as given in Equation 3.12.

$$R_T = R_{si} + R_{m1} + R_c \cos\beta R_{m2} + \cos\beta R_{so} \tag{3.12}$$

where

β = angle of the pitch of the roof
R_{m2} = resistance of materials in the plane of pitched part of the roof

In many constructions, components are arranged so that heat flow in parallel paths of different conductances. If no heat flow between lateral paths, heat flow in each part may be calculated using Equations (3.8) and (3.9). The average transmittance is then given by equation 3.13.

$$U_{av} = aU_a + bU_b + \ldots + nU_n \tag{3.13}$$

where, a, b … are respective fractions of a common basic area composed of several different paths with transmittance U_a, U_b, …, U_n.

TABLE 3.2
Surface Resistance (m² K/W)

Structure	Direction of Heat Flow	High Emittance $\varepsilon = 0.90$ R (m² K/W)	Low Emittance $\varepsilon = 0.05$ R (m² K/W)
	Inside		
Walls (vertical)	Horizontal	0.12	0.30
Ceilings or floors (horizontal)	Upward	0.11	0.23
	Downward	0.16	0.80
Ceilings or floors (pitched 45°)	Upward	0.11	0.24
	Downward	0.13	0.39
	Outside		
Walls or roofs (severe exposure)	Any	0.03	–
Walls or roofs (normal exposure)	Any	0.04	–

Source: Based on average values ASHRAE (2009, ch. 26, p. 26.1).

TABLE 3.3
Airspace Resistance (m² K/W)

	Direction of Heat Flow	High Emittance $\varepsilon = 0.90$ R (m²K/W)	Low Emittance $\varepsilon = 0.05$ R (m²K/W)
	Unventilated		
5 mm airspace	Any	0.11	0.18
≥ 25 mm airspace (wall)	Horizontal	0.18	0.44
≥ 15 mm airspace (roof)	Upward	0.16	0.34
	Downward	0.22	1.06
> 25 mm airspace (pitched 45°)	Upward	0.19	0.40
	Downward	0.20	0.98
Roof (attic)	Downward	0.33	1.14
	Ventilated		
Walls	Any	0.13	0.29
Roofs (attic)	Any	0.46	1.36

Source: Based on ASHRAE (2009, ch. 26, p. 26.1), ASHRAE (1997), ISO 6946 (2007).

If heat flow is two-dimensional, three methods, the parallel path method, isothermal-planes method, and zone or modified zone methods are available to compute thermal transmittance through building components, ASHRAE (2009, ch. 27, pp. 27.3–8).

An external wall consisting of 88.9 mm (3.5 inches) wood studs (20% of surface area) and insulation (80% of surface area) with 25.4 mm (1 inch) plyboard on the outside and 12.7 mm (0.5 inches) gypsum board on the inside. The outdoors and indoors are considered black bodies at 0°C and 20°C, respectively. The surface resistance to heat flow calculation includes combined effect convection and radiation heat transfer (Equation 3.10), and it is assumed that there is no lateral heat transfer. In this composite construction, thermal resistances of the insulation section and wood stud section of the wall are calculated separately (Equation 3.8); then the average U-value is calculated (Equation 3.12) (Table 3.4).

The incident solar radiation falling on a glazed surface is partly transmitted, partly reflected, and the remainder is absorbed within the body of the glass. The corresponding properties of glass are transmittance (t), reflectance (r), and absorption (a). The SHGC is given by Equation 3.14.

$$SHGC = \frac{\text{Solar heat transmitted}}{\text{Solar irradiance on the window surface}} \tag{3.14}$$

Thus, the window energy performance is measured in terms of U-value, SHGC, and VLT (Figure 3.22).

3.4.2 Sol-Sir Temperature (T_{sa})

Sol–air temperature, first introduced by Mackey and Wright (1944) and later modified by several researchers, combines the effect of solar radiation, ambient air temperature, and long-wave radiant heat exchange with the environment. A precise definition of sol–air temperature is

the equivalent outdoor temperature that will cause the same rate of heat flow at the surface and the same temperature distribution through the material, as results from the outdoor air temperature and the net radiation exchange between the surface and its environment. (Rao, Ballantyne 1970)

TABLE 3.4
Example Calculation of *U*-value of Wall Construction

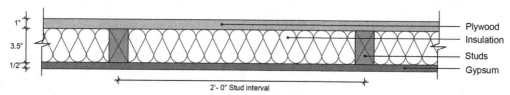

Element	Thickness (m)	Thermal Conductivity (w/m K)		Thermal Resistance (m² K/W)
Outside surface convection coefficient ($1/f_{cout}$)	–	–	1/20	
Outside surface radiative coefficient ($1/f_{rout}$)) = $1/4\sigma T^3$	–	–	$1/(4*5.67e{-}8*(273)^3) = 1/4.6$	
Outside surface resistance ($1/f_{cout} + 1/f_{rout}$)	–	–	$1/(20 + 4.6)$	0.04065
1 Plywood	0.0254	0.150	0.0254/0.15	0.16933
2 Insulation	0.09	0.035	0.09/0.035	2.57142
3 Gypsum board	.0127	0.200	0.0127/0.2	0.0635
1 Plywood	0.0254	0.150	0.0254/0.15	0.16933
2′ Wood	0.09	0.150	0.09/0.15	0.6000
3 Gypsum board	.0127	0.200	0.0127/0.2	0.0635
Inside surface convection coefficient ($1/f_{cin}$)	–	–	1/5	
Inside surface radiative coefficient ($1/f_{rin}$) = $1/4\sigma T^3$	–	–	$1/(4*5.67e{-}8*(293)^3) = 1/5.7$	
Inside surface resistance ($1/f_{cin} + 1/f_{rin}$)	–	–	$1/(5 + 5.7)$	0.09346
m²K/W				2.93836
m²K/W				0.96694
W/m² K			$(1/2.93836)* 0.8 +$ $(1/0.96694)*0.2$	$0.27226 +$ $0.20684 =$ 0.47909

Consider the situation of external surfaces of a building exposed to air temperature and solar and other radiation exchanges. The outdoor air temperature is T_o, the global radiation per unit area at short wavelength (solar) H_h, and the net radiation per unit area at long wavelength is $\in. \Delta I_l$. The heat from outdoors flows into the surface by means of radiation, convection, and conduction through the still air adhering to the surface, which can be represented by air film conductance f_o (reciprocal of resistance R_{os}). The heat flow per unit area into the building element can be represented as Q_{os} (Figure 3.23).

A sol–air equivalent temperature is that that will lead to the same surface temperature and heat flow; these latter are connected by the conditions within the surface, and matching surface temperature on a continuous basis will lead to the same heat flow in the two systems. The heat flow in Figure 3.23 (a) is,

$$Q_{os} = \alpha I_g + f_o \left(T_o - T_{os} \right) - \in \Delta I_l \qquad (3.15)$$

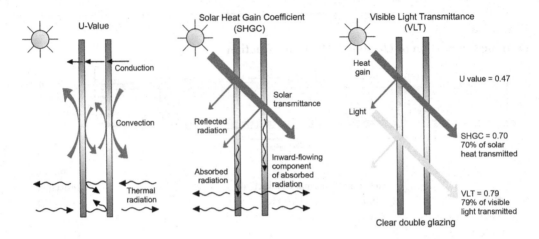

FIGURE 3.22 Window energy performance: *U*-value, SHGC and VLT, line work by Ar Vaibhav Ahuja.

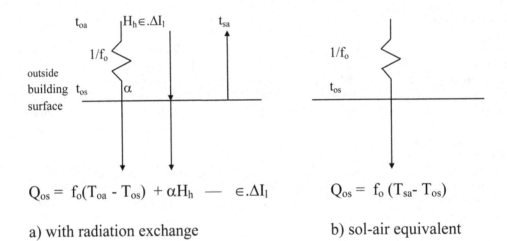

$$Q_{os} = f_o(T_{oa} - T_{os}) + \alpha H_h \quad — \quad \in.\Delta I_l \qquad\qquad Q_{os} = f_o (T_{sa} - T_{os})$$

a) with radiation exchange   b) sol-air equivalent

FIGURE 3.23 Energy balance at the exterior surface of outside building surface.

and in Figure 3.23 (b) is

$$Q_{os} = f_o \left(T_{sa} - T_{os} \right) \tag{3.16}$$

The resulting sol–air temperature is,

$$T_{sa} = T_o + \frac{1}{f_o} \left(H_h \alpha - \in \Delta I_l \right) \tag{3.17}$$

where
T_{os} = outside surface temperature (°C)
T_o = outdoor air temperature (°C)
H_h = total solar radiation (direct + diffuse) incident on the surface (W/m²)
α = absorptance of the material for short-wave or solar radiation (Table 3.5)
f_o = Outside film or surface conductance (W/m² K)

TABLE 3.5
Absorbance (α) and Emittance ($\in$)
of Surfaces (Average Value)

Surfaces	α	$\in$
Clean surfaces:	0.25	0.95
Brick	0.40	0.90
White glazed	0.80	0.90
Light colors		
Dark colors		
Roofs	0.90	0.96
Asphalt	0.65	0.85
Red tiles	0.40	0.50
White tiles	0.20	0.11
Aluminum (oxidized)		
Paint	0.30	0.95
White	0.96	0.96
Matt black		
Dirty building surfaces:	0.50	0.60
Light	0.80	0.85
Medium	0.90	0.95
Dark		

$\in$ = emittance of surface, for low-temperature radiation usually taken as 0.9 for most building surfaces, but only 0.05 to 0.2 for low-emittance surfaces such as polished metals (e.g., 0.2 for dull aluminum and 0.05 for polished aluminum).

ΔI_l = net long-wave radiation exchange between a black body at outside air temperature and the outside environment (W/m^2).

The term $(T_{sa} - T_o)$ is often referred to as 'sol–air excess' temperature, especially in British literature. For vertical walls, the long-wave exchange can be neglected, and the sol–air computation can be simplified,

$$T_{sa} = T_o + \frac{1}{f_o} I_g \alpha$$

Although there is abundant literature on the precise computation of the term ΔI_l, the degree of accuracy depends on the availability of data on sky cover and other atmospheric factors. Further, the associated computational efforts and complexity make the more sophisticated methods of limited use for practical purposes. However, a reasonable approximation was proposed by Loudon (1968) at BRS. For a horizontal surface, ΔI_l may be taken as 95 W/m^2 for a cloudless sky, 15 for an overcast sky, and intermediate values proportional to the cloud amount for partially clouded skies. Therefore,

$$\in .\Delta I_l = 0.9 \times (95 - 15) * m/8 \tag{3.18}$$

For horizontal surfaces that receive long-wave radiation from the sky only, an appropriate value of is about 63 W/m^2, so that if $\in$ = 1 and f_o = 17 W/(m^{2}K), the long-wave correction term is about 4 K (Bliss 1961). Since vertical surface receives long-wave radiation from the ground and surrounding buildings as well as from the sky, accurate ΔI_l values are difficult to determine. When solar radiation intensity is high, surfaces of terrestrial objects usually have a higher temperature than the outdoor air; thus, their long-wave radiation compensates to some extent for the sky's low emittance. Therefore, it is common practice to assume $\Delta I_l = 0$ for vertical surfaces (ASHRAE 2009, ch. 18, p. 18.23).

In the climate data tables (7.52 to 7.101 available on the publishers' web site) hourly values of solar irradiation on a horizontal plane are given for a typical summer and winter day (Wh/m²) for a period of one-hour beginning and ending at the hours indicated. The irradiance in W/m², at mid of the hours indicated, will be numerically the same.) The direct (beam-) and diffuse components are shown separately. The winter day data may be used for solar (primarily passive) heating design, the summer day data for the assessment of solar gain as a load for a cooling system.

Select the 'design time' irradiance is at its maximum (on the horizontal surface) at noon, but the temperature maximum usually occurs two to three hours later. Note the worst condition hour, say between 14:00 and 15:00 h, and the values of H_b (direct or beam component) and H_d (diffuse component) at this time from the climate tables.

Find the total irradiance at the selected time for all building surfaces.

$$H_h = H_b + H_d \tag{3.19}$$

For any vertical surface
INC = angle of incidence of beam radiation
ALT = solar altitude angle above the horizontal
ρ = reflectance of foreground, normally taken as 0.2

The solar position angles, solar altitude (ALT) and solar azimuth (AZI 0 to 360°), can be read from sun-path diagrams given in Chapter 7, or can be calculated as follows:
find solar hour angle: HRA = 15 × [HR-12] where HR = hour (0–24) then

$$ALT = \sin^{-1}\left[\sin LAT \times \sin DEC + \cos LAT \times \cos DEC \times \cos HRA\right] \tag{3.20}$$

$$AZI = \cos^{-1}\left[\frac{\cos LAT \times \sin DEC - \sin LAT \times \cos DEC \times \cos HRA}{\cos ALT}\right] \tag{3.21}$$

The above equation gives the result of AZI in the range 0–180°, i.e., for forenoon only, for afternoon AZI = 360° – AZI (as computed from the above equation).

The horizontal shadow angle is the azimuth difference between the sun and the building orientation (of the wall considered):

HSA = AZI –ORI (maximum 90°, beyond this the sun is behind)
if 90°<|HSA| < 270° then the sun is behind the vertical face of the building
if HSA > 270 then HSA = HSA –360°
if HAS ≤ 270° then HAS = HAS +360°

Furthermore, the angle of incidence for the vertical surface will be:

$$INC = arc\cos[\cos HSA \times \cos ALT] \tag{3.22}$$

The climate tables give standard clock time, and the time used for calculation of the hour angle in the Equations 3.20 and 3.21 is apparent local time (solar time). This can be obtained from the standard time observed on a clock by applying two corrections using Equation 2.11.

3.4.3 Space Heating Requirements

The design and selection of space heating systems aim to establish three quantities:

i) The size (capacity) of the system
ii) The annual (seasonal) or monthly heating requirement
iii) The energy requirements and the predicted fuel consumption

Space heating system sizing can be calculated using the steady-state method (CIBSE 1999) based on the simple model of the building heat loss under assumed design conditions, and the heating capacity will have to match that heat loss; the steps for calculation are:

i) Read the recommended 'outdoor design condition,' winter DBT (T_o), from the climatic data tables in Chapter 7.
ii) Establish 'indoor design temperature' (T_i) by comfort requirements to the lower limit of thermal neutrality for the coldest month given in the climatic data table so that the heating equipment will not be oversized.
iii) Calculate the 'specific heat loss rate' of the building (q) in W/K.

Conduction:

$q_c = \sum (A \times U)$ i.e., the sum of the products of the area and U-value of all enclosing elements (m^2.W/m^2K = W/K)

Lower U-value (i.e., higher insulation) would reduce the heat loss rate; thus both the installation and running costs of heating would be reduced. There are several cost–benefit analysis studies that have been carried out to optimize insulation versus heating expenditure.

Convection:

$q_v = 0.33 \times V \times N$, i.e., the volume of building times number of air changes per hour, times the volumetric specific heat of the air.

$$\left(Wh/m^3K \cdot m^3/h = W/K \right)$$

$$q = q_c + q_v$$

iv) Find the 'near-worst-case conditions', the difference between outdoor and indoor design temperatures to make sure that the system will cope with such conditions. Under less severe conditions, the system can be operated at partial capacity.

$$\Delta T = T_i - T_o \left(°C - °C = K \right)$$

and the required heating capacity will be

$$Q = q.\Delta T \left(W/K.K = W \right)$$

Heating load calculations ignore solar and internal gains, providing a built-in safety factor (ASHRAE 2009).

Take the example of a simple house (ignoring internal walls) (Table 3.6); 10×5 m on plan and 3 m high, with 40% windows/doors, vertical surfaces: $2 (10 + 5) \times 3 = 90$ m^2

TABLE 3.6

Construction, Occupancy Details and Winter Heat Loss Calculation of Example House in Denver, Colorado

S. No.	Building Element	Area (A) m²	U-Value W/m²K	Heat Loss (A × U) W/K
1.	Walls, cavity brick	90 × 0.6	1.7	54.0 × 1.7 = 91.8
2.	Windows, single glazed	90 × 0.4	5.4	36.0 × 5.4 = 194.4
3.	Floor, concrete on ground	10 × 5	0.62	50 × 0.62 = 31.0
4.	Roof, tiled, plaster ceiling	10 × 5	1.5	50 × 1.5 = 75
				q_c =3 92.2
5.	Ventilation: 3 air changes/h	Volume 50 × 3 = 150 m³		q_v = 0.33 × 150 × 3 = 148.5
6.	$Q = q_c + q_v$			540.7

In Denver the coldest month is January (Table 7.16)

$T_o = 1.9°C$
$T_i = 18.4°C$ (lower limit of thermal neutrality)
$\Delta T = 16.5$ K
$Q = 540.7 \times 16.5 = 8921.55$ W = 8.92 kW

which is the required heat output rate for the heating system.

Solar and internal gains need to be adjusted by the 'balance point temperature,' at which the heat loss rate (q) equals the internal (Q_i) plus solar (Q_s) heat gains. The term Q (= $q.\Delta T$) is plotted against outdoor temperature, as in the graph below (Figure 3.24).

Q_i is the heat output rate of human bodies, the total wattage of lighting (if used) plus any other heat-producing equipment.

Q_s can be evaluated as in Section 3.5.3.

The value of $Q_i + Q_s$ is plotted as a constant (a horizontal line), and where it intersects the $q.\Delta T$ line, the balance point temperature is determined.

The annual (or seasonal) heating requirements (D_h) can be accurately estimated as heating degree-hours (HDH) as summation can be carried out on an hourly basis above the balance point temperature (T_b).

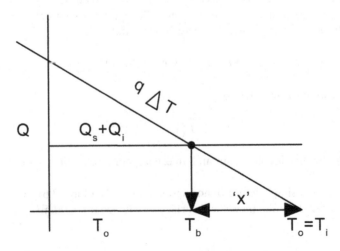

FIGURE 3.24   Balance-point temperature.

$$HDH = \sum (T_b - T_h)^+ \text{(from hour 1 to 8760)}$$

T_h = hourly temperature

Alternatively, HDH can be estimated from heating degree days (HDD) on a set base temperature of 18.3°C given in Chapter 7 as (HDH) = (HDD) × 24

$$D_h = q \times (\text{HDH}) \qquad (W/K \times Kh = Wh)$$

The result will be valid for continuous heating and should be adjusted by the factors for the duration of occupancy, building thermal mass, and system response.

3.4.4 SPACE COOLING REQUIREMENTS

The space cooling requirements can be calculated using the steady-state method (CIBSE 1999) at the 'design time' based on maximum use and worst-case climate, following the steps:

i) Read the climatic data tables for the recommended outdoor design condition, summer DBT (T_o).

ii) Establish indoor design temperature (T_i) the 'thermal neutrality' given in the climatic data table is valid for summer and can serve as a guide.

iii) Calculate conduction heat gains (Q_c in watts) due to transmission through the building envelope by summing the mean gains through the external opaque and glazed surfaces.

$$Q_c = \sum A_n U_n (T_{sa} - T_i)_n \tag{3.23}$$

where

A = surface area for each n element, in m²

U = U-value for each n element, in W/m²K

T_i = indoor air temperature (°C), taken as the summer neutrality

T_{sa} = sol–air temperature (°C), outdoor air temperature (T_o) taken as the summer design temperature given in the climate data and solar radiation different for each orientation element is facing

Note that, for glazing, T_o is used in equation 3.23 rather than T_{sa}, because the effect of solar radiation is included in the solar heat gain; see below.

Calculate solar heat gains (Q_s in watts) through glazed elements consist of solar radiation which is incident on the glazing and transmitted to the inside.

$$Q_s = \sum A_g H_n SHGC \tag{3.24}$$

where

A_g = area for each n glazing, in m²

H_n = the global solar irradiance for each n elevation, in W/m²

$SHGC$ = solar heat gain coefficient for each n glass

For the case of external shading devices, effective SHGC should be taken based on the procedure given in ASHRAE 90.1 (2010).

v) Calculate the convection heat gain (Q_v in watts) consists of the rate of heat flow between the interior of a building and the outdoor air, depends on the rate of ventilation, i.e., air changes. Ventilation rates must include air infiltration, natural ventilation due to open windows, and, where appropriate mechanical ventilation. The rate of ventilation can be given in m³/s. The recommended outdoor air supply rate for the sedentary occupant is 8 L/s/person (CIBSE 1999).

$$Q_v = \frac{c_p \rho N_v V}{3600}\left(T_o - T_i\right)$$. (3.25)

c_p = specific heat capacity of air (J/kg K)
ρ = density of air (kg/m³)
N = number of air changes per hour, Table 3.7
V = volume of space

For practical purposes $(c_p \rho/3600) = \frac{1}{3}$; therefore

$$Q_v = 0.33 NV\left(T_o - T_i\right)$$. (3.26)

Calculate the internal heat gain (Q_i in watts) from internal sources such as the heat output of occupants (Table 3.8), lighting, computers, and appliances is calculated by multiplying each load by its duration, summing over all sources and averaging the total over 24 hours. Hence,

$$Q_i = \frac{\sum q_{in}\theta_{in}}{24}$$ (3.27)

where
q_{in} = the instantaneous heat gain from internal heat source n (W)
θ_{in} = duration of internal heat source n (h)

vii) Calculate evaporation heat loss (Q_e, in watts) if it takes place on the surface of the building from, say, a roof pond, fountain or within the building from, say, human sweat, water in an aquarium, washing) and the vapors are removed; this can be computed as

$$Q_e = E_r L$$ (3.28)

TABLE 3.7

Recommended Values for Air Changes

S. No.	Space	Air Changes per Hour
1.	Banks, offices	4–8
2.	Bathrooms	6–10
3.	Bedrooms	2–4
4.	Cafes/restaurants	10–12
5.	Cinemas/theaters	10–15
6.	Classrooms	6–9
7.	Conference rooms	8–12
8.	Factories and workshops	8–10
9.	Hospital-wards	6–8
10.	Kitchens	6–9
11.	Laboratories	6–15
12.	Lecture theaters	5–8
13.	Libraries	3–5
14.	Living rooms	3–6
15.	Refectory	8–12

TABLE 3.8

Heat Emission (W) per Person for a Mixture of Men, Women, and Children in Different States of Activity

S. No.	Activity	Typical Application	Total Heat Emission (W)
1.	Seated at theater	Theater, matinee	95
2.	Seated at theater, night	Theater, night	105
3.	Seated, very light work	Offices, hotels, apartments	115
4.	Moderately active office work	Offices, hotels, apartments	130
5.	Standing, light work walking	Department store; retail store	130
6.	Walking, standing	Drug store, bank	145
7.	Sedentary work	Restaurant	160
8.	Light benchwork	Factory	220
9.	Moderate dancing	Dance hall	250
10.	Walking 4.8 km/h; light machine work	Factory	295
11.	Bowling	Bowling alley	425
12.	Heavy work	Factory	425
13.	Heavy machine work; lifting	Factory	470
14.	Athletics	Gymnasium	525

where

E_r = the rate of evaporation (kg/h)

L = the latent heat of evaporation (J/kg)

The rate of evaporation depends on many parameters, such as available moisture, the humidity of the air, temperature of the moisture itself, and the air and velocity of the air movement. It can be estimated from the number of people in the room, their activity, and thus their likely sweat rate (BS 5250 suggest a typical daily moisture production rate of 7 kg for a five-person family but clothes washing use of moisture-producing room heaters can increase this to 20 kg). Usually, evaporation heat loss is either ignored for calculations (except in mechanical installations), or it is handled qualitatively only: evaporative cooling will be utilized to reduce air temperature 'as far as possible.'

The space cooling requirement (Q), in watts, will be

$$Q = Q_c + Q_s + Q_v + Q_i - Q_e$$

Take an example of a simple office (ignoring internal walls), 5 × 5 m on the plan, and 3 m high, with windows/doors, located in Austin, Texas, latitude 26° 17′N and longitude 73°1′E (Figure 3.25).

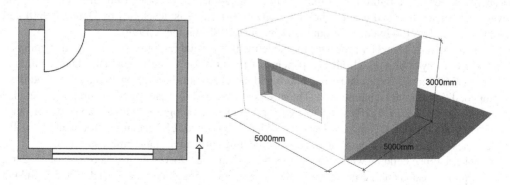

FIGURE 3.25 Plan and view of example office.

Constructional and occupancy details, along with surface areas and thermal transmittances, are given in Table 3.9. The calculation of steady-state design heat gain is based on the following assumptions:

- The window will be open during the day and closed at night.
- The thermal transmittance of the window frame is equal to that of the glass.
- There are no internal blinds; therefore, there will be solar gain.

Austin the hottest month is July.

$T_i = 26.8°C$ (Table 7.6) (thermal neutrality)

Design temperature is $T_o = 36.6°C$ (Table 7.6)

Direct solar irradiance $H_b = 656$ W/m^2, diffuse solar irradiance $H_d = 241$ W/m^2 at 12:30 h

Equation of time correction = 5 m 11 s

The local apparent time is 12.07 h for the standard time of 1230 h.

The above results obtained would be approximately valid for a building of negligible mass. For massive buildings, it would give an over-estimate. To predict the long-term effects of thermal mass, a detailed dynamic thermal model is required.

3.4.5 DYNAMIC MODELS

The energy requirements and the fuel consumption for space cooling and heating of the building must be estimated throughout the year based on non-steady-state or dynamic (heat flow) methods. It must take into account the hours of occupation of the building, the changes in external conditions throughout that time, the efficiency of the cooling and heating system, and the thermal performance of the building envelope.

There are necessarily two approaches to calculate the periodic thermal response of buildings (Milbank & Harrington-Lynn 1974), which has proponents in different parts of the world. In their more sophisticated forms, the two solutions give similar results to given design problems.

The first approach is a numerical method to solve the transient conduction equations; the response factor method and finite difference method are the most widely used in energy calculations. For use in design, it is common to feed this type of program with the same sequence of weather data for several days until the model reaches a stable condition. In other words, the hourly temperature and energy requirements are the same from day to day. The 'response factor' method is adopted by the ASHRAE (1972). The finite difference method treats storage effects by dividing the building structure into several layers and calculating the temperature or energy distributions at selected time intervals. The exterior and interior temperature variations are modeled with piecewise linear functions.

The second approach is the analytical or harmonic approach; the transient conduction equations are solved analytically in building thermal energy calculation with a periodic temperature boundary condition on one side of the slab and a constant temperature boundary condition on the other side. Therefore, this approach is called the 'harmonic' solution. This approach gives the temperature or energy pattern for a building, assuming it has reached the stable, or equilibrium, state for some given cycle of weather and usage. The 'admittance' procedure is developed by Danter (1960) and adopted by the IHVE (Guide, 1970) and later by the CIBSE (Guide A 1999).

There are several computer programs for modeling the thermal response of buildings that compute heat flow hour-by-hour through all components of the building using an annually hourly climatic database (8760 hours). These can predict hourly indoor temperatures or the heating/cooling load if set indoor conditions are to be maintained. The most sophisticated of these is ESPr (European reference program) and DOE-2/EnergyPlus (US reference program). The programs with a graphical user interface include eQuest (DOE-2 engine), Design Builder, Transys and IES Virtual Environment (IES VE). Nevertheless, describing the building in energy modeling software is quite time-consuming.

A mid-rise office building with a floor plat of 50 × 33 m, four-story and gross floor area of 6582 m^2, with a central core and four perimeter zones 6 m depth with 3.95 m floor to floor height and 2.74 m high ceiling, as shown in Figure 3.26 is modeled in eQuest (DOE 2016) with full air conditioning in Los Angeles, California. The building is modeled as a standard north-south orientation with 25% of glass.

TABLE 3.9
Construction, Occupancy Details, and Heat Gain Calculation of Example Office

Item	Details
External wall (opaque)	88.9 mm (3.5 inches) wood studs (20% of surface area) and insulation (80% of surface area) with 25.4 mm (1 inch) plyboard on outside and 12.7 mm (0.5 inch) gypsum board on inside. U-value 0.47909 W/m² K
Roof	Metal deck, topped with board insulation and membrane roofing. U-value 0.40 W/m² K
Window	South-facing, 3 m × 1.2 m SHGC 0.25, U-value 3 W/m² K
Door	25 mm plywood, north-facing 1.5 m × 2.0 m; U-value 2.7688 W/m² K
Ventilation	Two air changes per hour
Lighting	10 W/m² floor area; in use 0700–0900 h and 1700–1900 h
Occupancy	Occupied 0900–1700 h by three persons; 90 W sensible heat output per person
Electrical equipment	IT equipment generating 10 W/m²; in use 0900–1700 h

S. No.	Surfaces	Direct H_b (W/m²)	Diffuse H_d (W/m²)	Reflected (W/m²)	Solar Irradiance (W/m²)	Sol–Air Temperature
1.	Roof	656	241	—	897.0	68.89
2.	North wall	—	241 × 0.5 = 120.5	897 × 0.5 × 0.2 = 89.7	210.2	44.16
3.	East wall	0.23 × 656 = 150.88	241 × 0.5 = 120.5	897 × 0.5 × 0.2 = 89.7	361.08	49.59
4.	South wall	0.17 × 656 = 111.52	241 × 0.5 = 120.5	897 × 0.5 × 0.2 = 89.7	321.72	48.10
5.	West wall	—	241 × 0.5 = 120.5	897 × 0.5 × 0.2 = 89.7	210.2	44.16

S. No.	Building Element	U-value (W/m² K)	Area (m²)	$\Delta T = T_{sa} - T_i$ (K) or $\Delta T = T_o - T_i$ (K)	Heat gain – $A \times U \times \Delta T$ (W)
1.	Roof	0.40	25	68.89 − 26.8 = 42.09	420.9
2.	North wall	0.47909	15 − 3 = 12	44.16 − 26.8 = 17.36	99.8
3.	North door	2.7688	2.4	44.16 − 26.8 = 17.36	115.35
4.	East wall	0.47909	15	49.59 − 26.8 = 22.79	163.77
5.	South wall	0.47909	15 − 3.6 = 11.4	48.10 − 26.8 = 21.30	116.33
6.	South window	3	3.6	36.6 − 26.8 = 9.8	105.84
7.	West wall	0.47909	15	44.16 − 26.8 = 17.36	124.75
	Q_c (W)				1146.74

(Continued)

TABLE 3.9 (CONTINUED)

Construction, Occupancy Details, and Heat Gain Calculation of Example Office

S. No.	Building Element	Area (m²)	U-value (W/m² K)	$\Delta T = T_{sa} - T_i$ (K) or $\Delta T = T_o - T_i$ (K)	Heat gain – $A \times U \times \Delta T$ (W)
8.	Q_s (W) South window	3.6	SHGC = 0.25	$H_h \times A \times SHGC = 321.72*3.6*0.25$	**289.55**
9.	Q_v (W)	$V = 75\,m^3$	$N = 2$ air changes/h	$36.6 - 26.8 = 9.8$	$0.33NV\Delta T$ = **485.10**
10.	Lighting	20	10 W/m²	4 h	$20 \times 10 \times 4 = 800/24 = 33.33$
11.	Occupancy	3 persons	90 W/per	8 h	$90 \times 3 \times 8 = 2160/24 = 1080$
12.	IT equipment	20	10 W/m²	8 h	$20 \times 10 \times 8 = 1600/24 = 66.67$
	Q_i (W)				**1665.1**
	Q (W)				**3586.49**

The wall is steel-framed construction with insulation U-value 0.471 W/m² K (0.083 Btu/h ft² °F), roof is insulation entirely above deck U-value 0.273 W/m² K (0.048 Btu/h ft² °F) and glass is with metal framing U-value 3.407 W/m² K (0.6 Btu/h ft² °F) with SHGC 0.25. The light power density is taken as 0.9.

Cooling and heating set points: occupied cooling and heating setpoints of HVAC unit are determined as 24°C (75°F) and 21°C (70°F), respectively: that is, it is assumed that heating unit starts to work when zone temperature falls below 21°C in winter. In summer, the cooling process starts to work when the zone temperature exceeds 24°C. While as unoccupied cooling and heating points are set as 26.67°C (80°F) and 18.33°C (65°F) respectively. Cooling design temperature indoor is 24°C (75°F) and supply 12.78°C (55°F). Heating design temperature is 21°C (70°F) and supplies 32.22°C (90°F). Airflows minimum is 0.8495 m³/hour (0.5 cfm/ft²) and VAV minimum flow 30%. Indoor design temperatures are used by eQuest to size airflow requirements.

Occupancy: 09:00 to 18:00 for five days of the week throughout the year, weekends, and holidays are considered as unoccupied days. Hourly scheduled values of the internal loads (people, lights, and plug loads).

Table 3.10 and Figure 3.27 present the annual energy consumption by end-use.

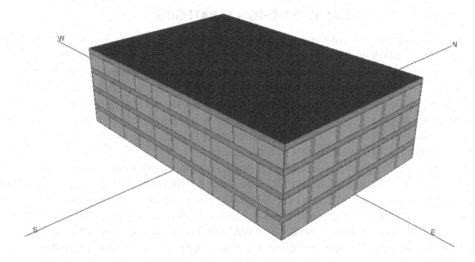

FIGURE 3.26   Building model for dynamic modeling (after Kalpana Tyagi).

TABLE 3.10
Annual Energy Consumption by End-Use

End-Uses of Energy	Electricity kWh (× 000)	Natural Gas MBtu
Space Cooling	162.61	0
Heat rejection	0	0
Refrigeration	0	0
Heat pump supp.	0	0
Water heating	0	0
Ventilation fans	19.35	0
Pumps & aux.	2.81	0
Exterior usage	0	0
Misc. equipment	163.47	0
Task lighting	0	0
Area lighting	171.46	0
Space heating	0	341.64
Total	519.7	341.64

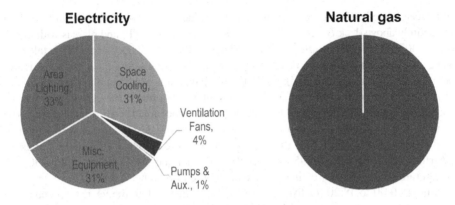

FIGURE 3.27   Results of energy modeling (after Kalpana Tyagi).

3.5   ENERGY-EFFICIENT ACTIVE DESIGN STRATEGIES

Generally, energy-intensive active strategies for heating, ventilation, and air conditioning (HVAC) are required to maintain desired environmental conditions when passive and hybrid strategies are insufficient. A primary role of passive and hybrid design strategies is to harness energy from nature to provide passive heating, cooling, ventilation, and lighting, as well as to reduce heating and cooling load. Thus the design of the energy-efficient active systems for a sustainable built environment is completed in consonance with the passive design strategies and hybrid-low energy strategies, as discussed in Sections 3.2 and 3.3. The active system should also be integrated with the design of the on-site renewable energy system, addressed in Chapter 5.

Very early in the design process, the HVAC design engineer must analyze and ultimately select appropriate systems. Occupant comfort (as defined by ASHRAE Standard 55), process heating, space heating, cooling, and ventilation criteria should be considered in selecting a system. Next, the production of heating and cooling is selected as decentralized or centralized. Table 3.11 presents a comparison of decentralized and centralized HVAC systems in terms of many criteria. Finally, the distribution of heating and cooling to the end-use space can be done by an all-air system or a variety of all-water or air/water systems and local terminals.

3.5.1   SPACE HEATING SYSTEMS

In ancient times, the radiant effect of an open fire (Figure 3.28) was the only method of heating after the sun. However, it has low efficiency and generates polluting CO, NO_2, and causes many health issues like throat and nose irritation. The space heating technology evolved considerably over one and a half century since the industrial revolution. The space heating systems can be classified into decentralized or direct and centralized or indirect heating systems. Decentralized or direct systems provide heating by conversion of some form of energy (usually the chemical energy of some fuel material) into heat within the space to be heated. Centralized or indirect systems, on the other hand, convert the fuel energy into heat in a central position (the boiler or furnace) from where it is distributed around the building and emitted to space by some conveying medium. This conveying medium is most often water or air, but it can be steam (or, theoretically any other liquid or gas). The level of centralization can vary, from central heating of a single flat or of a house, of a block of flats or offices, through a campus consisting of many buildings to an extensive 'district heating' scheme. Figure 3.29 illustrates the taxonomy of space heating systems.

TABLE 3.11

Comparison of Decentralized and Centralized HVAC Systems

Criteria	Decentralized Systems	Centralized Systems
Applicability	Most small to mid-sized installations, many built, expanded, and/or renovated buildings with equipment located in, throughout, adjacent to, or on top of the building.	All classes of buildings, but particularly in very large buildings and complexes or where there is a high density of energy use. Primary equipment located in a central plant (either inside or outside the building) with water or air required for HVAC needs to distribute from this plant.
Temperature, Humidity, Space pressure requirements	Fulfill any or all of these design parameters.	Fulfill any or all of these design parameters with greater precision and efficiency.
Capacity requirements	Each piece of equipment to be sized for zone peak capacity unless the systems are variable-volume.	HVAC diversity factors can be considered to reduce installed equipment capacity.
Air distribution	Constant air volume (CAV) perimeter unit ventilators are designed to respond to variations in thermal loads among different locations by varying the temperature of the air delivered to a given zone in the building.	Either CAV or VAV ventilation systems VAV systems attempt to respond to the differing thermal requirements by varying the quantity of air delivered to that zone.
Redundancy	No benefit of back-up or standby equipment.	Standby equipment for troubleshooting.
Space requirements	The equipment room may or may not be required; if required it can be located on the roof and/or the ground adjacent to the building due to space restrictions. Additional space may be required in the building for chillers and boilers. Duct and pipe shafts may or may not be required throughout the building.	The equipment room is generally located outside the conditioned area: in a basement, penthouse, service area, or adjacent to or remote from the building. An additional cost to furnish and install auxiliary equipment for the air and/or water distribution. The access requirements and physical constraints that exist throughout the building to the installation of the secondary distribution network of ducts and/or pipes and for equipment replacement.
Facility management	System performance can be maximized using proper business/ facility management techniques in operating and maintaining the HVAC equipment and systems.	System performance can be maximized using good business/ facility management techniques in operating and maintaining the HVAC equipment and systems.
Electric supply	Distributed electric supply required	Minimal distribution cost by centralized supply near the substation
Initial cost	The best first cost–benefit. This feature can be enhanced by phasing in the purchase of decentralized equipment as needed (i.e., buying the equipment as the building is being leased/occupied).	Significant initial cost, but longer service life may compensate for this A life-cycle cost analysis is critical
Operating cost	Energy consumption based on peak energy draw is high and less energy-efficient Operating costs can be saved by strategically starting and stopping multiple pieces of equipment	Larger, more energy-efficient primary equipment. Operational efficiency by the staging of the equipment operation to match building loads.

(Continued)

TABLE 3.11 (CONTINUED)
Comparison of Decentralized and Centralized HVAC Systems

Criteria	Decentralized Systems	Centralized Systems
Maintenance cost	Maintenance costs can be saved when equipment is conveniently located, and equipment size and associated components (e.g., filters) are standardized. When equipment is located outdoors, maintenance may be difficult during bad weather	There are usually fewer pieces of HVAC equipment to service. Access to occupant workspace is not required, no disruption to the space function. The equipment room for a central system provides the benefit of being able to maintain HVAC equipment away from occupants in an appropriate service work environment.
Reliability	Reliable equipment, although the estimated equipment service life may be less Equipment may require maintenance in the occupied space.	Centralized system equipment generally has a longer service life.
Flexibility	Very flexible because it may be placed in numerous locations	Flexibility can be a benefit when selecting equipment that provides an alternative or back-up source of HVAC.
Level of Control	Often use direct refrigerant expansion (DX) for cooling, and on/off or staged heat, greater variation in space temperature and humidity. Close control is not possible. Oversizing DX or stepped cooling can allow high indoor humidity levels and mold or mildew problems.	Generally, use chilled water for cooling and steam or hydronic heat. This usually allows for close control of space temperature and humidity where desired or necessary.
Noise and vibration	Noisy machinery is located close to building occupants, although equipment noise may be less.	Noisy machinery is sufficiently located remote from building occupants or noise-sensitive processes.
Constructability	Standardized in purchase and construction since multiple and similar-in-size equipment.	More coordinated installation required. However, consolidation of the primary equipment in a central location also has benefits.

Decentralized Heating

The available energy sources and the mode of energy delivery for such decentralized or direct heating may be:

   i) Solid fuel – (coal, coke, firewood) in batch (cans, bins, baskets)
   ii) Liquid fuel – (oil, kerosene) piped from an external tank or in batch (cans, bottles)
   iii) Gas – piped, from the grid or externally located bottle
   iv) Electricity – by cables

Solid fuel appliances, oil, and gas heaters should ensure air supply (oxygen) for the combustion of fuel and should exhaust the combustion gases through a chimney.

Electric resistant heaters have an enormous variety in terms of size capacity in kilowatts, heat output (radiant/ convective), and form. These devices have the disadvantage of using high-grade

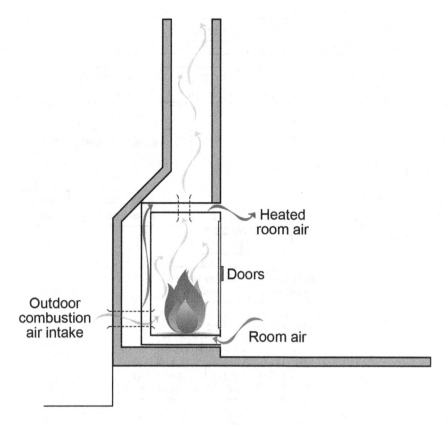

FIGURE 3.28   Fireplace, line work by Ar Vaibhav Ahuja.

energy to a low-grade task. Their advantages, however, are low first cost and individual thermostatic control so that each room can be made a separate heating zone. Electric heating depends on air velocity and electrical load.

A heat pump is an energy-efficient electrical device, which can deliver about three to four times more heat energy than the electrical energy it consumes. The efficiency of the heat pump is higher than the gas/electric resistant heater and results in energy-saving economy and also reduce greenhouse gas emissions.

By definition, a heat pump extracts heat from a source and transfers it to a sink at a higher temperature. In its most basic form, the heat pump consists of an evaporator, a compressor, a condenser, a throttling device which is usually an expansion valve or capillary tube, and the connecting tubing. The working fluid is the refrigerant (such as an organic fluoride or a hydrocarbon), which goes through a thermodynamic cycle in a closed loop. The heat pump thermodynamic cycle is shown schematically in Figure 3.30; four important processes take place during the cycle:

i)   Heat (Q_c) is transferred to the refrigerant in the evaporator from stations 4 to 1, where both its pressure and temperature are lower than a thermal source such as air, water, or the ground. Evaporation of the refrigerant occurs from a liquid to a saturated vapor theoretically at a constant pressure.

ii) Work (W) is done on the refrigerant as saturated vapor at low pressure, and temperature enters the compressor and undergoes adiabatic compression (from 1 to 2) – it becomes hot

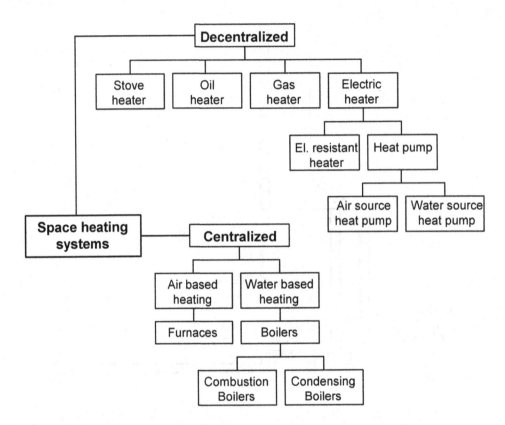

FIGURE 3.29 A taxonomy of space heating systems.

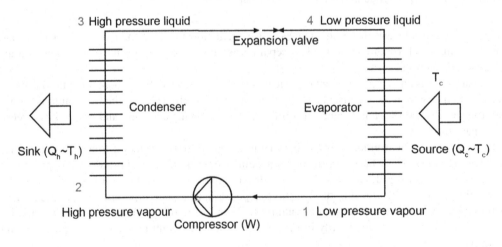

FIGURE 3.30 Thermodynamic cycle of a heat pump (or cooling machine).

and liquefies. The result is a compressed refrigerant at high pressure and temperature at the compressor outlet (point 2).

iii) Heat (Q_h) is transferred from the hot vapor in the condenser (from 2 to 3), where its pressure and temperature are higher than a thermal sink that is at a higher temperature than the source. Condensation from a vapor to a saturated liquid occurs in the condenser, theoretically at constant pressure. The refrigerant leaves the condenser as a saturated liquid.

iv) The throttling of the refrigerant occurs at an expansion valve or capillary tube from 3 to 4. During this process its pressure and temperature drop are adiabetic. The refrigerant reenters the evaporator at low pressure and temperature.

The performance of a vapor-compression cycle is usually described by a coefficient of performance (CoP), defined as the benefit of the cycle (amount of heat removed from the source or heat delivered to the sink) divided by the required energy input to operate the cycle.

$$CoP = \frac{\text{Useful refrigerating or heating effect}}{\text{Net energy supplied from external sources}}$$

If the purpose of using the heat pump is to gain heat, then the coefficient of performance (*CoP*) is defined as

$$CoP = \frac{Q_h}{W} = \frac{Q_c + W}{Q_h - Q_c} = \frac{\text{Heat delivered to sink}}{\text{Compressor work input}} \tag{3.29}$$

This CoP is higher for a small temperature increment (or step-up), but it reduces if the necessary step-up is large. In the ideal (Carnot efficiency) cycle the CoP is inversely proportionate to the temperature increment:

$$CoP = \frac{T_h}{T_h - T_c} \tag{3.30}$$

where T_h = sink temperature and T_c = source temperature (in °K).

However, a real cycle will give 0.82–0.93 (average 0.85) of the Carnot performance. This will be further reduced by the actual component efficiencies, such as electric motor 0.95, compressor 0.8, and, heat exchangers 0.9.

If the same machine is used for cooling, i.e., to remove heat, then the definition of CoP is slightly different:

$$CoP = \frac{Q_c}{W} = \frac{\text{Heat removed from source}}{\text{Compressor work input}} = \frac{T_c}{T_h - T_c} \tag{3.31}$$

Air-to-air heat pump either extracts thermal energy out of the atmosphere or transfers thermal energy from buildings to the atmosphere. Water-to-air heat pump relies on water as heat source and sink, and use air to transmit heat to or from the conditioned space. It may be 'grey' water discharged into a sump or a natural body of water (a river or the sea). A geothermal heat pump either extracts thermal energy out of the earth or transfers thermal energy from buildings to the ground (Figure 5.23).

Centralized Heating

Heat may be produced centrally in a building (or group of buildings), distributed to the occupied spaces by a heat transport medium and emitted to provide the required heating. The energy source may be non-renewable fossil fuels such as coal, oil or gas, electricity, produced from non-renewable fossil fuels, nuclear or hydropower, or renewable sources, such as solar-, wind-, tidal-, wave-, ocean-, biogas-, thermal-, or geothermal energy. Heating with coal, oil, gas, and solar energy requires building storage space. Heat can be distributed by transport medium water, steam, or air.

Any central heating system consists of three primary components:

1. the 'heat generating plant' (boiler for a water system or furnace for an air system)
2. the 'distribution network' (piping for a water system or ducting for an air system)
3. the 'heat emitter' units (radiators or convectors for a water system or grill or diffusers for an air system)

Architectural implications pertain to the accommodation of fuel storage, the heat production plant and its chimney, the distribution network as well as the placement of emitters for air and water-based heating systems are shown in Figure 3.31. The design of larger systems is the task of mechanical consultants.

Heating causes steep temperature gradient between areas of concentrated heat loss (e.g., windows) and zones of the heat output units generating strong convection currents (drafts), adversely affecting comfort conditions and causing, for instance, discoloration of surfaces. Heating also drops relative humidity of indoor air and may result in dryness.

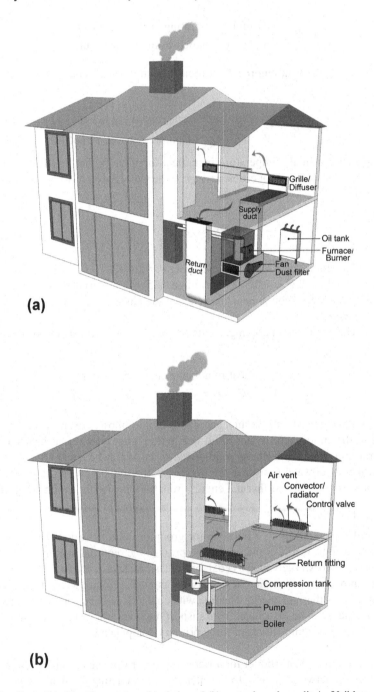

FIGURE 3.31 Centralized heating system: (a) air-based (b) water-based, credit Ar Vaibhav Ahuja.

Condensation can be caused indirectly. The warm indoor air will readily take on moisture from any available source: human exhalation (around 45 g/h per person), cooking, kettles, baths, etc. Its RH increases, consequently its dewpoint temperature is also increased. Air at 20°C DBT and 80% RH will have a dewpoint temperature of 16.5°C. If it comes into contact with a surface of 16°C and condensation will appear. Interstitial condensation may soak the wall material and increase its conductivity, thus lowering all the surface temperature, which in turn further increases the condensation.

3.5.2 Mechanical Ventilation

The major considerations for acceptable indoor air quality (IAQ) include the amount of outdoor air required to control moisture, carbon dioxide (CO_2), odors, and tobacco smoke generated by occupants as well as various additional pollutants that are not generated primarily by occupants. The outdoor air requirements are usually given as a function of occupancy density (in either volumetric or floor area terms) in L/s (person) by ASHRAE standard 62.1 (Table 3.12) or if no such information is available, then in terms of air changes per hour (number of times the whole volume of air must be exchanged every hour, Table 3.7).

The outdoor air requirements are ensured by ventilation; it may be composed of natural or mechanical ventilation, infiltration, suitably treated recirculated air, transfer air, or an appropriate combination. In free-running buildings, natural ventilation requirements can be stated in qualitative terms only, but for closed and conditioned buildings mechanical ventilation has the greatest potential to control fresh air requirements as per ASHRAE Standard 62.1. However, hybrid (or mixed-mode) buildings offer the possibility of saving energy in appropriate climates by combining natural ventilation systems with mechanical equipment (Emmerich 2006). The air-side economizer is one form of the hybrid ventilation control scheme and widely use in commercial, industrial, and institutional buildings in appropriate climates. The report of the International Energy Agency's (IEA) Annex 35 describes the principles of hybrid ventilation technologies, control strategies, design and analysis methods, and case studies (Heiselberg 2002).

Mechanical ventilation systems may be of three types:

i)   Supply (unbalanced) system
ii) Extract (unbalanced) system
iii) Balanced system

A supply system mechanically brings in filtered fresh outside air into space and create a positive pressure. Consequently, used air is forced out through vents. It is useful in hot or mixed climates, not appropriate for cold climates. A special form of this is the fire ventilation, which forces air at high pressure (some 500 Pa) into staircases and corridors, to keep the escape route free of smoke.

An exhaust system mechanically removes the used air from the space and creates a negative pressure. Consequently, fresh air finds its way in through grilles and openings (room under reduced pressure). Extract systems are useful in cold climates and typically not appropriate for hot, humid climates. It is also useful near a source of contamination, such as toilets, kitchen cooker hoods, laboratory fume cupboards.

A balanced system mechanically provides both supply and exhaust using two fans, one to bring in the fresh air and another to send indoor air out (Figure 3.32). The supply flow is usually kept higher than the exhaust, to keep a slight positive pressure and thus prevent unwanted dust entry. The balanced system is most dependable in all climates as it provides a great degree of control, but the most expensive. The balanced system with the 'heat recovery' ventilation (commonly known as HRV) and 'energy recovery' ventilation (commonly referred to as ERV) can yield improved indoor air quality and reduced energy consumption. HRVs transfer heat from exhaust air to incoming air during the heating season and from incoming air to exhaust air in the air-conditioning season to

TABLE 3.12
Minimum ventilation requirements

Outdoor Air Rate (L/s person)	Occupant Density Area/Person (m²)	Occupancy Category (Examples Only)
2.7	0.7	Auditorium seating area, lobbies
2.8	0.8	Multipurpose assembly, places of religious worship
2.9	1.4	Courtrooms
3.0	1.7	Telephone/data entry
3.1	2.0	Conference/meeting Legislative chambers
3.5	3.3	Dayroom, correctional facility
3.5	2.0	Break rooms, office building
3.5	3.3	Reception areas, Office building
3.5	4.0	Break rooms
4.0	0.7	Lecture hall (fixed seats)
4.0	5.0	Coffee stations, Barracks sleeping areas
4.1	1.0	Multiuse assembly, Transportation waiting
4.3	1.5	Lecture classroom
4.4	2.0	Booking/waiting, correctional facility
4.5	6.7	Guard stations, correctional facility
4.6	2.5	Museums/galleries
4.7	1.0	Cafeteria/fast-food dining, bars, cocktail lounges
4.8	3.3	Lobbies/prefunction area, hotels, motels, resorts, dormitories
4.9	4	Cell correctional facility
5.1	1.4	Restaurant dining rooms
5.3	2.5	Museums (children's)
5.5	10.0	Bedroom/living rooms, Main entry lobbies
5.9	2.9	Music/theater/dance
6.0	6.7	Banks or bank lobbies
6.7	2.9	Classroom (age 9 plus)
7.0	5	Kitchen (cooking)
7.4	4	Classroom (ages 5–8) Computer lab, media center
8.5	10	Photo studios, Libraries Laundry rooms, central, Laundry rooms within dwelling units
8.5	20	Office space, Bank vaults/safe deposit
8.6	4	Daycare (through age 4), Daycare sickroom, Science laboratories, University/college laboratories
9.5	5	Art classroom, wood/metal shop
10.0	25	Computer (not printing)
11.5	10	Pharmacy (prep. area)
12.5	14.3	Sorting, packing, light assembly
17.5	50	Occupiable storage rooms for dry materials
18.0	14.3	General manufacturing (excludes heavy industrial and process using chemicals
32.5	50	Occupied storage rooms for liquid or gels
35.0	50	Shipping/receiving

Source: Based on ASHRAE 62.1 (2013).

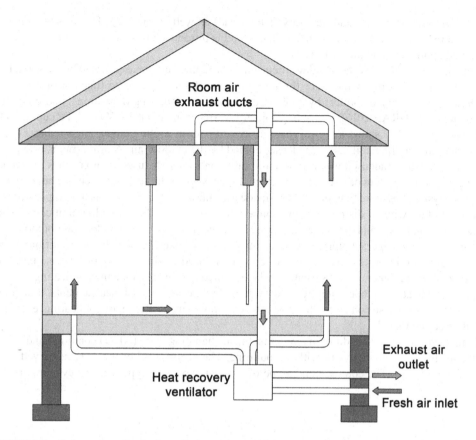

FIGURE 3.32 Balanced mechanical ventilation system with heat recovery, line work by Ar Shefali Ahlawat.

reduce the heating and cooling load and improve comfort. This would employ a heat exchanger (rotary, or a plate-type) or a heat transfer loop without mixing the two air streams. ERVs transfer heat and moisture between the exhaust air and incoming air. This provides additional savings in the summer by reducing the moisture content of the incoming air that would otherwise have to be dehumidified with the cooling equipment or a dehumidifier. ERVs also provide additional comfort in the winter by adding moisture from outgoing air to the incoming air to help avoid excessively dry indoor conditions.

The mechanical ventilation system may use either propeller (axial flow) fans or centrifugal (radial flow) fans. In case of supply or balanced system, the air will normally be filtered at the point of intake, by one of the four means: dry filters, wet filters (10 μm particle size), air washers, or electrostatic filters (0.01μm particle size). Ducts used to convey and distribute the air are usually made of sheet metal of rectangular cross section, but in recent times plastic materials are often used in circular or oval sections.

3.5.3 AIR CONDITIONING

Willis Haviland Carrier invented modern air conditioning; the US Energy Information Administration (EIA) defines air conditioning as cooling and dehumidifying the air in an enclosed space by use of a refrigeration unit powered by electricity or natural gas. Air-conditioning systems control the temperature (sensible load) and humidity (latent load) as well as the IAQ. The sensible loads are discussed in Sections 3.4.3 and 3.4.4. The latent load is introduced in space through people, equipment, and appliances and air infiltration.

The simplest system is the window room air conditioner: an encased assembly designed primarily for mounting in a window or through an external wall. The basic function of a room air conditioner is to provide comfort by cooling, dehumidifying, filtering or cleaning, and circulating the room air (ASHRAE 2016). Its cooling capacity varies from about 1 kW (~0.3 TR) up to 10 kW (~3.0 TR). Figure 3.33 shows a typical room air conditioner in cooling mode. It has a direct expansion evaporator cooling coil and a condenser cooled by the outdoor air. A drain tray is provided at the bottom for the condensed water. Room air conditioners may be used as auxiliaries to a central heating or cooling system or to condition selected spaces when the central system is shut down.

A split system has more than one factory-made assembly; an indoor unit has the evaporator cooling coil and fans with air distribution and temperature control, while the outdoor unit comprises of the noisier compressor and air-cooled condenser or condensing unit. The indoor and outdoor units are connected by refrigerant piping. The condensed water is taken away from the conditioned space using separate drain pipes. These units are commonly used in single-story or low-rise buildings. The room air conditioner or split system may be designed to provide supplemental heating in winter by air-source heat pump operation (reverse-cycle facility), electric resistance elements, or a combination of the two. These constitute the most effective way of using electricity for heating, even if the CoP is not more than 2.

In the centralized air-conditioning systems, the air-handling units (AHU) condition and distribute the air by insulated ductwork within a building. The heating coil of the AHU is served by hot water from a boiler. The cooling coil can be of a direct expansion type, i.e., the evaporator of the

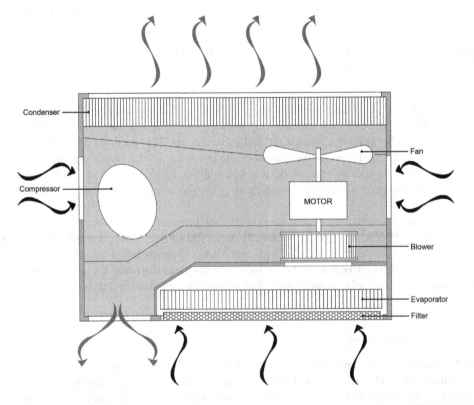

FIGURE 3.33 Schematic view of a conventional air conditioner. Source: adapted from ASHRAE (2016), line work by Ar Vaibhav Ahuja.

cooling machine itself, or the cooling machine can become a chiller (the evaporator shaped as a refrigerant- to water heat exchanger) supplying chilled water to the cooling coil. Figure 3.34 shows a typical central air-handling unit.

The chiller machine of an air conditioner may be a compressor type operated by electrical energy, such as that shown in Figure 3.30 or an absorption chiller operated by (waste or low-cost heat) thermal energy as shown in Figure 3.35.

The absorption chiller comprises a circulating refrigerant, an evaporator, a condenser, an expansion device, a chemical absorption process and generator, with a pump to provide the circulation and pressure change. For air-conditioning applications operating with evaporating temperatures above 0°C, lithium bromide solution is the absorbent, while water is the refrigerant. Below 0°C, the most common pairing is water as the absorbent and ammonia as the refrigerant. The 'weak' solution of lithium bromide and water in the absorber is pumped to a generator, where external heat is applied to boil off or vaporize the water from the solution. This results in the water (refrigerant) vapor leaving the generator and being condensed in a water or air-cooled condenser, back to liquid. Its pressure is then reduced before feeding back into the evaporator to continue the cooling process, where it picks up heat from its environment. Meanwhile, the now 'strong' solution in the generator is fed back to the absorber, also reducing in pressure as it goes and continuing the absorption process. Both chillers can be used to produce chilled water, which is then circulated to the air-conditioner unit cooling coil, or in a direct expansion coil, where the evaporator becomes the cooling coil.

The centralized air-conditioning systems can be classified based on the fluid media used in the thermal distribution system from the central plant to the conditioned space: all-air systems, all-water systems, and air–water systems. Figure 3.36 presents taxonomy of centralized air-conditioning system.

All-Air System

In an all-air system, the air is cooled or heated, humidity controlled, filtered, and freshened with outdoor air-all under controlled conditions at the central plant and the processed air is distributed to the desired zones by a network of either single or dual ducts using blowers and fans. Within the

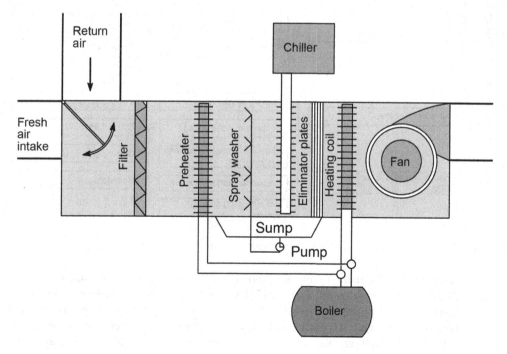

FIGURE 3.34 A typical central air-handling unit. Source: adapted from Szokolay (2008). Line work by Ar Shefali Ahlawat

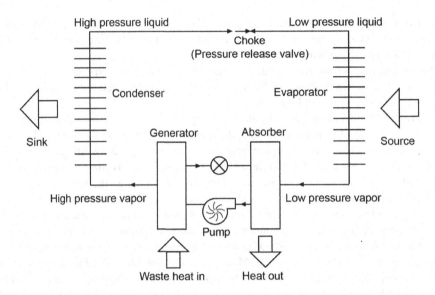

FIGURE 3.35   Absorption chiller.

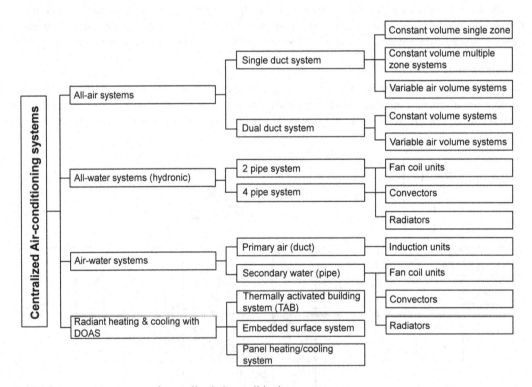

FIGURE 3.36   A taxonomy of centralized air-conditioning systems.

zones, terminal units, diffusers/registers and return grilles allow a well-planned stream of conditioned air to thoroughly permeate all work areas. The great advantage of an all-air system is that complete control over indoor air quality (IAQ). The main disadvantages are the large ducts for both supply and return air since air holds much less heat per unit volume than water. The air volume flow rate to each room is constant, and the required condition is set at the central plant. It may include a terminal re-heat facility, to provide some flexibility, but at a cost in energy. An energy-efficient alternative is the variable air volume (VAV) system, where the supply air condition is constant and the cooling requirement of each room can be matched by reducing or increasing the airflow at the diffuser. Figure 3.37 illustrates different configurations of centralized air-conditioning system for multi-story building.

All-Water System

In an all-water system, the water is chilled by the evaporator coil or heated by the boiler at the central plant and the processed water is supplied to each room or group of rooms by a network of either two-pipe or four-pipe system. Each room may have its local air-handling system (fan-coil units,

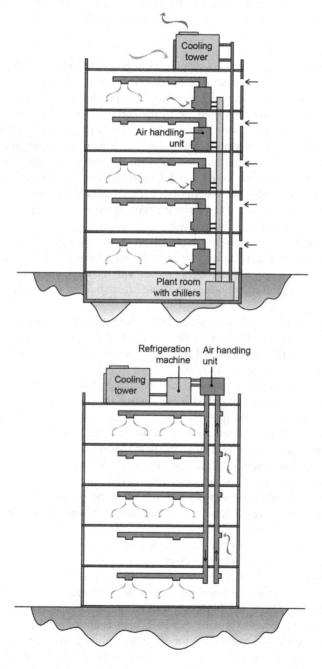

FIGURE 3.37 Configuration of air-conditioning system in multi-story buildings. Line work by Ar Vaibhav Ahuja.

convectors, or radiators) and controls. All water system is most appropriate for air-conditioning buildings with small zones (e.g., apartments, condominiums, motels, hotels, hospitals, and schools). Although the piping in the building takes up very little space, the fan-coil units in each room do require some space. Another advantage of the all-water system is the small amount of energy required by the pumps as compared with fans. However, air quality is dealt with either locally, using infiltration or windows; or by a separate fresh air supply system; or simply by fresh air from an adjacent system, such as a ventilated interior zone. Since air is handled locally, there is very little mixing of air from one zone to another, making this attractive where potential air contamination (or smoke from a fire) is a special concern. It is also an easy system to retrofit. However, maintenance is high; filters in each fan coil must be cleaned, and drain pans are potentially problematic.

Air–Water System

In the air–water system the central plant supplies both processed air and water to each zone of a building. The system consists of a central plant for cooling or heating of air and water, a ducting system with fans for conveying air, water pipelines, and pumps for conveying water and a room terminal. The in-room terminal units may be in the form of fan-coil units with supplementary air, an induction unit or a radiator panel. Almost 80–90% of the sensible heating and cooling in a building is accomplished by the water system due to the immense heat-carrying capacity of water as opposed to air. While as the remaining sensible load and indoor air quality-filtering, humidity, freshness is handled by air system, equal to total fresh air required. Fan-coil units for air–water systems are similar to that of all-water systems except that the centrally conditioned, tempered fresh air is brought to space in a constant volume stream; the fan moves both fresh and room air across a coil that either heats or cools the air, as required.

The over-cooled and very dry air and high-velocity constant volume fresh air is supplied at medium to high pressure to the induction unit in each room, induces a flow of air from the conditioned space and thus the supply air thoroughly mixes with the room air. Finally, the air is sensible cooled or heated as it flows through cooling or heating, which may be supplied from a central chiller or boiler.

All systems must provide fresh air supply for ventilation purposes as per ASHRAE standard 62.1. Acceptable indoor air in which there are no known contaminants at harmful concentrations as determined by cognizant authorities and with which a substantial majority (80% or more) of the people exposed do not express dissatisfaction.

Several energy conservation techniques can be integrated with air-conditioning systems. Building energy management system (BEMS) can control and coordinate all the building's energy-using equipment responsively, to optimize energy use. An *economy cycle* can increase out air supply without running the chiller plant when the outdoor air is cooler than the indoors. A *night flush* of (cool) outdoor air may be provided to remove the heat stored in the building fabric, thus reducing the following day's cooling requirement; see Section 3.3.3. An underfloor air distribution (UFAD) system or *displacement* ventilation using a large number of low-volume supply air outlets (floor diffusers) create laminar flow at the bottom of the room, where people are, instead of conditioning the whole space. The NOAA Daniel K. Inouye Regional Centre features displacement ventilation; see Section 6.2 for more information.

The dedicated outdoor air system (DOAS) decouples air conditioning of the outdoor air from the conditioning of the internal loads. The DOAS introduces 100% outdoor air, heats or cools it, may humidify or dehumidify it, and filters it, then supplies this treated air to each of its assigned spaces. DOAS can accommodate an exhaust or relief airflow for heat recovery between the outdoor and exhaust or relief airflows. Often, the DOAS serves multiple spaces and is designed not necessarily to control space temperature but to provide thermally neutral air to those spaces. A second application, the conventional system, is responsible for offsetting building envelope, and internal loads and DOAS are responsible for the condition or deliver outdoor air. A common example may be a large apartment building with individual fan-coil units (the conventional system) in each dwelling unit, plus a common building-wide DOAS to deliver code-required outdoor air to each housing unit for good indoor air quality and to make up a bathroom and/or kitchen exhaust.

3.5.4 RADIANT HEATING AND COOLING WITH DOAS

Radiant systems are active temperature-controlled indoor surfaces on the floor, walls, or ceilings, which act as the primary source of sensible heating and cooling in the conditioned space. The central idea of the radiant system is that maintaining lower or higher indoor surface temperatures will enhance radiant heat exchanges between the human body and indoor surfaces. The surface temperature is maintained by circulating water, air, or electric current through a circuit.

An actively controlled surface is considered a 'radiant system' if at least 50% of the design heat transfer is by thermal radiation (ASHRAE 2016) and is often part of a hybrid system that includes conditioning of ventilation air to address internal latent load (humidity) from occupants and infiltration, plus sensible and latent loads associated with outside ventilation using DOAS (Figure 3.38).

The radiant system is based on the principle of thermal radiation, a process by which energy, in the form of electromagnetic waves,

i)   Is emitted by a heated surface in all directions
ii) Travels directly to its point of absorption at the speed of light in straight lines and can be reflected
iii) Elevates the temperature of solid objects by absorption but does not noticeably heat the air through which it travels
iv) Is exchanged continuously between all bodies in a building environment

The rate at which thermal radiation occurs depends on the following factors:

i)   The temperature of the emitting surface and receiver
ii) Emittance of the radiating surface
iii) Reflectance, absorptance, and transmittance of the receiver
iv) View factor between the emitting and receiver surface (viewing angle of the occupant to the thermal radiation source)

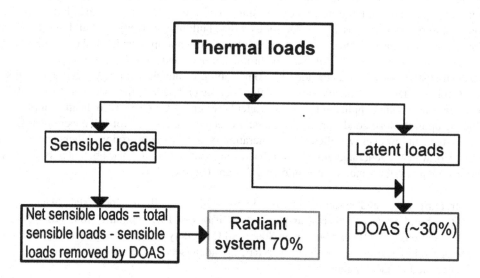

FIGURE 3.38   Thermal loads in radiant heating and cooling system in a building.

When considering radiant heating or cooling for occupant thermal comfort, the following terms describe the temperature and energy characteristics of the total radiant environment:

i) Mean radiant temperature (T_{MRT}) The temperature of an imaginary isothermal black enclosure in which the occupant exchanges the same amount of heat by radiation and convection as in the actual non-uniform environment.

ii) Ambient temperature (T_{air}) is the temperature of the air surrounding the occupant.

iii) Operative temperature (T_o) is the temperature of a uniform isothermal black enclosure in which the occupant would exchange the same amount of heat by radiation and convection as in the actual non-uniform environment.

For air velocities less than 0.4 m/s (1.3 ft/s) and mean radiant temperatures less than 50°C (122°F), the operative temperature is approximately equal to the adjusted dry-bulb temperature, which is the average of the air and means radiant temperatures.

$$T_{op} = \frac{f_r T_{MRT} + f_c T_{air}}{f_r + f_c} \qquad (3.32)$$

$$T_{op} \approx \frac{T_{MRT} + T_{air}}{2} \qquad (3.33)$$

The cooling capacity of the chilled slab:

$$Q = f_{rad+conv} \times \Delta T \times \text{Active area} \qquad (3.34)$$

$f_{rad+conv}$ = Combined heat transfer coefficient = 8–11 W/m²K
ΔT = Operative temperature-ceiling surface temperature = 25 K
Allowable surface temperature = 18°C
Cooling capacity per unit chilled surface area = 56–77 W/m²

Thus, the radiant systems lead to a more controlled, acceptable thermal condition in the space as well as are more energy-efficient for both space heating and cooling requirements for an entire building compared to the conventional all-air system (Table 3.13). Radiant heating and cooling systems are applicable in buildings, where sensible loads are more dominant than latent loads and indoor humidity control is possible; laboratories, office buildings, educational buildings, healthcare buildings, government facilities, and such other places. It is not applicable in places where indoor humidity control is not possible, kitchens, bathrooms, toilets, and similar places.

Radiant systems can have embedded hydronic tubing or attached piping in ceilings, walls, or floors. Chilled water is circulated through the floor/ceiling embedded pipes between 15°C and 17°C temperatures for cooling applications and hot water between 20°C and 35°C for heating applications depending upon the load conditions. Chilled water is generated either through conventional electric chiller systems or low energy chilled water generation systems like absorption chillers and desiccant chillers. Hot water is generated either through conventional boilers or solar hot water systems. The hydronic radiant systems include the following designs (Figure 3.39):

i) Structure integrated systems – tubing in the ceiling, walls or floors slab (Thermally active building system-TABS) as per the provisions of ISO standard 11855 (ISO 2012). Applications of TABS can be seen in Standford University Central Energy Facility (section 6.3.4) and Research Support Facility, NREL, Golden (section 6.5.4).

ii) Embedded surface system

iii) Panel systems (ceiling-suspended, wall-mounted). Applications of panel systems can be seen in Edith Green - Wendell Wyatt (EGWW) Federal building, Portland (section 6.4.4).

TABLE 3.13

Conventional Air-Conditioning vs Active Radiant System

S. No.	Parameters	Air System	Radiant Cooling
1.	Space cooling medium	Air (sensible + latent)	Radiant system (sensible load) + DOAS (latent load)
2.	Chilled water supply temperature	5–9°C	Radiant system: 14–18°C DOAS: 5–9°C
3.	Space air temperature	22–24°C	26–28°C
4.	Surface mean radiant temperature	26–28°C	22–24°C
5.	Air quantity handled by a fan	(Recirculated + fresh air)	Only fresh air-DOAS
6.	Indoor environment quality	Noise and drafts of air movement	Less noise due to less draft Even temperature distribution
7.	Spatial	Space for air-handling unit on every floor in addition to mechanical equipment room outside More floor to floor height to accommodate duct/pipework	Space savings per floor Less floor to floor height implies a reduction in building wall material usage
8.	Facility management	Trained facility managers are required	
9.	Building envelope	An efficient envelope with non-openable windows is required to reduce condensation	

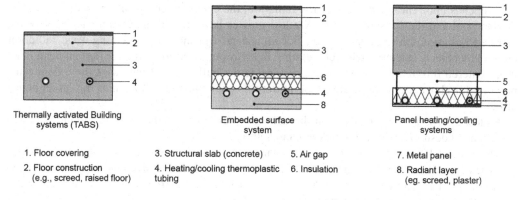

Thermally activated Building systems (TABS)	Embedded surface system	Panel heating/cooling systems

1. Floor covering	3. Structural slab (concrete)	5. Air gap	7. Metal panel
2. Floor construction (e.g., screed, raised floor)	4. Heating/cooling thermoplastic tubing	6. Insulation	8. Radiant layer (eg. screed, plaster)

FIGURE 3.39   Active Radiant heating and cooling systems. Line work by Ar Vaibhav Ahuja.

3.6   SOLAR CONTROL DESIGN

Windows are a multivalent and essential feature of nearly every type of building. The environmental attributes of windows determine the luminous, thermal, acoustical and psychological environment inside of a building (Figure 3.40), Le Corbusier

> one could say that the history of windows is the same as that of architecture or, at least, this is the trait which is the most characteristic in the history of architecture.

The thermal performance of any fenestration system (any light-transmitting opening in a building wall or roof) depends on the environmental factors, the thermal behavior of its components, and the comfort conditions required inside. Heat gain, or loss, through the fenestration system, is affected by many environmental factors. These include solar radiation, intensity and angle of incidence,

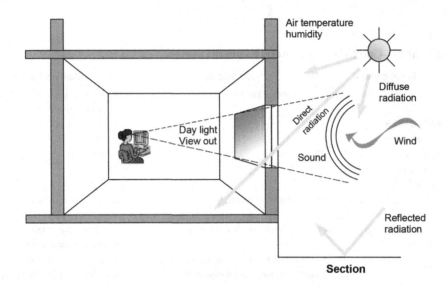

FIGURE 3.40 Environmental attributes of a window.

outdoor-indoor temperature differential, and velocity and direction of an airflow across the exterior and interior fenestration surfaces. However, the example in Section 3.4.4 (Table 3.9) shows that solar radiation entering through a window is the dominant source of heat gain.

According to the ASHRAE (1972), 'The ability of glazing materials to transmit solar radiation depends upon the wavelength of the radiation, the chemical composition and thickness of the material and the incident angle.' Window glasses have high transmittance for short-wave infrared radiation emitted by the sun, but almost no transmittance for long-wave radiation emitted by objects at a terrestrial temperature in the room. Consequently, the radiant heat, once it has entered through a window, is trapped inside the building and increases the indoor temperature far above the outdoor air temperature; this phenomenon is known as the 'greenhouse effect.'

Protection against solar penetration through openings is one of the most important means to prevent an undesirable increase in indoor temperature.

The well-established principle of thermal solar control is to allow the solar energy into the building when the weather is cool, conversely to intercept it when it is hot. Given the broad spectrum of buildings and the different climate zones, solar control has to be exclusively designed. There are four methods available to the designers for solar control:

i) Orientation and window size
ii) Internal shading devices
iii) High-performance glasses
iv) External shading devices

The proper orientation from the solar control point of view is that the building receives the maximum solar radiation in summer and the minimum in winter. The key considerations are:

- The horizontal surfaces receive the highest solar radiation in summer as well as in winter. It is, therefore, essential to restrict skylights, and the SRR (skylight roof ratio) should not exceed 5% as prescribed by ASHRAE 90.1 (2010). Also, consider shading the roof even if there are no skylights since the roof is a significant source of transmitted solar heat gain into the building.

- The east- and west-facing surfaces receive the second-highest intensities of solar radiation. Hence limit the amount of east and west glass. Direct solar radiation on the west side coincides with the highest temperature in the afternoon. Generally, the west side should accommodate services spaces not meant for occupants.
- The south wall (facing equator) receives the next highest intensities in the winter (when the altitude of the sun is less), but it receives comparatively less solar radiation in summer. The windows may be provided on the south wall for allowing winter sun but with solar control, devices to intercept the direct sun in summer.
- The north side of a building will receive solar radiation in summer. The windows may be provided on the north wall but with solar control, devices to intercept the direct sun in summer.
- The window to wall ratio (WWR) should not exceed 40% as prescribed by ASHRAE 90.1 (2010) (Figure 3.41).

The interior shading devices such as Venetian blinds or vertical louvers do not considerably reduce cooling loads since the solar heat has already been admitted into space. However, these interior devices do offer glare control and can contribute to visual comfort in the workplace. These devices absorb solar heat and can increase temperature. The absorbed heat will be partly convected to the indoor air and partly reradiated. Half of this reradiation is outward, but as it is of a long wavelength, it is stopped by the window glass. The usual narrow space between the window and the blind will thus be quite substantially overheated. The hot surface of the blind causes the indoor mean radiant temperature (MRT) to rise far above the air temperature.

3.6.1 HIGH-PERFORMANCE GLASSES

An ordinary glass transmits a large proportion of all radiation between 300 and 3000 nm, i.e., both visible light and short-wave infrared, but very little around and outside the 300 and 3000 nm range. Its transmittance is selective. This selective transmittance can be modified by varying the

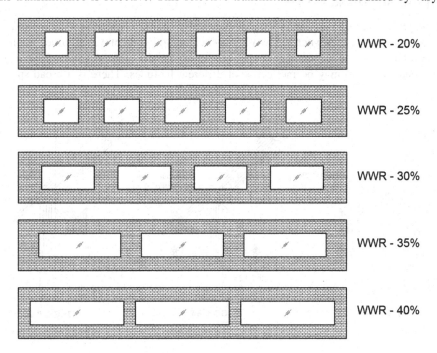

FIGURE 3.41 Window to wall ratio (WWR).

composition of the glass to reduce the infrared transmission substantially, while only slightly affecting the light transmission. The primary function of high-performance glass is to act as an efficient heat filter with little effect on the other functions of the window, such as view and contact and provision of daylight. These high-performance glass elevations have great current popularity. This is probably since a relatively simple solution to the problem of sun control can be integrated with purely abstract design without necessitating the reduction of glass area. Solar heat gain has been defined in general terms as 'transmitted and absorbed solar energy.'

A broad spectrum of high-performance glasses for solar control application of fenestration has been developed by the glass industry. The types of high-performance glasses are usually classified under the following broad categories:

With fixed performance

- Heat-absorbing glass
- Solar-reflecting glass
- Solar-control-reflective polyester-coated film glass
- Low-emittance (low-E) coating glass

With variable performance (smart glasses)

- Electrochromic glass
- Thermochromic glass
- Photochromic glass
- Gasochromic glass

Figure 3.42 compares the solar heat gain coefficient of plate glass, heat-absorbing glass, and heat-reflecting glass.

3.6.2   EXTERNAL SHADING DEVICES

The solar control devices have to satisfy the diametrically opposite functions to let the sun's energy into the buildings at all times when the weather is cool and to intercept it at all times when it is hot. A solution suitable for one season may not be satisfactory for the other. A solution that is effective for a given site (latitude) may be ineffective at different latitudes. There is a broad spectrum of

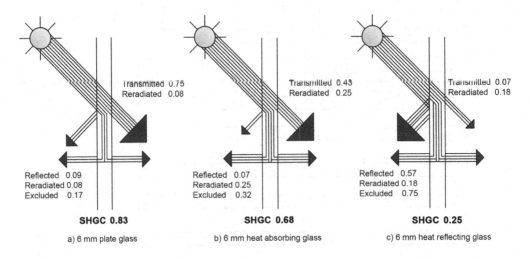

FIGURE 3.42   Comparison of solar heat gain coefficient of three types of glass.

design variations for shading devices, as can be seen in the work of modern architects Le Corbusier, Oscar Niemeyer, Richard Neutra, Marcel Breuer, Paul Rudolph, and others (Olgyay and Olgyay 1957). Figure 3.43 is an elegant application of an external shading device by the master architect Marcel Breuer at Saint John University, College Ville, Minnesota.

The external shading devices are characterized by horizontal and vertical shadow angles (Figure 3.44). These two angles are measured from a line perpendicular to the elevation and indicate the limit beyond which the sun would be excluded, but within which the sun would reach the point considered.

FIGURE 3.43 External shading devices, an architectural expression, Saint John's University, College Ville, Minnesota, Marcel Breuer, 1967. © C. Kabre

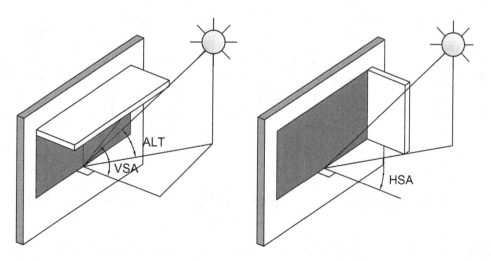

FIGURE 3.44 Vertical shadow angle (VSA) and horizontal shadow angle (HAS).

i) The horizontal shadow angle (HSA) characterizes a vertical shading device, and it is the difference between the solar azimuth and wall azimuth, the same as the horizontal component of the angle of incidence.

ii) The vertical shadow angle (VSA) characterizes a horizontal shading device, e.g., along with a horizontal projection from the wall, and it is measured on a vertical plane normal to the elevation considered.

There are three basic types of external shading devices (Figure 3.45):

i) Horizontal shading device
ii) Vertical shading device
iii) Egg-crate shading device

Horizontal shading devices are beneficial on south-facing windows during the summer since the sun is opposite to the building's face and at a high angle in the sky. Although less effective, the horizontal overhang may also work well on the east, southeast, southwest, and west orientations. Their performance will be measured by a VSA.

Vertical shading devices are required on north-facing windows in equatorial and low latitude climates because, during the summer, the sun rises north of east and sets north of west. Since the sun is low in the sky at these times, the horizontal overhangs are not effective, and small vertical fins work well on the north façade. The HSA measures their performance.

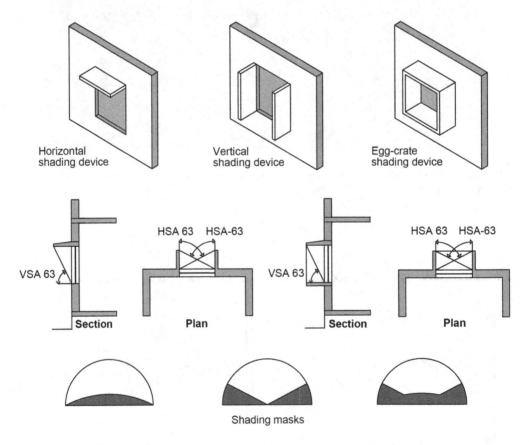

FIGURE 3.45 Three basic types of shading devices and corresponding shading masks.

Egg-crate shading devices are combinations of horizontal and vertical elements. This type of device is most effective when the sun is to one side of the elevation, such as a north-east or south-west elevation. These can also be effective for south orientation. The performance of shading devices is specified by two angles: the horizontal and the vertical shadow angles. East- and west-facing windows receive the low altitude angle of the sun in the morning and evening. A combination of horizontal and vertical shading devices is found useful in some cases depending on the latitude.

The process of design of external shading devices mainly involves three steps:

i) To delineate overheated and underheated periods
ii) To plot the overheated period on the sun-path diagram
iii) To find optimum shadow angles for given window orientation

Delineate Overheated and Underheated Periods

For any given location the bioclimatic analysis based on the temperature data and thermal neutrality delineates the 'underheated' (cold discomfort) and 'overheated' (hot discomfort) periods. The over-heated period is the one when $T_o > T_c$ (mean outdoor temperature is greater than thermal neutrality for the hottest month); the solar radiation is to be intercepted. The underheated period is the one when the mean outdoor temperature is below the lower limit of thermal neutrality of the coldest month ($T_o < T_c - 2.5°C$); the solar radiation is desirable in the building.

Plot the Overheated Period on the Sun-Path Diagram

Table 3.14 illustrates the climate analysis of Austin with the use of hourly temperature for an average day of each month; the thermal neutrality for the hottest month July is 26.8°C, and the lower limit of thermal neutrality for the coldest month January is 18.9°C. The dark color designates the overheated period, the mid-shade designates a comfortable period, and the light area shows underheated period. Three periods are delineated for 50 US cities are available on publisher's website, Table 7.52 to 7.101.

A sun-path diagram can be used as a chart to plot the overheated period, where the long east-west arcs represent the month and the cross curves are the hour lines. The diagram can be used to find the specific position of the sun during the overheated period. Sun-path diagrams for 50 US cities are available on publisher's website, Figures 7.2 to 7.51.

The overheated periods will differ according to city or place, and the sun position will differ according to the latitude. Except for the summer and winter solstice date, each sun-path arc is representing two dates; therefore, two sun-path diagrams must be used, one from December 22 to June 22 and the other from June 22 to December 22. The overheated period (when outdoor temperature greater than 26.8°C) is plotted on the sun-path diagram (Figure 3.46). It is relevant to mention that the overheated period is asymmetrical on both the sun-path diagrams; December 22–June 22 half-year has less overheated period than June 22–December 22 half (temperatures are lagging behind solar heating by a couple of months due to heat-retaining capacity of the earth). Hence the shading device design will have to be optimized between the two limits.

Determine Geometrical Parameters of Shading Devices

To determine the geometrical parameters of shading devices, draw a line across the center of the sun-path diagram, representing the plan of the window wall such that the normal to the wall is the orientation; Figure 3.47 shows four orientations, case 'a' is north, case 'b' is south, case 'c' is north-east, and case 'd' is southwest, and the sunrise and sunset for each wall is marked with small circles. The sun shines for the maximum period on the south and southwest wall.

The plot on the sun-path diagram the overheated period when shading is desirable. Overlay the shadow protractor (Figure 7.52 available on publisher's website) to find the horizontal shadow angle and (or) vertical shadow angle for the given window orientation (Figure 3.48).

TABLE 3.14

Overheated, Underheated Periods Austin N 30° 17′ W 97°44′

Months / Hours	Jan	Feb	Mar	Apr	May	Jun	Jul	Aug	Sep	Oct	Nov	Dec
0:01–1:00	9.4	11.2	14.5	18.8	19.8	23.2	25.7	23.2	21.6	19.5	15.5	10.1
1:01–2:00	9.0	10.9	14.0	18.4	19.4	22.5	25.3	23.0	21.4	19.4	15.2	9.5
2:01–3:00	8.8	10.6	13.8	18.3	19.3	22.5	25.1	22.7	21.0	19.0	15.1	9.3
3:01–4:00	8.5	10.3	13.4	17.9	19.1	22.3	24.6	22.5	20.7	18.9	15.0	9.4
4:01–5:00	8.1	9.8	13.2	17.9	19.0	22.2	24.2	22.1	20.3	18.7	14.9	9.0
5:01–6:00	7.7	9.7	13.0	17.9	18.9	22.1	23.8	22.0	20.2	18.7	14.8	8.9
6:01–7:00	7.5	9.8	13.0	18.3	20.0	23.7	24.9	23.0	20.4	18.7	14.8	8.8
7:01–8:00	7.5	10.0	13.6	19.2	21.4	25.3	26.5	25.2	22.2	19.6	15.2	8.9
8:01–9:00	8.6	11.5	15.0	20.7	22.9	26.9	28.3	27.3	24.0	20.7	16.3	10.1
9:01–10:00	10.0	13.2	16.4	22.1	24.3	28.1	30.0	29.1	25.8	22.0	17.3	12.0
10:01–11:00	11.4	14.7	17.8	23.4	25.7	29.3	30.7	30.6	27.3	23.3	18.9	13.4
11:01–12:00	13.0	15.7	19.3	24.6	26.8	30.5	32.1	31.6	28.3	24.5	19.9	14.6
12:01–13:00	14.0	16.7	20.3	25.5	27.9	31.6	33.2	32.4	29.3	25.0	20.8	15.9
13:01–14:00	14.8	17.7	21.4	25.9	28.6	32.0	33.5	33.2	29.9	25.6	21.4	16.9
14:01–15:00	15.3	17.9	21.9	26.2	29.1	32.2	33.7	33.6	30.1	25.9	21.6	17.5
15:01–16:00	15.6	18.2	21.9	26.2	29.1	31.9	34.0	33.5	29.8	25.7	21.3	17.1
16:01–17:00	14.7	17.7	21.6	25.6	28.7	31.7	33.8	32.7	29.1	25.0	20.5	16.4
17:01–18:00	13.5	16.8	20.7	24.7	28.0	31.1	33.1	31.5	27.8	23.9	19.1	15.1
18:01–19:00	12.2	15.1	19.0	23.2	26.5	29.8	31.8	29.5	25.9	22.7	17.9	13.9
19:01–20:00	11.5	14.3	17.7	21.8	24.0	27.6	30.1	27.5	24.8	22.0	17.4	13.1
20:01–21:00	11.0	13.6	16.6	21.0	22.6	26.1	29.0	26.2	24.0	21.4	16.7	12.6
21:01–22:00	10.5	12.9	15.9	20.3	21.8	25.3	28.0	24.7	23.2	21.0	16.6	12.0
22:01–23:00	10.0	12.7	15.3	19.8	21.1	24.3	27.0	24.3	22.6	20.5	16.1	11.3
23:01–24:00	9.5	12.3	14.7	19.4	20.6	23.8	26.3	23.7	22.1	19.8	16.0	10.8
Legend	Underheated		<18.9°C		Comfortable					Overheated		>26.8°C

Select the horizontal shadow angle and (or) vertical shadow angles which would cover as much as possible of the given overheated period while taking care not to cover too much of the underheated period, when solar radiation is needed. The efficiency of a shading device depends on the proportionate success with which it covers the overheated area without covering the underheated area.

A north-east facing window is illustrated for Austin in Figure 3.49 below the given combinations of vertical and horizontal shadow angles may give satisfactory results:

- A combination of VSA 48° and HSA 19° would provide complete shading of the over-heated period but still exclude the mid-winter sun after 12:00 h.
- A combination of VSA 50° and HSA 16° would also give the required shading but still exclude the mid-winter sun after 12:00 h.

A window facing the equator (south in the northern hemisphere and due north in the southern hemisphere) is the easiest to handle, it can give an automatic seasonal adjustment: full shading in summer but allowing solar heat gain in winter. For complete summer 6 months sun exclusion (for an equinox cut-off), the VSA will have to be 90° – LAT; e.g., for LAT = 30.28° it will be VSA = 90 – 30.28° = 59.72°.

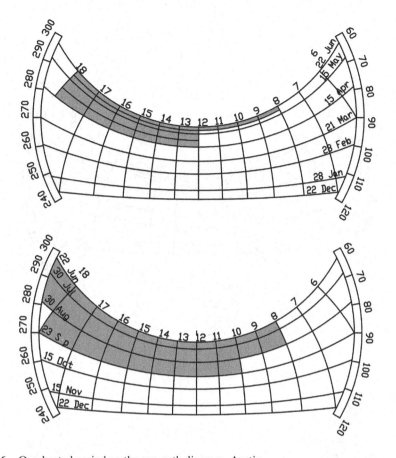

FIGURE 3.46 Overheated period on the sun-path diagram, Austin.

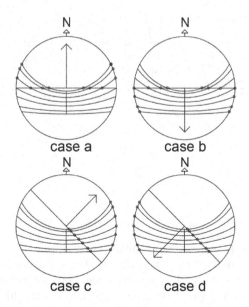

FIGURE 3.47 A window wall in four different orientations, (a) north, (b) south, (c) north-east, and (d) south-west.

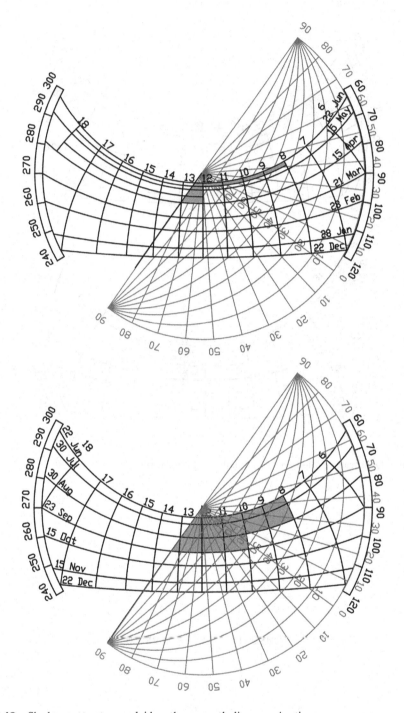

FIGURE 3.48 Shadow protractor overlaid on the sun-path diagram, Austin.

This shading mask exactly matches the equinox sun-path line. For other dates, the match is not so exact, but still quite similar to the sun-path line. For orientations other than due north the situation is not so simple. A combination of vertical and horizontal shading devices may be the most appropriate answer. During any period when the sun is behind this line, its radiation would not reach that wall; thus, it is of no interest. The illustration shows a north-east orientation (LAT 30.28°, ORI 45°).

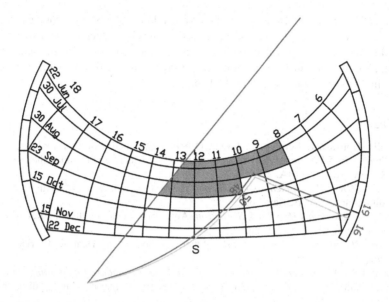

FIGURE 3.49 Optimum shadow angles, Austin.

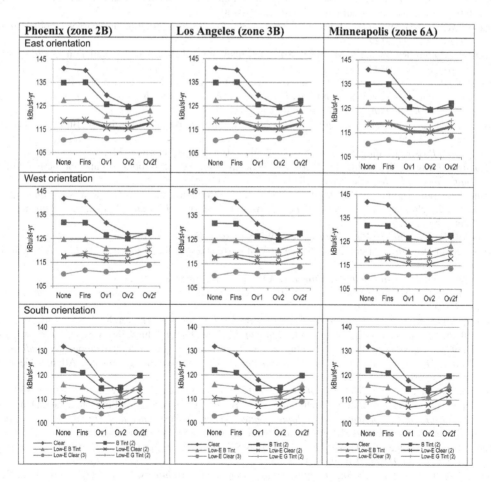

FIGURE 3.50 A parametric study of external shading devices and high-performance glass (WWR = 0.3). Source: after Carmody, Haglund (2006)

A study at the University of Minnesota (Carmody, Haglund 2012) shows that the exterior shading devices (overhangs and vertical fins) result in energy savings by reducing direct solar gain through windows, and sometimes it is possible to obtain performance equivalent to unshaded high-performance glazing. Figure 3.50 illustrates a parametric study for three cities, Phoenix, Los Angeles, and Minneapolis, in different climate zones in the United States.

REFERENCES

Arens E, McNall P, Gonzalez R, Berglund L, Zeren L (1980) A new bioclimatic chart for Passive Solar Design. *Proceedings of the 5th National Passive Solar Conference*, American Section of the International Solar Energy Society., Amherst, MA.

ASHRAE (1972) *ASHRAE Fundamentals Handbook (SI)*. American Society of Heating, Refrigerating and Air-Conditioning Engineers, New York.

ASHRAE (1997) *Fundamentals Handbook SI Edition*. American Society of Heating, Refrigerating and Air Conditioning Engineers, Inc., Atlanta.

ASHRAE (2009) *Handbook of Fundamentals*. American Society of Heating Refrigerating and Air Conditioning Engineers, Inc., Atlanta.

ASHRAE (2010) *ANSI/ASHRAE/IES Standard 90.1-2010, Energy Standard for Buildings Except Low-Rise Residential Buildings*. American Society of Heating Refrigerating and Air Conditioning Engineers, Inc., Atlanta.

ASHRAE (2016) *ASHRAE Handbook – HVAC systems and equipment (SI)*. American Society of Heating, Refrigerating and Air-Conditioning Engineers, Atlanta.

Bahadori MN (1985) An improved design of wind towers for natural ventilation and passive cooling. *Solar Energy*, 35(2), 119–129.

Bansal N, Mathur R, Bhandari M (1993) Solar chimney for enhanced stack ventilation. *Building and Environment*, 28(3), 373–377.

Bliss, RJV (1961) Atmospheric radiation near the surface of the ground. *Solar Energy* 5(3), 103–120.

Carmody J, Haglund K (2006) *External Shading Devices in Commercial Buildings, the Impact on Energy Use, Peak Demand and Glare Control*. Regents of the University of Minnesota.

CIBSE (1999) *Environmental Design: CIBSE Guide A*. The Chartered Institution of Building Services Engineers, London.

Cook J (1989) *Passive Cooling*. Solar Heat Technologies/MIT Press, Cambridge, MA.

Cristofalo S, Orioli S, Silvestrini G (1989) Thermal behavior of 'Scirocco rooms' in ancient Sicilian villas. *Tunneling and Underground Space Technology*, 4(4), 471–473.

Cunningham WA, Thompson TL (1986) Passive cooling with natural draft cooling towers in combination with solar chimneys. *Proceedings of the Passive Low Energy Architecture (PLEA)*, Pecs, Hungary, 1–5 September, pp 23–34.

Davies MG (2004) *Building Heat Transfer*. John Wiley & Sons, Chichester, England.

Ford B (2012) Downdraught cooling: An overview of current research and practice. *Architectural Science Review*, 55(4), 237–240.

Givoni B (1969) *Man, Climate and Architecture*. Elsevier Publishing Company, London.

Givoni B (1991) Performance and applicability of passive and low-energy cooling systems. *Energy and Buildings*, 17(3), 177–199.

Givoni B (1994) *Passive and Low Energy Cooling of Buildings*. Van Nostrand Reinhold, New York.

Givoni B (1998) *Climate Considerations in Building and Urban Design*. John Wiley & Sons, Inc, New York.

Hay H (1978) A passive heating and cooling system from concept to commercialization. *Proceedings of Annual Meeting of American Section of the International Solar Energy Society*, Atlanta, Georgia, 28–31 August, pp. 262–272.

Heiselberg, PK (ed.) (2002) *Principles of hybrid ventilation*. International Energy Agency Annex 35 Final report. Department of Building Technology and Structural Engineering, Aalborg University, Denmark.

ISO (2007) *Building Components and Building Elements-Thermal Resistance and Thermal Transmittance Calculation Method: 6946*. International Organization for Standardization, Geneva.

Koenigsberger OH, Ingersoll TG, Mayhem A, Szokolay SV (1974) *Manual of Tropical Housing and Building: Part I. Climatic Design*. Longman, London.

Lechner N (2001) *Heating, Cooling, Lighting Design Methods for Architects*. John Wiley & Sons Ltd., New York.

Loudon AG (1968) *Summertime Temperatures in Buildings.* Building Research Station, Garston, England.

Mackey CO, Wright LT (1943) Summer comfort factors as influenced by the thermal properties of building materials. *ASHVE Transactions Heating, Piping & AC Section,* 49, 148–174.

Milbank NO and Harrington-Lynn J (1974) *Thermal Response and the Admittance Procedure,* Building Research Station, Garston, England.

Milne MB, Givoni B (1979) Architectural design based on climate. In: Watson D (ed.), *Energy Conservation Through Building Design.* McGraw-Hill, Inc, New York, 96–113.

Muneer T (2004) *Solar Radiation and Daylight Models.* Elsevier Butterworth Heinemann, Amsterdam.

Olgyay V (1963) *Design with Climate - Bioclimatic Approach to Architectural Regionalism.* Princeton University Press, Princeton, NJ.

Olgyay V, Olgyay A (1957) *Solar Control and Shading Devices.* Princeton University Press, Princeton, NJ.

Pescod D (1976) *Energy Savings and Performance Limitations with Evaporative Cooling in Australia.* Techn. Report no. 5. CSIRO, Div. Mech. Eng., Highett, Vic., Australia.

Rao KR, Ballantyne ER (1970) *Some Investigation on the Sol-Air Temperature Concept.* Division of Building Research Technical Paper, no. 27. CSIRO, Melbourne, Australia.

Szokolay SV (2008) *Introduction to Architectural Science: The Basis of Sustainable Design.* Architectural Press, Elsevier, Oxford.

Watson D (1997) Bioclimatic design. In: Watson, D, Crosbie, MJ, Callender, JH (eds), *Time-Savers Standards for Architectural Design Data.* McGraw-Hill, New York, pp 23–34.

Watson D, Labs K (1983) *Climatic Design: Energy-Efficient Building Principles and Practices.* McGraw-Hill, New York.

Wooldrige MJ, Chapman HL, Pescod D (1976) Indirect evaporative cooling system. *ASHRAE Transaction,* 82(1), 146–155.

Yellott JI (1982) Passive and hybrid cooling research. In: Boer KW, Duffie JA (eds), *Advances in Solar Energy,* vol. 1. American Solar Energy Society Inc., New York, pp 241–263.

Zhao D, Aili A, Zhai Y, Xu S, Tan G, Yin X, Yang R (2019) Radiative sky cooling: Fundamental principles, materials, and applications. *Applied Physics Reviews,* 6(2), 1–40.

4 Luminous Environment Design Strategies

4.1 INTRODUCTION

Daylight is part of the architecture, in its historical, theoretical, and technical conception, with a unique capacity to inspire people and illuminate the elements of its design. The chronicles of architecture are synchronized with the evolution of the window from the earliest cave openings letting in light and air as well as heat and cold; the window has been the vehicle for the ingress of daylight, and ultimately to the dramatic interiors of the medieval cathedral, the Baroque churches, or the Renaissance buildings of the 18th century. The window has developed over the centuries and became an element of architectural expression during all periods. The infill of the window initially was made of thin slabs of stone, sheets of mica, or oiled paper; later with the advancement of technology, glass was introduced.

Many master architects, Le Corbusier, Alvar Aalto, Louis Kahn, and Paul Rudolph, spent their careers mastering the interaction of light and architecture. Weisman Art Museum is one of the museums designed by the architect Frank Gehry with an incredible sculptural form. The brick building is clad in panels of stainless steel. The form and the material of the building reflect the natural light and change its colors throughout the day. The interior also uses natural light. The most important quality of sunlight is its unique and dynamic properties, which can transform a stationary building into a dynamic element (Figure 4.1). Lighting is also a medium of architectural expression.

Indoor illumination was primarily provided through side lighting (windows) and top lighting (skylights) until the 19th century; electric lighting became mainstream with the development of many high rise buildings and underground spaces in the 20th century. However, in recent years it has been found that lighting is the second-highest energy end-use, at an average of 14% of the total building energy use, according to the Department of Energy's 2011 Building Energy Data Book. Lighting is an end-use consistent across commercial building types and all climate zones; daylighting can be considered as a major part of sustainable buildings for saving electric lighting and providing benefits such as health, visual comfort, and productivity of the occupants.

A symbiotic relationship between exterior illumination and interior illumination is the main concern of sustainable architecture. The objective is to harness the radiant energy from the sun, the sky, and the earth and external obstructions (reflected light) – in all its highly dynamic nature – to illuminate the interior space, while eliminating solar heat gain. The building envelope is the interface between the exterior illumination and the interior lighting requirement. Thus, the building envelope becomes the opening, the filter, the reflector, the diffuser, and the shaper of daylight. The quality of daylight varies across the earth and in its unique and dynamic properties adds visual beauty and complexity to architecture. The dynamic radiant energy exchanges characterize this relation and foster the luminous environmental performance of buildings and the visual comfort of their occupants and affect the energy use. Knowledge and understanding of the physical principles underlying these radiant energy exchanges, along with the concepts/strategies and computational tools to translate them into sustainable architecture, form the core of this chapter.

Daylighting design has taken up two approaches: architectural daylighting and technological daylighting (Guzowski 1999). Architectural daylighting is the simplest and concerns optimized building orientation and form, optimized building components (such as windows and skylights), and their placement in the given climate conditions. On the other hand technological daylighting concerns supplemental systems and components such as shading; high-performance glazing (fixed and

FIGURE 4.1   Weisman Art Museum, University of Minnesota, Minneapolis, Frank Gehry 1993. © C Kabre

variable performance); advanced glazing systems (holographic gratings, prisms, etc.); light pipes; solar concentrators; heliostats to gather, distribute, and control light; photovoltaic cladding; and electric lighting control systems. Daylighting requires an integrated design approach to be sustainable because it can involve architectural design decisions, integration of advanced technological components, lighting controls, and lighting design criteria.

This chapter is structured into three sections. Section 4.2 covers the fundamentals of light, Section 4.3 deals with daylighting strategies, and Section 4.4 discusses the daylighting prediction techniques.

4.2   FUNDAMENTALS OF LIGHT

4.2.1   PHYSICS OF LIGHT

The Illuminating Engineering Society of North America (IESNA) defines light as radiant energy that is capable of exciting the human retina and creating a visual sensation (IESNA 2013). As a physical quantity, the light is a narrow wavelength band of electromagnetic radiations (from about 380 to 780 nm) which is perceived by human eyes (Figure 2.4).

4.2.1.1   Attributes of Light

All forms of electromagnetic radiation are transmitted at the same speed in a vacuum; the velocity of light (c) is approximately 3×10^8 m/s (i.e. 299,793 km/s or 186,282 miles/s). Light has two major attributes, which are quality and quantity. Its quantitative aspects are discussed under the heading photometry. Its quality is characterized by wavelength (λ) and its reciprocal, the frequency (f): the product of these gives the velocity. The wavelength and velocity may be altered by the medium through which it passes, but the frequency remains constant, independent of the medium. Thus, through equation 4.1, velocity can be calculated.

TABLE 4.1

Wavelengths of Various Colors

Color	Wavelength Band (nm)
Red	780–660
Orange	660–610
Yellow	610–570
Green/yellow	570–550
Green	550–510
Blue/green	510–480
Blue	480–440
Violet	440–380

$$\text{Velocity} = \frac{\lambda \times f}{n} \tag{4.1}$$

where

n = index of refraction of the medium

λ = wavelength in a vacuum

f = frequency in Hz

4.2.1.2 Color of Light

The color of light is determined by its spectrum or spectral composition, Table 4.1. Light of a specific wavelength is described as monochromatic. White light contains all wavelengths, and it is described as polychromatic light. The color of broad-band light depends on the relative magnitude of its components and on its spectral composition. A continuous-spectrum white light can be observed when it splits through a prism which is perceived as colors.

The three-color theory of light distinguishes red, green, and blue as the primary colors, and any color can be defined in terms of its redness, greenness, and blueness.

4.2.1.3 Color of Surfaces

As the colored light from different sources is additive, the surface colors are subtractive. Surfaces appear in different colors because they absorb some colors (wavelengths) and reflect or transmit other colors. The colors perceived are the wavelengths that are reflected or transmitted. When a surface is painted blue, it appears to be this color, as it absorbs everything else and reflects only the blue component of the incident light. If a blue surface is illuminated by red light, it will appear to be black, because the red would be absorbed and the light imparted has no blue component to be reflected. The lighting would need to be of a continuous-spectrum white to reveal all colors, including the blue, thus producing good *color rendering*.

The international system for specifying opaque colors of surfaces is the Munsell color system, named after the American art instructor and painter Albert H. Munsell. It defines colors by measured scales of hue, value, and chroma. Figure 4.2 presents the Munsell color wheel.

a) **Hue:** it refers to color itself, a base-10 system for naming colors, using five principal hues and five secondary or intermediate hues, and subdividing each hue into 10 steps.

b) **Value (V) or lightness:** it is the unit of measure of reflectance, lightness, or darkness; it is measured on a scale from 0 (absolute black) to 10 (the perfect white). Practically values from 1 to 9 are encountered. The value (V) and reflectance (R) are correlated with

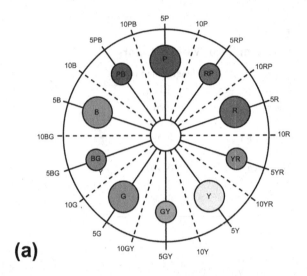

(a)

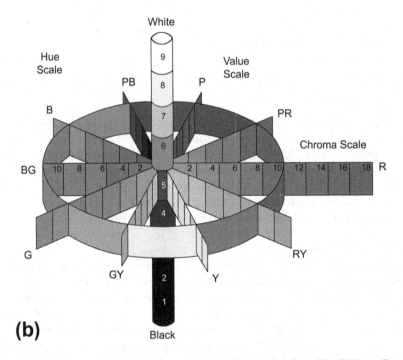

(b)

FIGURE 4.2 Munsell color wheel system. Legend Principal hues: R = Red; Y = Yellow; G = Green; B = Blue; P = Purple. Intermediate hues: YR = Yellow-Red (orange); GY = Green-Yellow; BG= Blue-Green; PB = Purple-Blue; RP = Red-Purple. Source: after munsell.com, line work by Ar Rishabh Jangra

Equations 4.2 and 4.3 (Longmore and Petherbridge 1961) and compared with CIE luminous reflectance factors in Table 4.2:

$$V = \sqrt{R} + 0.5 \qquad\qquad\qquad\qquad (4.2)$$

$$R = V(V-1) \qquad\qquad\qquad\qquad (4.3)$$

c) **Chroma:** the saturation or brilliance (intensity) of color. All colors have at least 10 or more classes from neutral gray to highly saturated.

TABLE 4.2

Surface Reflectance (*R*) Estimated for Various Munsell Values (*V*) vs CIE Luminous Reflectance Factor

Munsell Value (*V*)	*R* = *V* (*V* – 1)%	Luminous Reflectance Factor (%) CIE
1	0	0
1.5	0.75	2.0
2.0	2.0	3.0
2.5	3.0	4.5
3.0	6.0	6.4
3.5	8.75	8.8
4.0	12.0	11.7
4.5	15.75	15.2
5.0	20.0	19.3
5.5	24.75	24.0
6.0	30.0	29.3
6.5	35.75	35.3
7.0	42.0	42.0
7.5	48.75	49.4
8.0	56.0	57.6
8.5	64.0	66.7
9.0	72.0	76.7

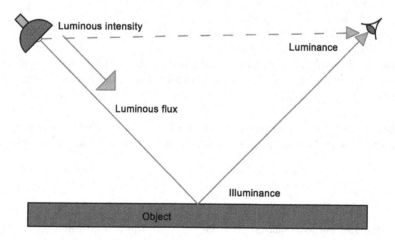

FIGURE 4.3   A simple luminous system. Source: after Szokolay (2008)

Photometry

It is defined as the science of measuring the light, in terms of the brightness or an illuminated surface to the human eye perceiving it from the source. The luminous system is best understood as a combination of a light source (lamp), an object illuminated, and an eye perceiving the light, both from the source and reflected by the object (Figure 4.3). There are four measurable photometric quantities:

a) **Luminous intensity (*I*)** is a characteristic of a light source, measured in units of candela (cd), which is the international standard candle, defined as the luminous intensity, in a given direction, of a 555.016 nm source that emits monochromatic radiation of frequency

540×10^{12} Hz and that has a radiant intensity in that direction of 1/683 W/sr (NBS 1991). The definition expresses the candela in terms of the watt (W) and the steradian (sr). The steradian is defined as the solid angle subtending an area on the surface of a sphere equal to the square of the sphere's radius. The unit of power, the watt, is defined as 1 J/s (energy per unit time) or in base units (1 $m^2 \times$ kg $\times$ s^{-3}).

Two important derived units based on the candela are those of luminous flux and illuminance.

b) **Luminous flux (φ, phi)** or flow of light, measured with the unit lumen (lm) in both SI and I-P systems, which is defined as the flux emitted within 1 steradian (sr) by an isotropic point source of I = 1 cd that radiates light uniformly in all directions. Thus 1 cd emits a total of 4π, or 12.57 lumens. In physical terms, the lumen is a unit of photometric power and is often a criterion of light bulb comparison.

c) **Illuminance (E)**, the measure of the illumination of a surface (illumination is the process; illuminance is the outcome); the amount of luminous flux incident on the surface, per unit area. The SI unit is the lux (lx), which is the illuminance caused by 1 lm incident on 1 m^2 area (i.e., the incident flux density of 1 lm/m^2). The I-P unit is foodcandle (fc), which is the illuminance caused by 1 lm incident on 1 ft^2 area (i.e., the incident flux density of 1 lm/ft^2). Table 4.3 presents some typical illuminance values.

d) **Luminance (L)**, the measure of the perceived brightness of a surface, that is, the visual effect that illumination produces when looked at from a given direction. Luminance depends not only on the illuminance on an object and its reflective properties but also on its projected area on a plane perpendicular to the direction of view. There is a direct relationship between the luminance of a viewed object and the illuminance of the resulting image on the retina of the eye. The unit of luminance is candela per square meter (cd/m^2 sometimes referred to as a *nit*, rarely used in English), which is the unit intensity of a source of unit area (source intensity divided by its apparent area viewed from the nominated direction).

(a 1 cd point source enclosed in a 1 m radius spherical diffuser has a projected area of πm^2; therefore, its luminance will be $1/\pi cd/m^2$)

e) **Luminous efficacy (F)** of a light source is defined as the ratio of the total luminous flux (in lumens) to the total power input (in watts), measured in lumens per watt (lm/W). As per the surface area of the sphere, i.e., $4.\pi.r^2$, the center of the sphere has 4π steradians.

Luminous flux (lm) is of the same physical dimension as watt (W), and illuminance (lux) is the same as irradiance (W/m^2), but the latter are energy units and the previous are luminous quantities.

TABLE 4.3
Some Typical Photometric Values

Total Flux Output of Some Sources	lm	Typical Illuminances	lux
Bicycle lamp	10	Bright sunny day, outdoors	100,000
40 W incandescent lamp	325	Overcast day, outdoors	10,000
40 W fluorescent lamp	2800	Moderately lit desk	300
140 W sodium lamp	13,000	Average general room lighting	100
400 W mercury lamp	20,000	Full moonlit night, outdoors	0.1
Typical Luminance Values	**cd/m^2**		
Sun (1600 Mcd/m^2)	1,600,000,000		
Filament in a clear incandescent lamp	7,000,000		
Fluorescent lamp (tube surface)	8000		
Full moon	2500		

Figure 4.4 explains the relationship between candelas, lumens, lux, and food candles. Figure 4.5 shows the luminous effect of the light of different wavelengths.

4.2.1.4 Transmission of Light

The light travels in a straight line in a vacuum or a transparent homogeneous medium (air). The *inverse square law* states that the illumination E at a point on a surface varies directly with the luminous intensity I of the source and inversely as the square of the distance d between the source and the point. Figure 4.6(a) illustrates how the same quantity of light flux is distributed over a greater area, as the distance from the source to the surface is increased, while the flux density (illuminance) is reduced. If the surface at the point is normal to the direction of the incident light, the law may be expressed by Equation 4.4.

$$E = I \times d^2 \tag{4.4}$$

The cosine law, also known as Lambert's law, states that the illuminance on any surface varies as the cosine of the angle of incidence. The angle of incidence, θ, is the angle between the normal to

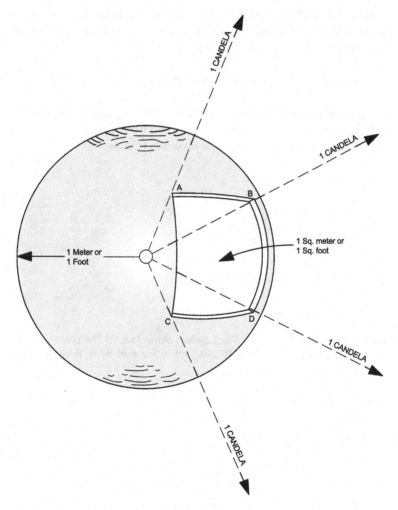

FIGURE 4.4 Relationship between candelas, lumens, lux, and food candles. Source: after IESNA (2013), line work by Ar Rohan Bhatnagar

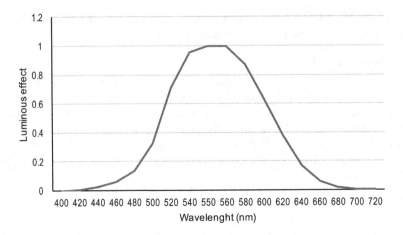

FIGURE 4.5   Relative luminous effect of light of different wavelengths. Source: based on data in Lim et al. (1979)

the surface and the direction of the incident light. Figure 4.6(b) shows that the light flux striking a surface at angles other than normal angle is distributed over a greater area, and the inverse square law and the cosine law can be combined as per Equation 4.5.

$$E_2 = I \times d^2 cos\theta = E_1 \times cos\theta \qquad (4.5)$$

Light behaves in different ways with various mediums. A pane of glass, plastic, acrylic, crystal, and so forth are transparent mediums; a sheet of wax paper and frosted glass are translucent mediums; and a wooden panel and a ceramic tile are opaque mediums (Figure 4.7).

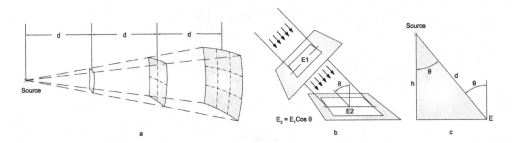

FIGURE 4.6   (a) The inverse-square law. (b) The Lambert cosine law. (c) The cosine cubed law explaining the transformation of the formula. Source: after IESNA (2013), line work by Ar Rohan Bhatnagar

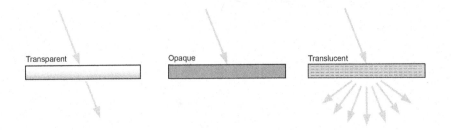

FIGURE 4.7   Transmission of light.

Light incident on an object can be distributed in three ways: transmitted, reflected, or absorbed. The corresponding properties are transmittance (t), reflectance (r), and absorptance (a), and in all cases $t + r + a = 1$ (as discussed in Sections 3.6.1 and 3.8.3 in relation to solar radiation). All three terms are functions of radiation wavelengths, and when applied to the visible wavelengths (light), they may be referred to as 'optical,' e.g. optical transmittance, optical reflectance, or optical absorptance.

Light reflected from a matt surface will be diffused. Most often a mixture of the two kinds of reflections will occur, termed as 'semi diffused' or 'spread' reflection, depending upon the relative magnitude of the two components.

Reflection is the process by which a part of light falling on a surface leaves that surface from the incident side. Reflection may be specular (a plane mirror or a polished metal), spread (corrugated, etched, or hammered), diffuse (ordinary building surfaces) or compound, and selective or nonselective.

Specular reflection happens when parallel rays of incident light remain parallel after reflection from a surface; the surface is a 'plane mirror.' The rules of geometrical optics apply to such surfaces: the angle between the reflected ray and the normal to the surface will equal the angle between the incident ray and the normal; from a convex mirror the reflected rays will be divergent, and from a concave mirror they will be convergent. Spread reflection spreads parallel rays into a cone of reflected rays. Diffuse reflection reflects the light at many angles. In diffuse reflection each ray falling on an infinitesimal particle obeys the law of reflection, but as the surfaces of the particle are in different planes, they reflect the light at many angles. Most common materials are compound reflectors and exhibit all three reflection components (specular, spread, and diffuse) to varying degrees (Figure 4.8).

The term 'transmission' is the movement of electromagnetic waves (whether visible light, radio waves, ultraviolet, etc.) through a material. If the light is reflected off the surface and absorbed by the molecules in the material, this transmission can be reduced or stopped.

The term 'absorptivity' is a property of the material, indicating the absorption per unit thickness, while 'absorptance' is the property of a body of given thickness or surface quality.

4.2.2 VISION

Perhaps the most important communication channel of the human with his or her environment is vision, and light is a prerequisite of vision. Thus, the objective of lighting is to provide visual

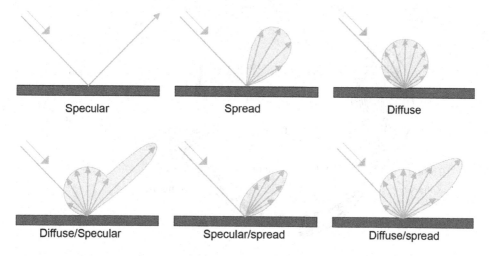

FIGURE 4.8 Taxonomy of reflective surfaces.

conditions in which occupants can function effectively, efficiently, and comfortably. To predict human behavior as a function of the lighting conditions, it is important to understand the physical, physiological, and perceptual characteristics of the visual system. This section highlights some of the basic relationships between light and vision.

4.2.2.1 The Eye and Brain

The visual system involves the eye and the brain working together to process the image and interpret the visual environment. The optical elements of the eye form an image of the world on the retina. At the retina, photons of light are absorbed by the photoreceptors and converted to electrical signals. These signals are transmitted by the optic nerve to the lateral geniculate nucleus (LGN) and then to the visual cortex for visual processing. In addition to the neural pathways from the eye to the visual cortex, there are many other pathways leaving the optic nerve shortly after it exits the eye that control pupil size, eye movements, and circadian rhythms.

The structure of the eye can be divided into three distinct parts (Figure 4.9):

- The ocularmotor components consist of three pairs of eye muscles, which position the lines of sight of the two eyes so that they are both pointed toward the same object concerned. The line of sight of the eye passes through the part of the retina used for discriminating fine detail, the fovea.
- The optical components (the cornea, crystalline lens, pupil, and intraocular humors) form an image of the target on the retina. For this to occur, light has to be transmitted through the eye without excessive absorption and scattering, and the image of the target has to be focused on the retina.
- The neurological and supportive components (the retina and optic nerve). The posterior 80% of the eye is enclosed by three layers of tissue: the outermost sclera (protects and shapes the eye), the choroid (nourishes the eye), and the innermost retina (transduces light into electrical signals that are sent to the brain). Together, the choroid and the retina constitute the fundus.

The human retina incorporates two main classes of photoreceptors: 75–150 million rods (dim light vision) and 6–7 million cones (bright light vision), which are distinguished by their morphology and

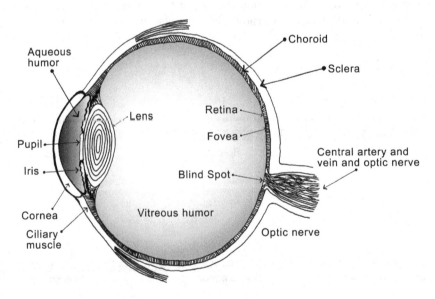

FIGURE 4.9 Section of the human eye. Source: adapted from IESNA (2013), line work by Ar Kapil Grover

by the spectral sensitivity of the photopigments which they contain. Rods, which are absent in the fovea, contain the same photopigment (rhodopsin, visual purple), which has a peak spectral sensitivity at approximately 507 nm, can detect luminances from 1/1000 cd/m² to approximately 120 cd/m² (scotopic vision), and lacks color sensitivity. Cones are concentrated in or around the fovea, which is sensitive to both quantity and quality (color) of light. Cones are divided into three known classes, each characterized by the photopigments that it contains: erythrolabe, chlorolabe, or cyanolabe. All three cone types acting together have a peak spectral sensitivity at approximately 555 nm and detect luminances in the range of 3 to 1000 cd/m² (photopic vision) and are capable of color vision. The light levels where both are operational are called mesopic vision.

The pupil's response is practically instantaneous; the iris constricts and dilates in response to increased and decreased levels of retinal illumination. Iris constriction has a shorter latency and is faster (approximately 0.3 s) than dilation (approximately 1.5 s). The typical range in pupil diameter for young people is from 3 mm for high retinal illuminances to 8 mm for low retinal illuminances. Photochemical adaptation, the second adaptation mechanism of the eye, is the variation of the retina's sensitivity by varying the photochemical compound present (e.g. of the visual purple). While the pupils' response to changed lighting conditions is almost instantaneous, the adaptation of the retina to dark conditions may take up to 30 min, as more visual purple is produced. Adaptation to brighter light is no more than about 3 min, as the visual purple is being removed. Both adaptation mechanisms respond to the average luminance of the field of vision, starting from the darkness.

4.2.2.2 Threshold Visual Performance

Threshold visual performance measurements depend on the task or the object, the lighting conditions, and the visual system of the observer. The primary factors associated with the task or object are visual acuity, contrast sensitivity, including color contrast and exposure time needed or given.

The visual size of an object or task can be measured in a plane of two dimensions as a visual angle (VA), approximated by equation 4.6.

$$\text{Visual angle} = 2\arctan\frac{d \times \cos\theta}{2l} \tag{4.6}$$

where d is the physical dimension of the object, θ is the angle of inclination, of the target from normal to the line of sight and l is the distance from the viewer (Figure 4.10).

Visual acuity (a) or sharpness of vision depends on illuminance. The word 'acuity' is the ability to resolve fine details; there are different kinds of acuity. For example, vernier acuity is the ability to identify misalignment between two lines, and resolution acuity is the ability to distinguish between two points, such as two nighttime stars. Visual acuity is represented by the reciprocal of

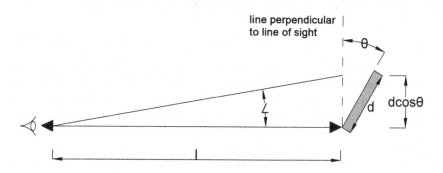

FIGURE 4.10 Dimensions required for calculating a visual angle.

visual angle (expressed in minutes of arc), subtended at the eye by opposite extremes of the least perceptible detail:

if VA = 2′ then $a = 1/2 = 0.5$
or
if VA = 4′ then $a = 1/4 = 0.25$

Luminance contrast is defined as the ratio of luminance difference to the lower of the two luminances (Equation 4.7).

$$C = \left| \frac{(L_T - L_B)}{L_B} \right| \tag{4.7}$$

where L_T and L_B are the luminance of the task and background, respectively, in any units. Thus, this equation results in luminance contrasts that range between 0 (for no contrast) and 1 (for the maximum contrast) for all objects, whether brighter or darker than their backgrounds. Since luminance is the product of illuminance (lux) and reflectance, contrast can also be expressed as equation 4.8.

$$C = \left| \frac{(R_T - R_B)}{R_B} \right| \tag{4.8}$$

where R_T and R_B are the reflectances of the task and background, respectively.

The ability to detect a target against a background can be quantified by its threshold contrast.

Visual acuity and threshold contrast separately define two aspects of a target that defines its visibility. Visual acuity sets the minimum size for a target to be seen, and threshold contrast sets the minimum luminance contrast that is required for a target of a given size to be seen. The contrast sensitivity function combines these two measures by showing the minimum contrast required for targets of different sizes to be seen. The contrast sensitivity is the ability to detect the presence of luminance differences. Quantitatively, it is equal to the reciprocal of the brightness contrast threshold. The contrast sensitivity of the eye depends on the lighting conditions. In bright daylight, a luminance difference between surfaces as little as 1% can be distinguished, but under overcast conditions, surfaces with up to 10% luminance difference may be perceived as the same. Contrast is the single-most important factor in visual acuity.

4.2.2.3   Lighting Requirements

The adequacy of lighting is a quantitative requirement, which depends on the visual task: the contrast, the fineness of details, and the speed at which the view changes. To set the required lighting level the risk of possible errors must be judged and balanced against the affordability of lighting.

The suitability of lighting is a qualitative requirement and has at least four component factors (Szokolay 2008):

i)   Color appearance and color rendering
ii) Color appearance of an environment is associated with mood and the expected 'atmosphere'
iii) The directionality of light must suit the functional as well as the psychological requirements of a visual task. More diffuse light is normally judged as more 'pleasant,' but it will cast little or no shadows, so it may create a hazy or even eerie atmosphere. Where 3-D perception is essential, more directional lighting is necessary, as shadows will reveal form and texture.
iv) Glare should be avoided, but the extent of acceptable (desirable?) glare must suit the visual task.

4.2.2.4 Glare

Glare is the sensation produced by luminances within the visual field that is sufficiently greater than the average luminance of the visual field to which the eyes (both the pupil and the retina) are adapted. Glare occurs in two ways, by extremely bright light sources (direct or reflected) causing saturation effect or by large luminance variations causing strong brightness contrasts in the visual field. Glare can be distinguished as disability glare and discomfort glare, depending on the effects.

Disability glare can be caused when the average luminance of the field of vision is over 25,000 cd/m², such as full sunlight (100,000 lx), bright sky, and white clouds or surfaces with the specular reflection of the sun (e.g., white sandy beach, $\rho = 0.9$). High luminance produces a simple photophobic response, in which the observer squints, blinks, or looks away. Disability glare is caused due to the intraocular light scattered in the eye, reducing the luminance contrast of the retinal image whereby visibility and visual performance are reduced. The magnitude of disability glare can be estimated by calculating equivalent veiling luminance, which is mimicked by adding a uniform 'veil' of luminance to the target. Reflection or veiling glare is caused when reflections on computer display screens or other task materials (e.g., glossy photo or paper) reduce the contrast between background and foreground for the visual task and thus reduce readability.

Discomfort (direct) glare has been characterized as exacerbation or pain or distraction as a consequence of light exposure to high luminance contrasts in the field of view; which does not necessarily impair visual performance or visibility. When the luminance ratio (L_{max}/L_{min}) within a visual field is greater than 15 or 10, then the visual efficiency is reduced and discomfort may be experienced. Laboratory studies have related discomfort glare to pupillary and facial muscle activity since, on the one hand, the highly stimulated parts of retina receiving illumination from the glare source demand to close the pupil while, on the other hand, the less stimulated parts of the retina demand the reverse.

Discomfort glare indoors are influenced by the full visual environment, including windows, reflections (especially specular), external surroundings, and/or interior surfaces – luminaries that are outside the visual task or region being viewed. In North America, the empirical prediction system is called the visual comfort probability (VCP) system. This system is based on assessments of discomfort glare for different sizes, luminances, and numbers of glare sources; their locations in the field of view; and the background luminance against which they are seen, for conditions likely to occur in interior lighting.

Discomfort glare can be mitigated by 'contrast grading.' All potential glare sources should be directly surrounded by areas of relatively high luminances. This can be achieved if the luminance of the visual task on the working plane is 100%; its immediate surrounding should not be less than 50%, and the rest of the visual field should not be less than 20%.

The visual field (central foveal vision) is the locus of objects or points in space that can be perceived when the head and the eyes are fixed, approximately a vision angle of 2°, while the field of view is objects perceived when the head is fixed, but the eyes are moving. Individually, human eyes have a horizontal field of view of about 135° and a vertical field of view of just over 180°. The monocular fields of view are stitched together by the brain to form one binocular field of view, which is around 114° horizontally and is necessary for depth perception. The peripheral vision makes up the remaining 60–70° and has only monocular vision because only one eye can see those sections of the visual field.

While in daylighting/sunlighting it is difficult to quantify glare and it is best to tackle the problem in qualitative terms. This approach is supported by the fact that under 'natural' lighting conditions people seem to be more tolerant and adaptable. Shading devices such as venetian blinds, awnings, vertical blinds, and roller blinds are suitable for reducing the glare, but the specific material characteristics should be taken into consideration.

4.2.3 DAYLIGHT AVAILABILITY

Exterior day illumination is generally referred to as natural light, and it is distinguished by its unique, changing spectra and distribution. The source of all-natural light is the sun. The light that arrives directly from the sun is referred to as *sunlight* (or *beam sunlight*). As sunlight passes through the atmosphere, varying fractions are scattered by gases, water vapor, and particulates, and its spectral composition changes; it is referred to as *sky luminance* or *daylight*. These effects depend very much on the condition of the sky. If the sky is absolutely clear, scattering is relatively small, the quality of light strongly differs from that of the sun, and the sky appears as blue. If the sky is overcast, scattering is very strong and the quality of light is only slightly different from that of the sun: colder. Light reflected from the surfaces of the earth, natural obstructions (plants, the terrain), and man-made obstructions (other buildings, constructions) is also part of exterior illumination and may be important in daylighting design. Consequently, the sun, the sky, the surfaces of the earth, and external obstructions may all serve as parts of the 'natural luminaire.'

4.2.3.1 Sky Conditions

The term 'daylight availability' refers to the amount of light from the sun and the sky for a specific location, time, date, and sky conditions. Over the last seven decades, daylight illuminance measured in different locations all over the world has have resulted in similar values. The empirical equations, therefore, derived from the daylight availability data are used to calculate mean values of daylight illuminance. The empirical equations provide the best fits to data average over time and measurement sessions. For this reason, measured instantaneous luminances and illuminances may differ widely from those determined by calculation methods based on daylight availability. It is not unusual for the instantaneous values to be more than twice or less than half the mean design values.

The daylight availability is determined by sky conditions. The illuminance distribution of the sky depends on weather and climate, and it changes during a day with the position of the sun. Commission Internationale de l'Eclaiage (CIE 2003, ISO 2004) defines a set of 15 standard general sky types that model different sky luminance distributions under a wide range of conditions from the heavily overcast sky to cloudless weather and with varying levels of direct sunlight. This standard defines relative luminance distributions: the luminance of the sky at any point is given as a function of the zenith luminance. For daylighting calculation purposes it may be used with values of zenith luminance or horizontal illuminance to obtain absolute luminance distributions. This is a mathematical model for calculating representative spatial distributions of daylight over the sky dome that can functionally simulate a range of different weather conditions. There is an additional 16th type sky based on a much simpler mathematical model for overcast skies defined and used by CIE prior to this standard.

Among the 15 standard skies, CIE standard overcast sky and CIE clear sky are two particular sky conditions that are commonly used in daylight analyses to define an appropriate range of natural lighting conditions.

The standard CIE overcast sky acts as a diffuse light source having illuminance of about 40–50 klx. The standard CIE overcast sky has a luminance distribution defined as a function of altitude angle (ALT) by the empirical Moon–Spencer (1942) equation. The sky luminance (L_{ALT}) of an overcast sky at any altitude angle (*ALT*) is expressed in terms of the zenith luminance (L_z), and the illuminance over the horizontal plane (E_h) is expressed as per equation 4.9.

$$L_{ALT=}\frac{L_z}{3}\times\left(1+2\sin\left(ALT\right)\right)=\frac{3}{7}E_h(1+2\sin ALT) \tag{4.9}$$

that is, the zenith luminance is three times that at the horizon and it gradually increases from horizon to zenith. It should be noted that the relationships refer purely to the relative luminance of different sky zones and not to any absolute value.

The overcast sky is used when measuring daylight factors. It can be modeled under an artificial sky. Although the distribution of luminance from the overcast sky is symmetrical about the zenith and independent of the position of the sun, the illuminance as received from the overcast sky on a horizontal plane depends on the solar altitude angle (*ALT*) behind the clouds; Equation 3.20 may be used to find the solar altitude angle for any given day and time. An indication of the average illuminance (*lux*) from the overcast sky is given in equation 4.10.

$$E_{avg} \approx 215 \times ALT \tag{4.10}$$

The luminance of the standard uniform sky does not change with altitude or azimuth and is taken as uniform. It has been used when calculations were done by hand or with tables. Today, it is still used for the rights of light cases.

The luminance of the standard CIE clear sky varies over both altitude and azimuth of the sun. It is the brightest around the sun and the dimmest opposite to it. The brightness of the horizon lies in between these two extremes. Under clear sky conditions, direct sunlight can give an illuminance of 100 klx (1 kilo-lux = 1000 lux), but if the sunlight itself is excluded, the sky can give 40–50 klx diffuse illuminance.

The standard CIE intermediate sky is a somewhat hazy variant of the clear sky. The sun is not as bright as with the clear sky, and the brightness changes are not as drastic. The average illuminance produced by such a sky (excluding direct sunlight) can be estimated as equation 4.11.

$$E_{avg} \approx 538 \times ALT \tag{4.11}$$

Figure 4.11 shows some examples of CIE standard sky, generated with a web app using the updated Perez All-Weather sky model as defined in ISO 15469: 2004 (E) and CIE S011/E:2003. Annual hourly weather data for Los Angeles are used to simulate sky luminance and radiance distribution on March 21 at noon.

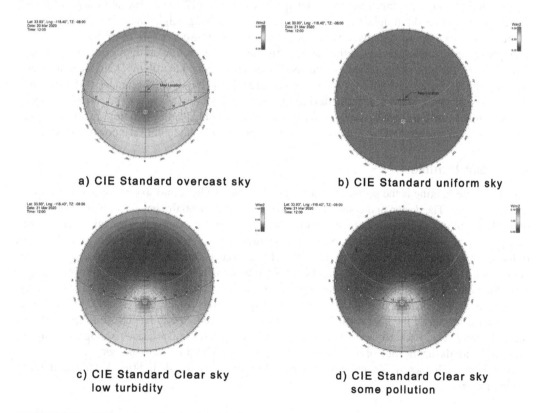

a) CIE Standard overcast sky   b) CIE Standard uniform sky

c) CIE Standard Clear sky
low turbidity

d) CIE Standard Clear sky
some pollution

FIGURE 4.11 CIE standard sky simulated for Los Angeles on March 21 at 12 noon. Source: http://andrewma rsh.com/software/cie-sky-web/

There is another way to estimate the illuminance from the measured solar irradiance data using the luminous efficacy values defined in Equation 4.12.

$$\text{Luminous efficacy}(F) = \frac{\text{illuminance}}{\text{irradiance}} = \frac{\text{lux}}{\text{W/m}^2} \frac{\text{lm/m}^2}{\text{W/m}^2} = \frac{\text{lm}}{\text{W}} \qquad (4.12)$$

4.3 DAYLIGHTING DESIGN STRATEGIES

A daylighting design strategy involves techniques that manipulate and control the delivery of daylight to enhance its performance or to distribute it a specific way or to a specific space. Daylight enters the building through openings in the building envelope. The efficiency of the opening is the ratio of the luminous flux entering the interior and the luminous flux incident on the outer surface of the transparent infill material of the opening. This indicates the efficiency or the resultant transmittance of the opening.

A complete matrix of daylighting systems is provided in *Daylight in Buildings, A Sourcebook on Daylighting Systems and Components* (Aschehoug et al. 2000). The determination of a viable daylighting strategy in a particular building entails a year-round analysis based on local climatic conditions.

This section presents commonly used daylighting strategies: side lighting, a perimeter daylighting system, and top lighting, followed by advanced daylighting strategies, light-guiding systems, and light transmission systems.

As per ASHRAE 90.1 standard side-light spaces, the daylight zone is considered equal to the head height of the window from the floor, and in case of top lighting daylight zone extends 0.7 times the height of the roof opening around the roof aperture (Figure 4.12). Thus, the application of traditional side lighting and top lighting methods is limited by building height and design.

Advanced daylighting systems are appropriate devices to transport daylight from exteriors into deep plan rooms. Advanced daylighting devices are classified according to their primary operating principles and the type of light (direct or diffused) they are designed to utilize. There are many benefits to using advanced daylighting systems. It increases usable daylight for climates with predominantly overcast skies, increases usable daylight for very sunny climates where the control of direct sun is required, increases usable daylight for windows that are blocked by exterior obstructions and therefore have a restricted view of the sky, and transports usable daylight to windowless spaces.

4.3.1 SIDE LIGHTING

Unilateral side lighting is the standard strategy of admitting daylight into space through windows in an exterior wall. The design and construction of windows are essentially governed by their functions other than lighting, such as visual connection, thermal insulation, and noise insulation. The direct sunlight entering the window creates excessive-brightness ratios as well as overheating, and the view of the sky is often a source of direct glare. Side lighting strategy, therefore, should reflect light to the ceiling or diffuse light to optimize daylight quality while reducing solar heat gain. An important feature of the illuminance distribution of side lighting is that it is the highest just adjacent to the window and rapidly decreases away from the window to inadequate levels for most visual tasks. Side lighting area as a percentage of the floor area should generally not exceed 20% to balance summer overheating and winter heat losses. However, in temperate, cold, or cloudy climates, side lighting area may be increased with the application of movable shading systems and high-performance glass.

The lighting characteristics and the efficiency of illumination of side lighting depend on the following factors:

- The orientation of window
- The location, form, relative size, and slope (60°–90°)

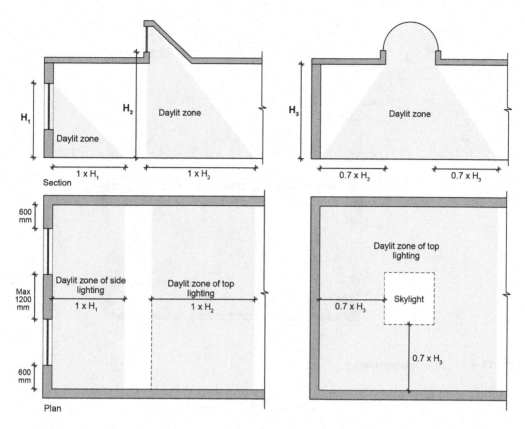

FIGURE 4.12 Daylight zone of side lighting and top lighting. Source: adapted from ASHRAE (2010), line work by Ar Vaibhav Ahuja

- The infill material, transparent or translucent glazing (clear, opal, sandblasted, or ornamental glass)
- The number of layers (double, triple)
- Cleanness of the infill material
- The construction
- The thickness of the wall surrounding the window and the manner of connection

The efficiency of ordinary, clear 2–5 m² double-glazed windows set in 350 mm walls is between 0.4 and 0.5.

The orientation of the window has a pronounced impact on the effectiveness of side lighting. East and west orientations are difficult for solar shading because of low altitude angles of the morning and evening sun. The west orientation can contribute to undesirable heat gains during the evening hours. The south and north are appropriate orientations for side lighting. The north orientation (in the northern hemisphere) provides a diffuse, high-quality light and does not need extensive solar shading or control. The direct sun in the south orientation can be controlled by installing shading and light-reflecting devices. It is rare to find that daylight alone would be sufficient beyond a depth of 2 1/2 times the window head height (from the work plane); side lighting may bring daylight about 4572 mm (15 ft) into a building (Figure 4.13). Placing windows on two walls (bilateral lighting) provides better light distribution and glare control compared to windows placed on only one wall (unilateral lighting) (Figure 4.14). Placing windows on adjacent walls is effective in reducing glare since the windows on each wall illuminate the adjacent walls and, therefore, reduce the contrast between each window and its surrounding wall. Placing windows adjacent to interior walls also

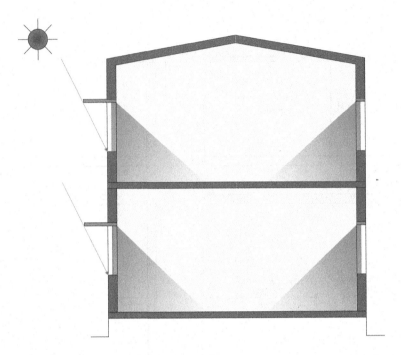

FIGURE 4.13   Side lighting strategy.

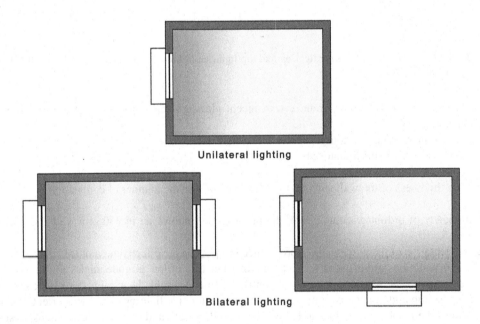

FIGURE 4.14   Plan diagram of unilateral and bilateral side lighting.

reduces glare and strong directionality of daylight since these walls act as low-brightness reflectors. Splayed or rounded edges of the window create a transition of brightness and therefore reduce the potential glare.

External daylight filters such as trees, vines, trellises, and screens may modify and soften daylight. Internal daylight filters such as venetian blinds, drapes, or translucent glazing diffuse direct sunlight but can cause glare.

4.3.2 TOP LIGHTING

The illumination of the sun falling on the horizontal plane of the roof may be many times more than that which falls on the vertical plane of a window even under an overcast sky. Unilateral top lighting is a strategy to bring daylight into the top floor of a building from the openings on the roof; top lighting can also contribute to lower floor levels through the use of core daylighting systems, atriums, light wells, or other devices. The roof openings can be of two forms: linear and localized. Linear openings are elongated rectangles with a constant vertical section, while localized openings are a square, a circle, or a polygon or have some other form. Top lighting can include glazed units that are horizontal, vertical, or tilted. Top lighting provides high-quantity and high-quality illumination over a large area year-round since a large part of the unobstructed sky is visible. However, the control of direct sunbeam and solar heat gain in summer is especially important in top lighting applications. In environments where visual acuity is critical, such as classrooms, libraries, and offices, the top lighting directly viewable from below may produce excessive brightness, causing glare and disabling veiling reflections on tasks below. Horizontal glazing systems may be effective in climates with a predominantly overcast sky. Tilted glazing has an advantage over horizontal glazing as sun penetration may be eliminated with polar orientation. Top lighting applications often utilize translucent and insulated glazing units or skylights to diffuse the daylight and prevent heat gain or heat loss. In the case of clear glazing applications, reflecting and shading elements (baffles) can be incorporated for solar control and light distribution. Figure 4.15 presents different configurations of top lighting, shaded roof monitors, and sawtooth roofs. The placement of top lighting determines both light distribution and efficiency.

A properly designed top lighting system with daylight and heat transfer controls will prove viable on a year-round basis in almost all climates. Much like a lighting fixture, the innovative skylights from small apertures with diffusers may be seen in the NOAA Daniel K Inouye Regional Center, Honolulu (see Section 6.2). Another major application of top lighting is the Louvre museum, Paris (Figure 4.16).

The vertical or near-vertical glazing of clerestories have many of the attributes of top lighting except that they occur in the vertical rather than the horizontal plane and, therefore, are exposed to

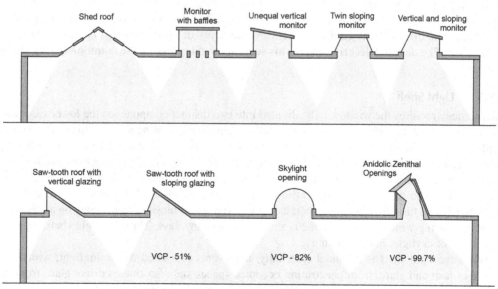

Daylit zones for Diferent rooflight profiles

FIGURE 4.15 Top lighting strategies. Line work by Ar Vaibhav Ahuja

FIGURE 4.16　Pyramid at the Grand Louvre, I. M. Pie, 1989. © C. Kabre

less quantity of direct daylight than are top lighting. Clerestories having equatorial orientation are desirable for collecting more sunlight in the winter than in the summer. The ingress of direct sun through clerestories can be eliminated with the addition of overhangs and/or horizontal louvers on the interior or exterior.

4.3.3　Light-Guiding System

Light-guiding systems can redirect both direct and diffuse natural light to the core of the building up to 8–10 m, through reflection, refraction, or deflection. These systems can be grouped as a direct and diffuse light-guiding system. A taxonomy of light-guiding systems is shown in Figure 4.17. The detailed descriptions in this section include some of the common light-guiding systems.

4.3.3.1　Light Shelf

This strategy requires the window to be divided into two distinct components: the lower view window and the upper daylight window. A light shelf is a classic daylighting system; this is a horizontal baffle positioned inside and/or outside of the window at a height of about 2.1 m with a reflective upper surface, which redirects/reflects incoming sunlight through daylight window onto the interior ceiling in a fairly high room (3 m), minimizing glare and enhancing light levels in the space (Figure 4.18). The interior reflective ceiling causes the distribution of daylight more evenly and more deeply into the interior. A light shelf can extend the useful range of daylighting on a building's equatorial facing windows to about 7620 mm (25 ft) on sunny days. Thus, the light shelf improves the quality of daylight, not the quantity.

When the light shelf is mounted externally, it prevents unwanted direct sunlight, which is a source of heat and glare, from penetrating occupied spaces and also shields direct glare from the sky. Glare from the upper daylight window can be controlled by setting a reflective louvered device into the entire frame or glazing unit. The louvers are parabolic in shape and specular to reflect light deep into space; the spacing of the louver blades creates a cutoff for direct sunbeam penetration.

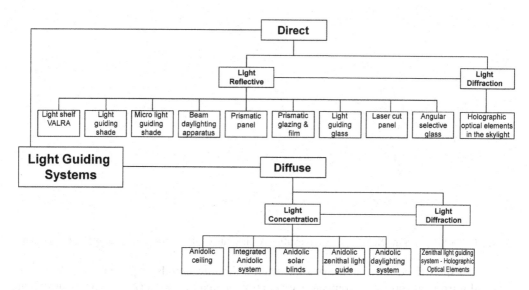

FIGURE 4.17 A taxonomy of light-guiding systems. Source: adapted from Aschehoug (2000) and Nair et al. (2014)

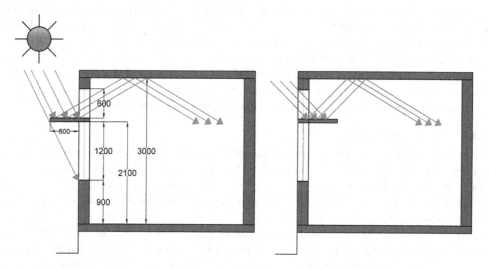

FIGURE 4.18 Light shelf, external and internal.

The Research Support Facility, NREL, Golden makes use of light shelf and LightLouver™ on the southern facade windows; see Section 6.6 for detailed discussion.

The light shelf performs better with highly reflective (as with polished metal), which needs to be maintained to enhance the daylight reflectance. The light shelf works well at high solar altitude angles, but at lower angles, the shelves need to extend deeper into the room to catch the sunlight. In the early 1980s, researchers tried to increase the performance of light shelves by developing a reflective light shelf that responds to the altitude angle of the sun.

The sun-tracking light shelf is an active system. A variable area light-reflecting assembly (VALRA) reflects light into a building for all sun angles by using a reflective plastic film surface (silver-coated Mylar film) over a tracking roller assembly within a 'V'-shaped fixed light shelf (Figure 4.19). One edge of the reflective film is fixed, and the opposite end is attached to a spring-loaded take-up roller. As the sun moves across the sky, daily and seasonally, the take-up roller automatically adjusts the angle and surface area of the film. At high solar angles, the roller retracts

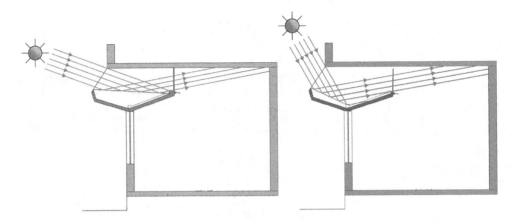

FIGURE 4.19   Sun-tracking light shelf. Source: after Aschehoug et al. (2000), line work by Vaibhav Ahuja

and forms a short tilted mirror. At low solar angles, the film unrolls to produce a wide, almost horizontal mirror. Motor-driven computerized control automates the film's movement. Dr. Wayne Place of North Carolina State University installed light sensors in a model room to test VALRA's effectiveness. Experiments have shown that a VALRA can direct sunlight up to 30.48 m (100 ft) into a building at solar noon on a day with full sun. However, because solar conditions vary, the more practical goal would be to count on a daylight penetration to a depth of 12.2 m (40 ft). Model studies and computer simulation show that a VALRA system could reduce electricity used for lighting by 71% in Phoenix and 44% in New York City.

An innovative patented idea (Howard 1986), VALRA was not cost-effective for commercial use. A simpler variant of a light shelf that can be adjusted to compensate for summer–winter differences in solar altitude angles or the sky luminance is movable (pivotable) external light shelf.

4.3.3.2   Light-Guiding Shades

In the tropics, windows are almost always shaded by external and internal shading devices and reflecting or absorbing glasses to reduce radiant heat gain through windows. Consequently, the average daylight level in a room with a strongly shaded window is less than 50 lux. A light-guiding shade is an external shading system that redirects all direct and diffuse natural light onto the ceiling to enhance the daylighting of buildings in tropical climates. A light-guiding shade consists of a diffusing glass aperture and two reflectors usually made of bright-finish aluminum; the upper reflector is an inclined plane, while the lower reflector is parabolic (Figure 4.20). The input-aperture-to-output-aperture ratio is usually 1:2 so that the output light lies within an angular range from 0 to 60°, thus maintaining the uniform nature of light. The shade is a source of diffuse light, which is non-luminous when viewed by the occupants of the room and is therefore entirely free of glare. The light-guiding shades are installed over the upper one-third or one-half of a window system; it shades the window from direct sunlight as a normal external shade does.

There is a considerable energy benefit from light-guiding shades in sunny tropical climates; for example, under clear sky conditions a light-guiding shade can produce a work plane illuminance of more than 1000 lux at a distance of 5 m in a room, while under overcast sky conditions the average illuminance obtained would be about five times smaller, i.e. 250 lux. The performance of light-guiding shade will depend on the shape and size of the window; the slope and reflectance of the ceiling, walls, and floor; the type of glazing on the window; and the ambient conditions.

4.3.3.3   Prismatic Panel

Prismatic panel works on the principle of refraction; it is a thin planar device made of a clear acrylic or glass having ridges on one side that bend light to specified angles. Prismatic panels can be applied

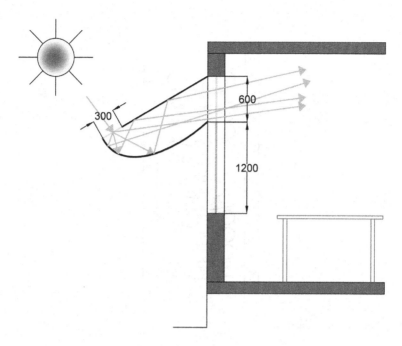

FIGURE 4.20   Light-guiding shade. Source: adapted from Edmonds and Greenup (2002)

in many different ways, in fixed or sun-tracking arrangements, to side lighting and top lighting in temperate climates. Prismatic glazing can be used to exclude sunlight from a space or redirect light, usually to the ceiling. Prismatic glass obscures the view out, so it is usually placed at the top one-third of a window to divert the beam sunlight (by refraction) upward, to the ceiling, which will then diffuse light deep into the interior of the room as illustrated in Figure 4.21.

For clear sky climates, the panels can direct sunlight at certain solar angles into a room and provide a relatively uniform daylight distribution. As the sun moves beyond the critical angles, sunlight passes through the panels and can cause glare. Prismatic panels have limited applications in climates dominated by overcast sky conditions.

4.3.3.4   Light-Guiding Glass

The main component of the light-guiding glass system is a double-glazed sealed unit that holds concave acrylic elements (Figure 4.22). These elements are stacked vertically within a double-glazed unit and redirect direct sunlight from all angles of incidence onto the ceiling. The sealed unit is normally placed above the view window. A sinusoidal pattern on the interior surface of the window unit can be used to spread outgoing light within a narrow horizontal, azimuthal angle. A holographic

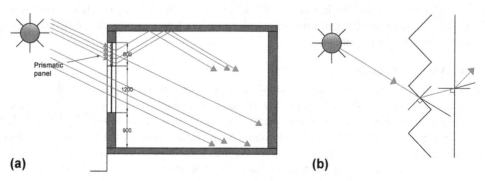

FIGURE 4.21   Prismatic panel. (a) application in top one third of the window, (b) detail section.

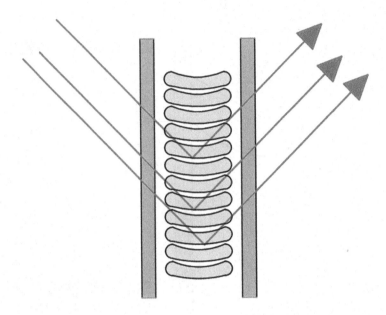

FIGURE 4.22   Light-guiding glass.

film on the exterior glass pane can also be used to focus incoming daylight within a narrow horizontal angle.

The system is designed for use in direct sunlight. The best orientation on a facade is south (in the northern hemisphere) or north (in the southern hemisphere) in temperate climates. On the west or east faces, it is useful only in the morning or afternoon. The system also deflects diffuse light, but the illuminance level achieved is much lower than that with direct sunlight. For north facades, these elements have to be larger.

4.3.3.5   Laser-Cut Panel

The laser-cut acrylic panel is a daylight-redirecting system produced by making an array of laser cuts to some 75% of the depth in a thin sheet of clear acrylic (polymethyl methacrylate or PMMA). The surface of each laser-cut rectangular element becomes a small internal mirror; light is deflected in each element of the panel by refraction, then by total internal reflection, and again by refraction (Figure 4.23).

Laser-cut light-deflecting panels can be applied as:

- A fixed sun-shading system for windows
- A light-redirecting system (fixed or movable in the upper half of windows) or angular-selective skylight in a pyramid or triangle configuration (Figure 4.24)
- A sun-shading/light-directing system for windows (in louvered or venetian form); they may be adjusted to the open, summer position to reject light or to the closed, winter position to admit light (Figure 4.25)

4.3.3.6   Anidolic Ceiling

Anidolic ceiling systems use the optical (non-imaging) properties of compound parabolic concentrators to collect diffuse daylight from the sky; the upward-looking anidolic optical concentrator at the outer end is coupled to a 3–4 m long specular light duct in a suspended ceiling (Figure 4.26), which transports the light to the back of a room. At the interior end of the duct, light is 'deconcentrated' by a second anidolic device to direct the flux toward the work plane, avoiding any back reflection. Anidolic ceilings can be used in both clear and cloudy skies as long as proper shading blind is provided to control sunlight over the entrance glazing. An anidolic ceiling system is designed to be located on a vertical facade above a view window. Anidolic ceilings can be used

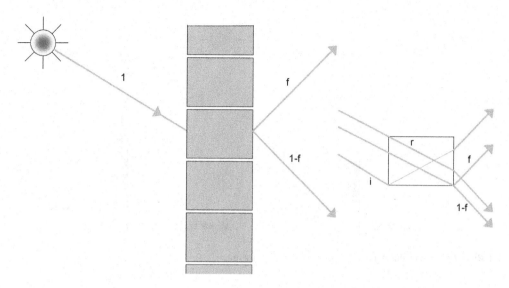

FIGURE 4.23   Laser-cut acrylic sheet.

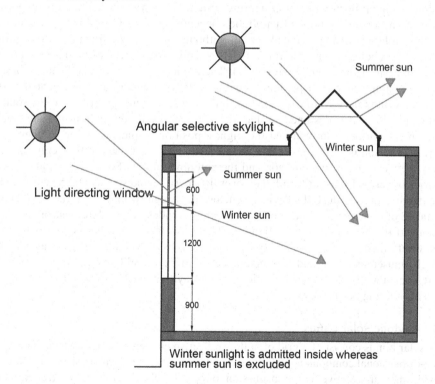

FIGURE 4.24   Laser-cut panel application as light-directing window and angular-selective sky light.

in commercial, industrial, or institutional buildings. The primary objective is to provide adequate daylight to rooms under predominantly overcast sky conditions as it 'sees' the upper part of the sky, which is of a greater luminance. Anidolic ceiling can provide about 30% yearly lighting savings in an office building for a given required desk illuminance (viz. 300 lux) assuming fully automatic control of the electric lighting (perfect daylight-responsive dimming). Anidolic ceiling improves overall light levels as well as uniformity from front to back; a daylight factor of more than 4% at a distance of 4–6 m from the facade under overcast conditions can be achieved.

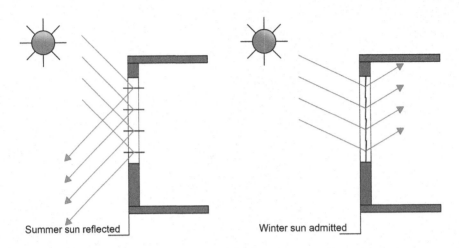

FIGURE 4.25 Laser-cut panels in louver or venetian form.

4.3.3.7 Anidolic Zenithal Openings

The anidolic zenithal opening is a top lighting system. The anidolic zenithal opening system is composed of an optical 'compound parabolic concentrating' (CPC) element to collect diffuse daylight from a large portion of the sky vault without allowing the direct sun to penetrate and a 'compound parabolic deconcentrating' (CPD) or emitting element to guide the daylight flux toward the bottom of the room (Figure 4.27). The anidolic device roof collector is based on a linear, non-imaging compound parabolic concentrator whose long axis is oriented east–west. The opening is tilted northward for locations in the northern hemisphere and to the south in the southern hemisphere. It is designed so that the sector where it admits light includes the whole sky between the northern horizon and the highest position of the sun in the southern sky during the year. The sun never comes inside the admission sector, except at the beginning and end of the day, between the spring equinox and the autumn equinox. Solar protection is completed with a series of vertical slats uniformly laid over the aperture and spaced at 0.5 m. To prevent the reflectors from gathering dust, the device is enclosed between two layers of glazing (visible light transmittance of 0.9). The anidolic zenithal opening provides better glare control and improved visual comfort than conventional top lighting. This form of top lighting system is best utilized at places where there are clear indoor spaces in which visual comfort is essential, for instance, sports halls, museums, atria, and markets. Under clear sky conditions and in a semi-temperate climate (Geneva) with a 20% opening ratio, interior daylight levels reach 500 lux during 79% of the hours of building occupancy.

4.3.3.8 Anidolic Solar Blinds

Anidolic solar blinds consist of a grid of hollow reflective elements, each of which is composed of two three-dimensional compound parabolic concentrators. The blinds are designed for side lighting and provide angular-selective light transmission to reject high solar altitude rays from direct sun in summer but to transmit lower altitude diffuse light or winter sunlight.

The anidolic solar blinds would typically be placed between two panes of glass for protection against dust. The anidolic solar-blind system either can be applied as a fixed louver to window openings that were principally designed to collect daylight (i.e., the view through them is blurred) or can be placed in the upper part of a normal window if the view to the outside must be maintained through a lower portion of the window.

The innovative feature of anidolic solar blinds compared to other anidolic systems (anidolic ceilings, anidolic zenithal openings) is their use of three-dimensional reflective elements and their small scale.

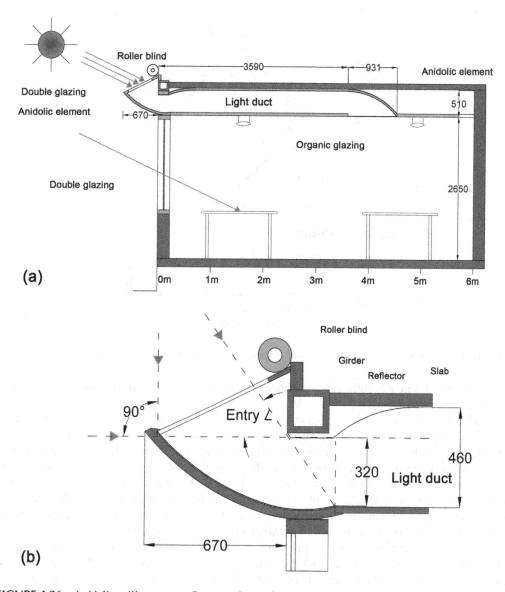

FIGURE 4.26 Anidolic ceiling system. Source: adapted from Linhart et al. (2010)

4.3.3.9 Zenithal Light-Guiding Glass with Holographic Optical Elements

Zenithal light-guiding glass redirects diffuse skylight into the depth of a room from the zenithal region of the sky. The main component is a polymeric film with holographic diffraction gratings, which are laminated between two glass panes. The system may cause color dispersion when hit by direct sunlight; it should only be used on facades that do not receive direct sunlight. The glass can be integrated into a vertical window system or attached to the facade in front of the upper part of a window at a sloping angle of approximately 45°. Since zenithal light-guiding glass slightly distorts the view, it should be applied only to the upper portion of a window.

4.3.4 Light Transmission System

Light transmission systems are mechanically complex daylighting devices, designed to transmit light (generally, the direct component of sunlight) at a long distance from the building exterior

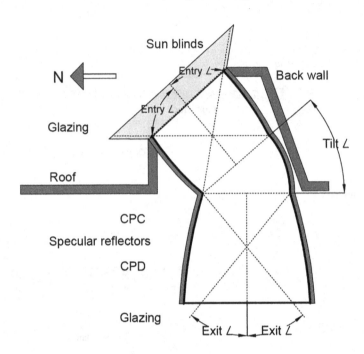

FIGURE 4.27   Anidolic zenithal opening. Source: adapted from Aschehoug et al. (2000)

(facade or roof) where it is collected to the interior and distributed either in the building core or at a rear place in a room adjacent to the facade.

A light transmission system consists of three major components (i) a passive or active light collector, which redirects or concentrates diffuse and/or direct sunlight to a specific point or direction; (ii) a light transmission element, which allows light propagation inside itself; and (iii) a light distributor that extracts the light and emits within the space. A taxonomy of the three components of light transmission systems is shown in Figure 4.28. Because they depend on direct sunlight and are relatively expensive to install, they will be cost-effective only in regions where blue skies and clean air can be guaranteed for much of the year. Energy-efficient backup lamps may be fixed at the head of the shaft to substitute for sunlight during infrequently overcast conditions.

Passive light collectors are fixed in one position outdoors to enhance light collection from sky illuminance and occasional direct sunlight incident on a collector's entrance. Light is collected and conveyed further to the transmission element directly for the flat glass or dome, by reflection for anidolic or hyperbolic concentrator, by refraction for the prismatic film in compound parabolic concentrator (CPC), by deflection for laser-cut panels, and by absorption for a fluorescent fiber solar concentrator or a luminescent solar collector.

Active collectors have a fixed position, but their light concentrators are motorized systems tracking the apparent movement of the sun by computer to efficiently convey the light; tracking mechanism is either single or double axis. Light is collected and conveyed further to the transmission element by light redirection for heliostats and mirror sunlight systems and light concentration for the Fresnel lens and parabolic concentrator. Heliostat and the Fresnel lens are shown in Figure 4.29. Power-operated tracker is energy-intensive and is expensive, while the thermo-hydraulic tracking systems powered by solar cells or renewable energy are cost-effective.

The light transmission elements guide light to a remote place where it is to be exported. Light propagates in the guide by (i) multiple specular reflections (mirror light pipe) (ii) total internal reflections (solid guides made of polymethyl methacrylate – PMMA, glass or liquid in the form of optical fibers, rods, and hollow pipes), or (iii) light convergence (system of lenses and mirrors). Figure 4.30 shows different configurations of light transmission elements.

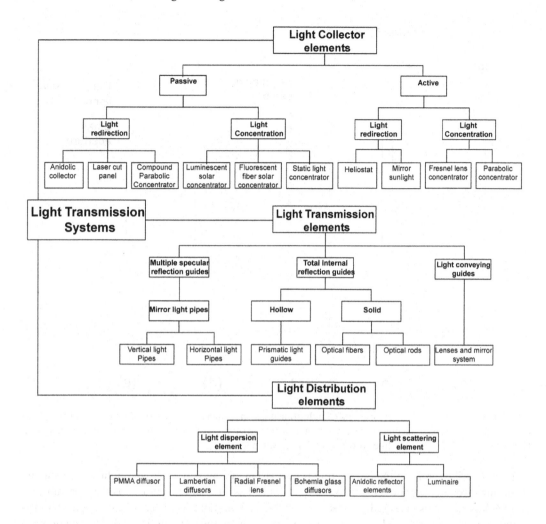

FIGURE 4.28 A taxonomy of light transmission system. Source: adapted from Obradovic and Matusiak (2019)

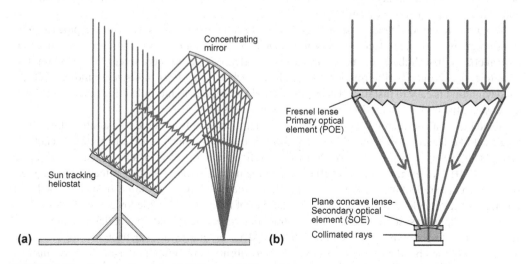

FIGURE 4.29 Active collectors: (a) Heliostat, (b) Fresnel lenses.

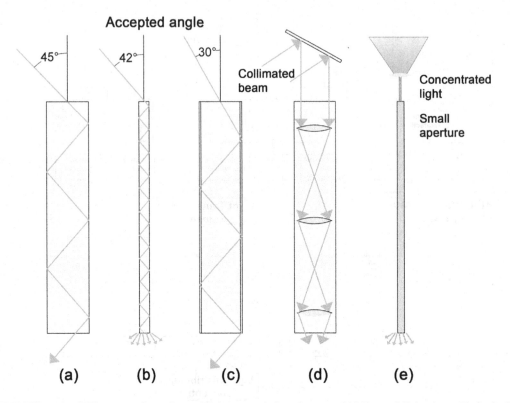

FIGURE 4.30 Different configurations of light transmission elements. (a) Mirrored light pipes, (b) Optical rods, (c) Hollow prismatic pipes, (d) Lenses and mirror system, (e) Optical fibers. Source: adapted from Hansen and Edmonds (2003), line work by Ar Sakshi Singhal

Light distribution from the daylighting transmission system is done by the light dispersion and scattering through the emitters, which can be produced as circular or square (prismatic or transparent) PMMA diffusers or as radial lenses (Fresnel) in a luminaire form. There can be just one distribution point at the end and many distribution points along the light guide, or the distribution can happen continuously along with the guide. If the light is transmitted by a wider guide, as light pipes or duct, diffusers are bigger, while for highly concentrated light from fibers, emitters are smaller and in a spots or downlights form.

Tubular Daylight Guidance System (TDGS) utilizes an inner mirror surface of the pipe to transmit sunlight by multiple reflections. A typical TDGS consists of an outside collector (dome or laser-cut panel), a mirror light pipe, and a luminaire that releases light into the interior. TDGS could be used to enhance interior illumination for buildings such as schools or industrial buildings. TDGS for daylighting is easy to install and cost-effective. However, due to multiple reflections, light loss is serious and the transmission distance is only a few meters (0–4 m).

Solatube® is a commercially available TDGS that uses a rooftop dome to capture natural light (Figure 4.31). The rooftop dome of high-transmittance material filters ultraviolet light and captures daylight, which is transmitted through a highly reflective sun tube that can transfer light up to 12.2 m (40 ft). However, the daylight efficiency of the transmission process is affected by the angle of incidence, the reflectance of the coating inside the tube, and the ratio diameter/length. The lower the angle of the incident light, the higher the number of reflections and the lower the efficiency.

Fiber daylighting system consists of a sun-tracking system, lens, and optical fibers (6 mm wide): silica fiber (SiO_2), polymethyl methacrylate (PMMA), or quartz and distribution fixtures. The advantages of the fiber daylighting system include flexibility in the installation and longer transmission distance. The disadvantage of the fiber daylighting system is its transmission loss due to fiber

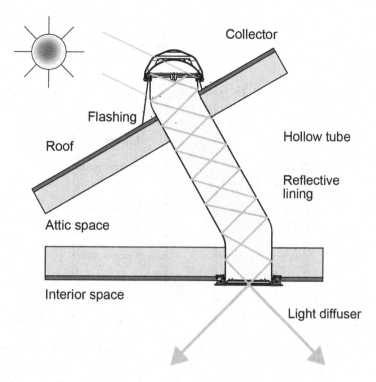

FIGURE 4.31 Tubular daylight guidance system. Source: adapted from the Solatube®

attenuation, which is closely related to wavelength. For example, silica fiber has high transmission loss in the visible range, with its transmission loss being 0.015 dB m⁻¹ at 600 nm, while the attenuation of commonly used PMMA fiber is about 0.2 dB m⁻¹ at 650 nm. Besides, optical quartz fiber bundles are too expensive to be popularly applied to daylighting systems. Therefore, the light traverses for 20–50 m before losing half of the intensity. Fibers are only 6 mm wide and can be 40 m in length.

For large buildings higher than 50 m, it is difficult for optical fiber daylighting systems to realize inner natural lighting. Heliostat mirror is suggested for fiber optical systems as sunlight redirecting and concentrating devices are placed on a south facade. Most heliostat daylighting systems reported adopting small mirrors with an aperture of less than 7 m². For a roof-mounted heliostat with an area of 22.95 m², the light transmission distance is more than 70 m, and the system can provide a level of 20–80 klx daylighting illuminance in the daytime (Song et al. 2018). Figure 4.32 illustrates the principle of a light transmission system in a high rise building.

4.4 DAYLIGHT PREDICTION METHODS

Daylight prediction methods are useful in comparing alternative daylight design strategies or considering the limits of daylight utilization for various systems under a wide spectrum of lighting conditions. Daylight prediction can be accomplished in three ways by analytical, computer modeling, and analogue methods.

Analytical methods for predicting illuminance at a point involve mathematically modeling the sun, sky, and interreflected components of daylight. The sun and sky components are computed from the flux that reaches an aperture or points directly. The interreflected component results from sunlight and skylight that have initially reached other surfaces and have then been reflected to the point of interest. Average illuminance on a work plane can be calculated using the lumen method for a simple rectangular room with a flat ceiling. Prediction of point illuminance on a work plane

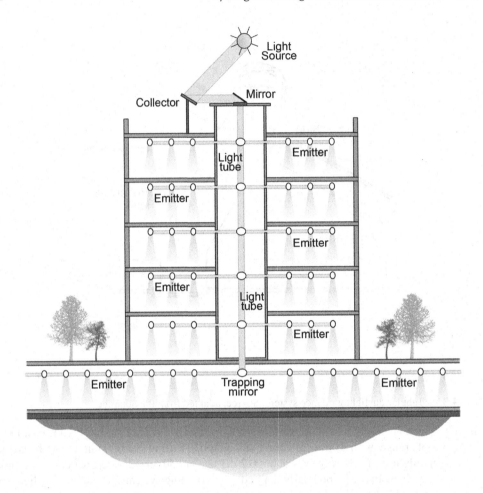

FIGURE 4.32 Light transmission system for a high rise building. Credit Ar Vaibhav Ahuja

can be made using the daylight factor method for a uniform or an overcast sky condition. Climate-based daylight modeling is an alternative to the daylight factor method since it can predict daylight based on the actual climatic data of a given location. Standard references give the detailed procedure for the calculation of interior day illuminances using the lumen method and daylight factor method.

Precise prediction of point illuminances is usually made with computer modeling because of the capacity to model daylighting and calculate lighting energy savings. Comparable physical scale model of complex built form can be measured and evaluated.

4.4.1 IESNA Lumen Method

The lumen method provides a simple way to predict interior daylight illumination through side lighting and top lighting (IESNA 2013). It assumes the building (or a room in the building) as a cuboid with simple apertures that will admit a light flux. By definition, the interior illuminance (E_i) reaching a prescribed point is a function of the amount of the exterior illuminance (E_x) available on the aperture plane, expressed by a simple equation 4.13.

$$E_i = E_x \times NT \times CU \tag{4.13}$$

where
 Ei = interior illuminance in lx
 Ex = exterior illuminance in lx
 NT = net transmittance of glazing
 CU = coefficient of utilization

The lumen method consists of four basic steps:

1. The exterior illuminances (E_x) at the window or skylight are determined in terms of day-light availability (sunlight and skylight).
2. The net transmittance is the product of the transmittance (T) of the glazing: a light loss factor (LLF) representing dirt accumulation; the net-to-gross aperture area ratio (R_a), representing such elements as mullions and glazing bars; and a factor T_c, representing the transmittance of other solar control elements such as louvers, shades, and drapes, which reduce the transmittance of the window.
3. Coefficients of utilization are ratios of the interior to exterior horizontal illuminances. For the lumen method for top lighting, the coefficients provide the average daylight illuminance on the work plane. For the lumen method for side lighting, coefficients give the illuminance at five predetermined points located on a line perpendicular to the window wall across the center of the room at the same height as the window sill (Figure 4.33).
4. The interior illuminance is calculated by taking the product of the factors determined in the first three steps.

4.4.2 DAYLIGHT FACTOR METHOD

The daylight factor method is a simple procedure for determining the illuminance at any point in an interior space produced by a sky with known luminance distribution. Direct sunlight is excluded. The method is generally used with uniform or CIE overcast skies.

Daylight factor (DF) is the ratio of the illuminance (E_i) at a point within an interior space, generally, the horizontal work plane, produced by luminous flux received directly or indirectly at that point from a sky whose luminance distribution is known to the illuminance on a horizontal plane produced by an unobstructed hemisphere of this same sky (E_o) (equation 4.14).

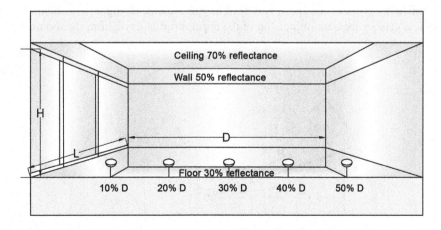

FIGURE 4.33 Standard conditions in a room for calculating side lighting. H, L, and D represent room height, width, and depth, respectively. Source: adapted from IESNA (2013), line work by Ar Sakshi Singhal

$$DF = \frac{E_i}{E_o} \times 100\% \qquad (4.14)$$

The higher the daylight factor, the more natural light is available in the room.

There are three ways in which daylight may reach a point on a horizontal plane within an interior space (Figure 4.34).

a) The sky component (*SC*) is due to daylight received directly at the point from the sky vault.
b) The externally reflected component (ERC) is due to daylight received directly at the point from external reflecting surfaces (buildings/walls), which obscure the view of the sky.
c) The internally reflected component (IRC) is due to daylight reaching the point after one or more interreflections from interior surfaces.

The daylight factor is the sum of the three components as Equation 4.15.

$$DF \ = \ SC \ + \ ERC \ + \ IRC \qquad (4.15)$$

Rooms with an average DF of 2% or more can be considered daylit, but electric lighting may still be needed to perform visual tasks. A room will appear strongly daylit when the average DF is 5% or more, in which case electric lighting will most likely not be used during daytime (CIBSE 2002). The key building properties that determine the magnitude and distribution of the daylight factor in space are:

a) The size, distribution, location, and transmission properties of the facade and roof windows
b) The size and configuration of the space
c) The reflective properties of the internal and external surfaces

The daylight factor can be determined by hand calculation methods using tables of pre-calculated components for typical simple geometries, usually rectangular rooms with a wall of windows (IESNA 2013). The Building Research Establishment (BRE) has developed a set of protractors to provide percentages of direct reading of the sky component. Daylight factors from individual windows can be added to produce the daylight factor due to all the windows. In most cases, daylight factor levels in rooms are measured at work plan height (e.g. 0.85 m above the floor), leaving a 0.5 m border from the walls around the perimeter of the work plane.

4.4.3 COMPUTER MODELING

Computer modeling provides the speed and flexibility to evaluate daylight performance and building design and gets an accurate impression of the appearance of daylight in the rooms in terms of

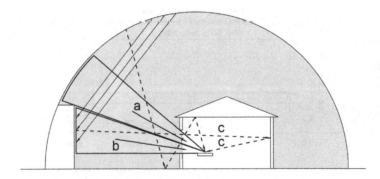

FIGURE 4.34 Components of daylight entering a room. (a) Sky component, (b) externally reflected component, and (c) internally reflected component.

work plane illuminance, daylight factors, surface luminance, a variety of glare and illumination quality, and performance metrics, and often can produce realistic color renderings. Most CAD visualization programs used today are capable of generating images that look realistic, but they do not provide information about the quantity and quality of daylight in the rooms.

There are two basic approaches to the computer modeling of daylight: radiative transfer and ray tracing. If illuminances at points are the only requirement, a radiative transfer procedure is usually sufficient. The advantage of radiative transfer is that one computation allows for all views of the room to be readily redisplayed without additional computations, facilitating the simulation of a walk-through of the space. Several commercially available programs use this technique. If accurate and realistic visualization of space is required, ray tracing can be the better technique. The advantage of ray tracing is that non-diffuse surfaces and greater geometric complexity are inherently more accurately and easily calculated. Programs using this technique are also available. The software packages most successful at solving real-world problems usually employ a hybrid of these methods.

Computer modeling is a practical method to parametrically evaluate and compare the performance of design alternatives and provide guidelines for design modification. Daylight availability inside a building is influenced a number of variable parameters including the window size, placement, geometry, the glazing type, the design of shading devices, the external obstruction, the interior layout and finishes. Figure 4.35 provides the summary of the results of a parametric study to observe the effect of the size, shape, and position of the window on daylight distribution in a room at Los Angeles, California. It can be observed that the height of the window determines the depth of daylight penetration, while the width influences the sideways spread of daylight. This is the result of using ECOTECT interfaced with Radiance. The software employs the BRE 'split flux' method for daylight factor calculation and uses CIE standard overcast sky or clear sky for its illuminance distribution model. Most of the recent daylight prediction programs include all six models of CIE standards and climate-based data, which consider not only the diffused daylight entering the room but also beam sunlight and its internal lighting effects.

4.4.4 CLIMATE-BASED DAYLIGHT MODELING (CBDM)

Climate-based daylight modeling relies on the annualized simulation of work plane illuminance using local sun and sky conditions derived from standardized annual meteorological data sets (Mardaljevic et al. 2012). It is sensitive to building orientation, geographical location and daily or seasonal variations in the sun and sky conditions. Work plane illuminance metric is given by daylight autonomy (DA) and useful daylight illuminance (UDI).

Daylight autonomy is a daylight availability metric that corresponds to the percentage of the occupied time when the target illuminance at a point in space is met by daylight (Reinhart 2001). A target illuminance of 300 lux and a threshold DA of 50%, meaning 50% of the time daylight levels are above the target illuminance, are values that are currently promoted in the Illuminating Engineering Society of North America (IESNA 2013).

Useful daylight illuminance is a daylight availability metric that corresponds to the percentage occupied time when a target range of illuminances at a point in space are met by daylight. Daylight illuminances in the range of 100–300 lux are considered effective either as the sole source of illumination or in conjunction with artificial lighting. Daylight illuminances in the range of 300 to around 3000 lux are often perceived as desirable.

Figure 4.36 provides the summary of the results of a parametric study to observe the effect of size, shape, and position on spatial daylight autonomy (sDA), the percentage of floor area that receives over 300 lux for at least 50% of 3650 annual hours. It can be observed that the sDA is related to the position and size of windows; the window located centrally is achieving the highest sDAs for each percentage of wall-window ratio, while the wide and high windows give the next highest performance. This is the result of using REVIT 2016, AutoDesk®.

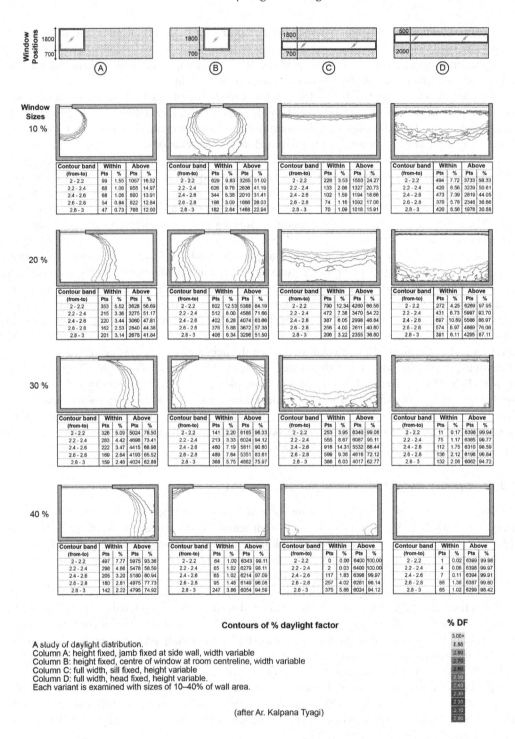

Contours of % daylight factor

A study of daylight distribution.
Column A: height fixed, jamb fixed at side wall, width variable
Column B: height fixed, centre of window at room centreline, width variable
Column C: full width, sill fixed, height variable
Column D: full width, head fixed, height variable.
Each variant is examined with sizes of 10–40% of wall area.

(after Ar. Kalpana Tyagi)

FIGURE 4.35 A parametric study of daylight factor distribution, Los Angeles, California. Source: Computer modeling by Ar Kalpana Tyagi, line work by Vaibhav Ahuja

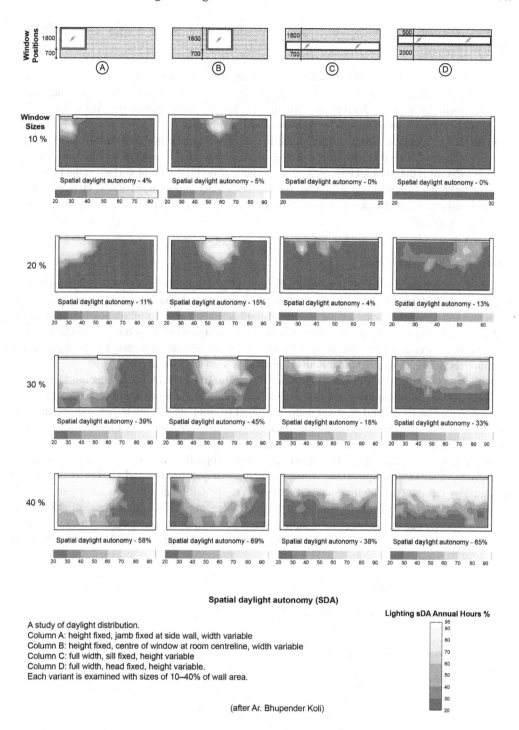

Spatial daylight autonomy (SDA)

A study of daylight distribution.
Column A: height fixed, jamb fixed at side wall, width variable
Column B: height fixed, centre of window at room centreline, width variable
Column C: full width, sill fixed, height variable
Column D: full width, head fixed, height variable.
Each variant is examined with sizes of 10–40% of wall area.

(after Ar. Bhupender Koli)

Lighting sDA Annual Hours %

FIGURE 4.36 A parametric study of spatial daylight autonomy (sDA), Los Angeles, California. Source: Computer modeling by Ar Bhupender Koli, line work by Vaibhav Ahuja

Further analysis can be done adjusting two important parameters that influence daylight availability inside a building:

i) Visible Light Transmittance: Visible Light Transmittance (VLT) of the exterior glazing is the percentage of the daylight that falls within the visible band of the electromagnetic spectrum (i.e. between 380 nm and 780 nm wavelength) that is transmitted through a glass. Higher the VLT more the daylight that will be transmitted.

ii) Surface reflectance: Surface reflectance is a property of the material that describes the percentage of light that is reflected off the surface compared to a perfectly black body that absorbs all radiation incident on it. A hypothetical perfectly white surface reflects all of the light that is incident on it, and has the Surface reflectance value of 100%. Hence lighter colored interior surfaces reflect more light and facilitate a greater amount of daylight in the interior space.

4.4.5 PHYSICAL MODELING

The analytical methods or even currently available computer modeling can predict the daylight for simple geometrical spaces, but complex geometrical spaces and fenestration systems can be realistically investigated using physical scale models and actual simulation of artificial skies. Physical models provide a simple means of parametric study to change window geometry, shading system, surface reflectance, and other parameters to find optimum design solutions. The performance of scale models can be evaluated under both clear and overcast sky conditions.

Smaller models (1: 50 and smaller) are used for qualitative evaluations of shading angles, sunlight penetration, glare, building massing, and siting issues. Scale models can be tested under outdoor conditions under natural sunlight using a sundial to simulate sun angles for specific days and hours or under similar natural overcast sky conditions, if available. The qualitative study is usually documented visually, photographically, or on video.

Larger scales (1:20 and larger) have internal surface properties close to the actual area required for interior evaluations and quantitative measurements. For quantitative studies, the sky condition is simulated by reproducing the luminance distribution of the CIE standard overcast sky or clear sky inside a model testing room, known as the artificial sky (Figure 4.37). The artificial sun can be created with a nearly parallel beam electric light source; the variation in solar angles can be reproduced either by moving the 'sun' in an arc around the model or by inclining the model on a tilting heliodon

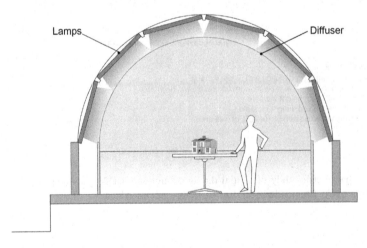

FIGURE 4.37 An artificial sky using lamps and diffuser. Source: after Szokolay (2008), line work by Ar Vaibhav Ahuja

table. A range of sun positions can be tested, for example, at 4-min intervals from 8 a.m. to 6 p.m., 4 min being the interval in which the sun changes position by 1°.

The climate of the site will determine which of the two critical sky conditions must be utilized in testing the model. In most parts of the United States, a model must be tested under both an overcast sky and a clear sky with the sun. The overcast sky determines whether minimum illumination levels will be met, and the clear sky with sun indicates possible problems with glare and excessive-brightness ratios.

The process of scale model studies is simple: first, measure the 'outdoor' illuminance in the center of the artificial sky and then place the model, measure the illuminance 'indoor' at points on a grid 1 × 1 m located inside the model, and calculate the daylight factor at each of these points as $DF = (E_i/E_o) \times 100$. Alternatively, the measurement can be taken with a series of photocells placed on the grid points and connected to a lighting logger or computer. The measurements are used to calculate DF values at grid points and generate DF contours. They are also used to convert these into illuminance (isolux) contours and also to estimate potential electric lighting energy savings from daylighting strategies.

Artificial skies and suns have limitations. Illuminances can be determined only for the sky conditions under which the measurements are made; they may not be representative of average conditions. The measured sky conditions can be compared with daylight availability measurements.

4.5   ELECTRIC LIGHTING AS A SUPPLEMENT TO DAYLIGHTING

Lighting design for buildings requires an appropriate balance between daylighting and electric lighting, depending on the type, function, and form of the building and the visual activity performed by the users. The integration of daylight and electric light starts with an assessment of the contribution of daylight across space at different times of the day and year. This may include evaluating daylight surface luminance ratios, illuminance levels and contours, day light autonomy, useful daylight illuminance, daylight zones, temporal variations in daylight availability (direction and intensity), and how daylight distribution changes with adjustable shading and fenestration elements.

Daylight and electric light are designed to function together in a task/ambient approach, where daylight provides the ambient lighting (supplemented by electric light as necessary), and electric light provides the task and accent lighting. In this case, the ambient electric lighting system is designed to reduce daylight gradients and balance luminances in the space. The ambient electric lighting must be circuited and zoned to follow the daylight zones. This is accomplished by aligning the electric lighting circuits parallel to the daylight contours. Ideally, the ambient electric light is delivered to the same surfaces as the daylight so that the transition between daylight and electric light is smooth.

4.5.1   ELECTRIC LIGHTING CONTROL

If ambient electric lighting has been designed to correspond with daylight zones, an automatic control system can be implemented to turn off or dim the electric lights in response to daylight availability. With advanced lighting controls, it is now possible to adjust the level of electric light when sufficient daylight is available. Three types of controls are commercially available (Ander 2016):

i) **Switching controls**: on-and-off controls that simply turn the electric lights off when there is sufficient daylight.
ii) **Stepped controls**: control individual lamps within a luminary to provide intermediate levels of electric lighting.
iii) **Dimming controls**: continuously adjust electric lighting by modulating the power input to lamps to complement the illumination level provided by daylight.

Any of these control strategies can, and should, be integrated with a building energy management system (BEMS) to take advantage of the system's built-in control capacity. Automatic control is accomplished by a photocell (sensor) that evaluates the light level and sends a signal to a control unit to dim or switch the electric lights to maintain a preset target level. For side lighting, the photocell (sensor) is usually placed on the ceiling looking down at a representative task area; for top lighting, it is frequently placed in a skylight looking up at the available daylight.

Automatic control provides more predictable performance than manual control, but control zones must be sufficiently small and matched well with daylight availability to achieve light-level consistency. It is critical that the lighting circuits and switching schemes are planned in relation to fenestration to effectively use available daylight and avoid dark zones.

In addition to daylight controls, other electric lighting control strategies should be incorporated where they are cost-effective (Ander 2016), including the use of:

i) **Occupancy controls**: using infrared, ultrasonic, or microwave technology, occupancy sensors respond to movement or object surface temperature and automatically turn off or dim down luminaries when rooms are left unoccupied. Typical savings have been reported to be in the 10–50% range depending on the application.

ii) **Timers**: these devices are simply time clocks that are scheduled to turn lamps or lighting off on a set schedule. If spaces are known to be unoccupied during certain periods of time, timers are extremely cost-effective devices.

Commissioning can be defined as a systematic process that ensures that all elements of the daylighting system and light control system perform interactively and continuously according to the documented design intent and needs of the building owner. Generally, commissioning should be carried out in spaces that are furnished and ready for occupancy.

Commissioning of the daylighting system primarily concerns fenestration controls; it can be divided into three steps, each with its specific requirements and procedures: (i) systems with fixed elements that may need adjustment after installation, such as some light shelf; (ii) systems with manually operated controls, such as interior venetian blinds, and (iii) systems with automated controls, such as exterior motorized louvers.

Commissioning of the lighting control system involves adjustment or calibration of any electrical or mechanical sensors such that they produce the desired control signal to the system under the specific conditions of the room design and over a wide range of incidents.

The specifics of this procedure vary with control system type (open versus closed-loop control) and hardware selection. This often requires special skills and equipment, such as a photometer to set proper light levels.

Commissioning must be completed for each unique physical zone or control system zone and each building orientation.

REFERENCES

Ander GD (2016) *Daylighting*. Whole Building Design Guide. National Institute of Building Sciences, Washington, DC, https://www.wbdg.org/resources/daylighting.

Aschehoug O, Christoffersen J, Jakobiak R, Johnsen K, Lee E, Ruck N, Selkowitz S (eds) (2000) *Daylight in Buildings, a Sourcebook on Daylighting Systems and Components. A Report on IEA SHC Task 21/ECBCS*. Annex 29. International Energy Agency.

ASHRAE (2010) *Energy Standard for Buildings except Low-Rise Residential Buildings*. ANSI/ASHRAE/IES Standard 90.1. American Society of Heating, Refrigerating and Air Conditioning Engineers, Atlanta.

CIBSE (2002) *Code for Lighting*. Chartered Institution of Building Services Engineers, Oxford.

CIE (2003) *S 0003 CIE Spatial Distribution of Daylight – CIE Standard General Sky, CIe S 011/E*. Commission Internationale De L'Eclairage, Vienna.

Edmonds IR, Greenup PJ (2002) Daylighting in the tropics. *Solar Energy*, 73(2), pp 111–121.

Guzowski M (1999) *Daylighting for Sustainable Design*. McGraw-Hill Professional, New York.

Hansen VG, Edmonds T (2003) Natural illumination of deep-plan office buildings: Light pipe strategies. In: *ISES Solar World Congress*, 14–19 June 2003, Gothenburg, Sweden.

Hopkinson RG (1963) *Architectural Physics - Lighting*. Her Majesty's Stationery Office, London.

Howard TC (1986) Variable area light reflecting assembly. United States, https://www.osti.gov/servlets/purl/866085.

IEA (2001) Application guide for daylight responsive lighting control. IEA SHC Task 21. International Energy Agency, https://www.iea-shc.org/Data/Sites/1/publications/8-8-1%20Application%20Guide.pdf.

IESNA (2013) *The IESNA Lighting Handbook: Reference and Application, Illuminating Engineering Society of North America, National Bureau of Standards (1991) The International System of Units (SI)*, 6th edition, NBS Special Publication 330. National Bureau of Standards, Gaithersburg, MD.

ISO (2004) *Spatial Distribution of Daylight – CIE Standard General Sky*. ISO 15496. International Organization for Standardization, Geneva.

Iversen A, Roy N, Hvass M, JØrgensen M, Christoffersen M, Jonsen K (2013) *Daylight Calculation in Practice: An Investigation of the Ability of Nine Daylight Simulation Programs to Calculate the Daylight Factor in Five Typical Rooms*. Statens Byggeforskningsinstitut (Danish Building Research Institute), Aalborg University, Copenhagen, SBi 2013:26, .

Larson GW, Shakespeare R (1998) *Rendering with Radiance: The Art and Science of Lighting Visualization*. Morgan Kaufman, San Francisco.

Lawrence Berkeley Laboratory, Windows and Daylighting Group (1985) *Superlite 1.0 Program Description Summary*, DA 205. Lawrence Berkeley Laboratory, Berkeley, CA.

Lim BP, Rao KR, Tharmaratnam K, Mattar AM (1979) *Environmental Factors in the Design of Building Fenestration*. Applied Science Publishers, London.

Linhart F, Wittkopf SK, Scartezzini JL (2010) Performance of anidolic daylighting systems in tropical climates – Parametric studies for identification of main influencing factors. *Solar Energy*, 84(7), pp 1085–1094.

Longmore J, Petherbridge P (1961) Munsell value/surface reflectance relationships. Letters to the editor. *Journal of the Optical Society of America*, 51(3), pp 370–371.

Majoros A (1998) *Daylighting. Passive and Low Energy Architecture International, Design Tools and Techniques, PLEA Note 4*. http://www.plea-arch.org/wp-content/uploads/PLEA-Note-4-Daylighting-lo wre.pdf.

Mardaljevic J (2015) Climate-based daylight modelling and its discontents. *CIBSE Technical Symposium*, 16–17 April, London.

Moon P, Spencer DE (1942) Illumination from a nonuniform sky. Illumination. *Engineering*, 37(12), pp 707–726.

Nair MG, Ramamurthy K, Ganesan AR (2014) Classification of indoor daylight enhancements systems. *Lighting Research and Technology*, 46(3), pp 245–267.

NBS (1991) *The International System of Units (SI). National Bureau of Standards*, 6th edition, NBS Special Publication 330. Gaithersburg, MD.

Obradovic B, Matusiak B (2019) Daylight transport systems for buildings at high latitudes. *Journal of Daylighting*, 6(2), pp 60–79.

Ruck N, Aschehoug O, Aydinli S, Christoffersen J, Edmonds I, Jakobiak R, Kischkoweit-Lopin M, Klinger M, Lee E, Courret G (2000) *Daylight in Buildings-A Sourcebook on Daylighting Systems and Components*. Lawrence Berkeley National Laboratory, Berkeley, CA.

Song J, Luo G, Li L, Tong K, Yang Y, Zhao J (2018) Application of heliostat in interior sunlight illumination for large buildings. *Renewable Energy*, 121, pp 19–27.

Szokolay SV (2008) *Introduction to Architectural Science: The Basis of Sustainable Design*. Architectural Press/Elsevier Science, Oxford.

Whang AJW, Yang TH, Deng ZH, Chen YY, Tseng WC, Chou CH (2019) A review of daylighting system: For prototype systems performance and development. *Energies*, 12(2863), pp 1–34.

5 Renewable Energy

5.1 INTRODUCTION

America's greatest inventor said:

> "I'd put my money on the sun and solar energy. What a source of power! I hope we don't have to wait till oil and coal run out before we tackle that. I wish I had more years left!"

Thomas Alva Edison, in conversation with Henry Ford and Harvey Firestone, March 1931

In the design of new buildings and major renovations of existing buildings characterized by a long life span, the achievement of energy efficiency and the appreciation of finite resources will play a major role. Errors made today in the design of the built environment, as a result of either want of expertise or well-formulated design philosophies, will place long-term environmental cost not just on the owners and users of buildings but also on the global society at large. The requirement to be sustainable when it comes to the built environment is extensive and exacting.

The synergistic design of the sustainable built environment offers an opportunity to architects or designers to synthesize the potentials of passive cooling, passive heating, natural ventilation, and daylighting with the hybrid (low-energy) and active design strategies – depending upon the climate and other site conditions – to meet occupants' comfort demand. Consequently, sustainable built environments need to integrate energy efficiency and renewable energy generation such that they consume only as much energy as can be produced onsite through renewable resources over a specified period. There is a wide spectrum of alternatives for renewable energy integration in a sustainable built environment. The potential and suitability of any renewable energy system depend on the geographical, topographical, climate, site characteristics, and resources. It is imperative to realize that the implementation of renewable energy will not be a case of choice but of necessity to be a carbon-neutral society. Also to meet the federal requirement, Energy Independence and Security Act (EISA 2007) requires new buildings and major renovations of federal buildings to reduce fossil fuel consumption relative to 2003 by:

    i. 55% by 2010
   ii. 65% by 2015
  iii. 80% by 2020
  iv. 100% by 2030

EISA 2007 requires that 30% of the hot water demand in new federal buildings (and major renovations) be met with solar hot water equipment, provided it is life-cycle cost-effective.

It needs to be emphasized that the design and implementation of renewable energy systems are highly specialized, requiring services of competent, experienced consultants and integrators in the project delivery team. This chapter aims to provide general guidance on renewable energy applications in the sustainable built environment both onsite and off-site at the neighborhood, district, and utility scale. The next section presents essential concepts of energy – forms and sources of energy, cogeneration, or combined heat and power systems and plug loads. The third section discusses solar energy applications – solar thermal and photovoltaic systems. The fourth section explains wind energy systems. The fifth section covers other renewable energy systems – biomass, geothermal, hydrogen and fuel cells, and hydropower. The final two sections pertain to energy storage and smart grid for renewable energy.

5.2 ENERGY

Albert Einstein defined the relationship of energy and matter by equation $E = mc^2$, which means energy is equal to matter multiplied by the speed of light squared (300,000 km/s).

A system possesses energy if it has the capacity for producing an effect or performing work. Energy is a scalar quantity; in the International System of Units (SI), it is measured in J (joule), named after James Prescott Joule. One joule is equal to the force of one Newton acting over one meter distance i. e. $N \times m$. The energy flow rate is measured with the unit W (watt), which is the flow of 1 J per 1 second (J/s). An accepted energy unit is the Wh (watt-hour), i.e., the energy that would flow if the rate of 1 W were maintained for 1 h. As there are 3600 seconds in an hour, 1 Wh = 3600 J or 1 kWh = 3600 kJ = 3.6 MJ.

5.2.1 FORMS OF ENERGY

Energy can be neither created nor destroyed (except in subatomic processes), but only converted from one form to another. This principle is known as the conservation of energy or the first law of thermodynamics; for any system, open or closed, there is an energy balance as (in the absence of a nuclear or chemical reaction):

[Net amount of energy added to the system] = [A net increase of stored energy in the system]
or
The energy in – Energy out = Increase of stored energy in the system

The second law of thermodynamics differentiates and quantifies processes that proceed only in a certain direction (irreversible) from those that are reversible. This is also known as the law of entropy. This law also states that heat flows from higher temperature to lower temperature.

Energy can exist in many different forms; all forms of energy can be classified as either stored or transient forms.

Stored Energy

Stored energy is of two types: potential energy and kinetic energy.

Potential energy (PE) is caused by attractive forces existing between molecules or the elevation of the system to a reference level:

$$PE = m\,g\,h \tag{5.1}$$

where

m = mass (kg)
g = local acceleration of gravity, 9.81 m/s^2
h = elevation above horizontal reference plane (m)

Chemical energy, mechanical energy, gravitational energy, and nuclear energy are all types of potential energy.

Chemical energy is stored in the bonds of atoms composing the molecules. Chemical processes requiring energy (heat) supply are termed 'endothermic,' those that release energy are termed 'exothermic,' and those that neither require heat nor release energy are termed 'thermoneutral.' Fuels like biomass, coal, natural gas, and petroleum are examples of high chemical energy content that can be released by combustion (an exothermic process). The process starts with little heat input (ignition), but then the process is self-sustaining, for example, when the wood is combusted in a fireplace or gasoline is burned in the car's engine. The chemical energy in coal is converted into electrical energy at a power plant. The chemical energy in a battery can also supply electrical power using electrolysis.

Gravitational energy is energy stored in an object's height. The higher and heavier the object, the more gravitational energy is stored. When a person rides a bicycle down a steep hill and picks up speed, the gravitational energy is converting to motion energy. Hydropower is another example of gravitational energy, where gravity forces water down through a hydroelectric turbine to produce electricity.

Nuclear (atomic) energy derives from the cohesive forces holding protons and neutrons together as the nucleus of an atom. Large amounts of energy can be released when the nuclei are combined (fusion) or split apart (fission).

Kinetic energy (KE) is the energy caused by the velocity of molecules and is expressed as

$$KE = \frac{1}{2}mV^2 \tag{5.2}$$

where
V is the velocity of a fluid stream crossing the system boundary.

Mechanical energy, thermal energy, electrical energy, electromagnetic radiation, and sound are all types of kinetic energy.

Mechanical energy is the sum of the kinetic energy and the potential energy in a system because of the position of energy stored in objects by tension. Compressed springs and stretched rubber bands are examples of potential mechanical energy. Swinging pendulum is an example of kinetic mechanical energy.

Thermal (internal) energy is caused by the movements of molecules of atoms and molecules in a substance and/or intermolecular forces. Heat increases when these particles move faster. Geothermal energy is the thermal energy of the earth.

Sound is the movement of energy through the system in longitudinal (compression/rarefaction) waves. Sound is produced when a force causes a system to vibrate. The energy is transferred through the system in a waveform. Typically, the energy in sound is smaller than in other forms of energy.

Electrical energy is delivered by tiny charged particles called electrons, typically moving through a wire. Lighting is an example of electrical energy in nature. The presence of free electrons in a body represents a charge, an electric potential. These electrons tend to flow from a higher potential zone to a lower one. The unit of electric charge is the coulomb (C). The rate of electricity flow (current) is the ampere (amp, A):

$$A = C/s \text{ conversely } C = A \times s$$

A potential difference or electromotive force (EMF) of 1 volt (V) exists between two points when the passing of 1 coulomb (C) constitutes 1 Joule (J) (Equation 5.3).

$$V = J/C \tag{5.3}$$

Electric current will flow through a body if its material has free electrons. For example, metal conductors (silver, copper, and aluminum) have only one electron in the outermost electron skin of the atom. In gases or liquids, electricity may flow in the form of charged particles called ions.

However, the electron flow has to withstand some resistance even in the best conductors. The unit of this resistance is the ohm (Ω), the resistance that allows the flow (current) of 1 ampere driven by 1 volt (Equation 5.4).

$$\Omega = V/A \text{ conversely } A = V/\Omega \tag{5.4}$$

The rate of energy flow in the current (or electric power) is the watt (W), a unit that is used for all kinds of energy flow (Equation 5.5).

$$W = V \times A \tag{5.5}$$

The above is valid for direct current (DC), i.e., when the current flows in one direction. Alternating current (AC) is produced by rotating generators, where the resulting polarity is reversed 50 times per second, i.e., at the frequency of 50 Hz (in the United States it is 60 Hz). With AC the above relationship is influenced by the type of load connected to the circuit: resistive load, capacitive load, or inductive load. For this reason, the power of AC is referred to as VA or kVA (rather than W or kW).

Energy in Transition

Two systems having stored energy will interact by transferring heat or work. Once the transfer has been accomplished, it will again convert into stored energy inside the system. Thus the system on the receiving end will not have increased stored energy compared to what it had before, and the donating system will have its stored energy decreased. Thus, there is energy in transition, heat, and work, i.e., energy in the process of transfer from one system to another. After it has transferred, energy is always designated according to its nature. Hence, heat transferred may become thermal energy, while work done may manifest itself in the form of mechanical energy.

Heat is the mechanism that transfers energy across the boundaries of systems with differing temperatures, always toward the lower temperature. Heat is positive when energy is added to the system.

Work is the mechanism that transfers energy across the boundaries of systems with differing pressures (or forces of any kind), always toward the lower pressure. The total effect produced in the system can be reduced to the raising of weight, then nothing but work has crossed the boundary. Work is positive when energy is removed from the system.

Mechanical or *shaft work* is the energy delivered or absorbed by a mechanism, such as a turbine, air compressor, or internal combustion engine.

Flow work is the energy carried into or transmitted across the system boundary because a pumping process occurs somewhere outside the system, causing fluid to enter the system. It can be more easily understood as the work done by the fluid just outside the system on the adjacent fluid entering the system to force or push it into the system. Flow work also occurs as fluid leaves the system.

$$\text{Flow work (per unit mass)} = pv \tag{5.6}$$

where p is the pressure and v is the specific volume or the volume displaced per unit mass evaluated at the inlet or exit.

5.2.2 SOURCES OF ENERGY

Energy sources can be categorized as renewable or nonrenewable. Nonrenewable energy sources are petroleum products, hydrocarbon gas liquids, natural gas, coal, and nuclear energy. Coal, crude oil, and natural gas are called fossil fuels because they were formed over millions of years by the action of heat from the earth's core and pressure from rock and soil on the remains (or fossils) of dead plants and creatures such as microscopic diatoms. Nuclear energy is produced from uranium, a nonrenewable energy source used through nuclear fission to generate heat and, eventually, electricity.

Energy Policy Act 2005 defines 'renewable energy' as electric energy generated from solar, wind, biomass, landfill gas, ocean (including tidal, wave, current, and thermal), geothermal, municipal solid waste, or new hydroelectric generation capacity achieved from increased efficiency or additions of new capacity at an existing hydroelectric project. Renewable and nonrenewable energy sources can be used as primary energy sources to produce useful energy such as heat or used to produce secondary energy sources such as electricity.

Figure 5.1 shows the energy sources used in the United States. In 2019, renewable energy sources accounted for about 12% of US energy consumption. Biomass, which includes wood, biofuels, and biomass waste, is the largest renewable energy source, and it accounted for about 43.5% of all renewable energy consumption and nearly 5% of total US energy consumption.

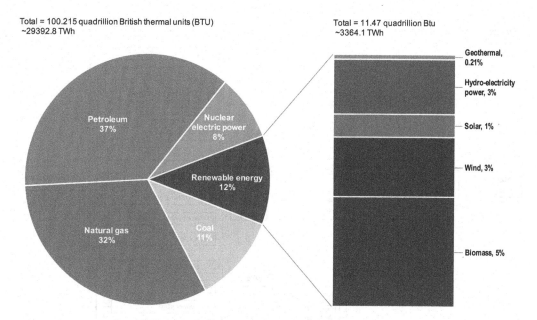

FIGURE 5.1 U.S. energy consumption by energy source, 2019. Source: U.S. Energy Information Administration, *Monthly Energy Review*, Table 1.3, March 2020, preliminary data

A taxonomy of renewable energy is shown in Figure 5.2. Three primary sources of renewable energy are gravity, solar energy, and geothermal. Solar energy systems can be further distinguished as direct and indirect forms of utilization.

5.2.3 Cogeneration or Combined Heat and Power (CHP) Systems

Cogeneration power plants use the heat of an engine or a power station to simultaneously generate both electricity and useable heat. In simple terms, it is a more efficient use of final energy than in exclusive power plants or exclusive thermal energy plants. In such 'pure' traditional power plants, the achievable performance efficiency is around 35% – in other words, only 35% of power input is being converted to electricity, with the rest being lost in the form of heat energy.

To achieve more efficiency, many years ago a large number of power plants were converted to heat power plants. In such power plants, the heat loss as a byproduct of electricity generation is used for thermal energy generation, with a resulting increase in efficiency of up to 90% of the energy input. In principle, CHPs are small power plants with combustion engines driving a generator to generate electricity. The excess heat of the combustion engine's cooling water and its exhaust gas heat energy is used for heating applications. Alternatively, such excess heat may be transformed into cooling energy in absorption chiller plants. A plant producing electricity, heat, and cooling energy is sometimes called a trigeneration or, more generally, a polygeneration plant. The results are very high year-round performance coefficients for such cogeneration power plants.

Concerning the energy input, CHPs achieve an efficiency of 30–40% for electricity generation and approximately 55% thermal yield, resulting in a total of around 85–95% efficiency.

Cogeneration plants should reach or exceed an annual minimum operation time of 4500 h/a to be economical. If large building complexes require emergency power systems, it is better to use a cogeneration power plant than diesel generators to provide continuous operation. Cogeneration power plants are today typically fueled by light fossil heating oil or natural gas, but the future fuels for such systems belong to the family of renewable fuels such as biodiesel and biogas to decouple the plants from fossil energy use.

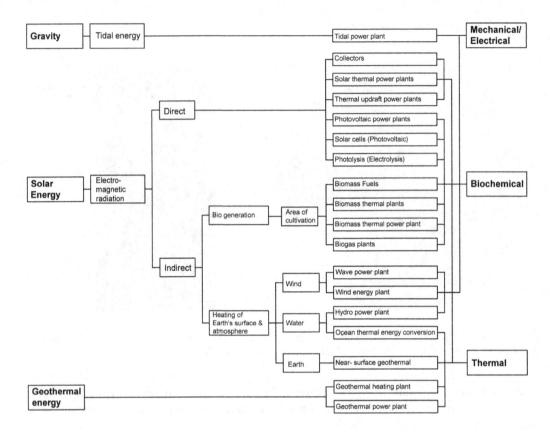

FIGURE 5.2 A taxonomy of renewable energy. Source: after archplus 184, October 2007, line work by Himanshi Jaglan

5.2.4 PLUG LOAD

Plug and process loads are the energy loads distinguished from energy load related to heating, cooling, lighting, and ventilation, but they typically account for about one-quarter of US commercial building energy use. These loads are specific to the function of the building and include items such as ATMs, appliances, chargers, computers, electrical devices, elevators, office equipment, data centers, server rooms, coffee machines, vending machines, and water coolers. Certain building types have specialized and heavy process loads, such as hospitals, manufacturing facilities, and buildings with foodservice operations.

The sustainable and low energy buildings, therefore, should be designed not only to effectively reduce energy related to heating, cooling, lighting, and ventilation but also to effectively manage the remaining plug and process load, which can grow to a high percentage of the building's energy use. Table 5.1 explains the basic electrical loads (power/electricity needs) by the short calculation method for an architect's studio; the plug load accounts for about 50% of the total energy requirement.

A comprehensive plug and process load management has the additional benefit of lowering internal heat gains, which have the potential of greater energy savings on space cooling.

5.3 SOLAR ENERGY

Solar radiation is an enormous amount of energy and is the driving force of all terrestrial energy systems. The theoretically available energy of $1,524,240 \times 10^9$ MWh/year as a result of solar radiation, assuming an average coefficient of the performance of solar technology of 15%, can be harnessed to the order of $2,28,636 \times 10^9$ MWh/year to fulfill all global energy requirements. The direct solar

TABLE 5.1
Worksheet for Determining the Electric Plug Load

Appliances	Watts	Number of Hours Used per Day	Watt-Hours Per Day
Coffee maker	200	1	200
Laptop computers (2)	200	8	3200
Plotter (1) 220 volts and 15 amps	3300	2	6600
Microwave	1000	1	1000
Television	200	2	400
Refrigerator[†]	500	6	3000
Total plug load (Wh/day) 35%			14,400
Lights (3)	60 + 6 = 66	4	792
Air conditioner	3000	4	12,000
Water heater	1500	1	1500
Total HVAC and lighting loads (Wh/day) 65%			14,292
Total energy load (Wh/day)			28,692

Total electric plug load per day = 28,692 × 1.5 = 43,038 Wh ~ 43.04 kWh

Total electric plug load per annum = 43,038 × 365 = 1,570,880 Wh ~ 1,57,08.9 kWh

Typical appliance wattage			
Devices	Wattage	Devices	Wattage
Incandescent lights (wattage of the lamp)	–	Vacuum cleaner	500
Fluorescent lamps (wattage of lamps plus 10% for ballast)	–	Clothes dryer (uses gas)	350
Coffeepot	200	Furnace blower	500
Microwave	1000	Air conditioner (central)	3000
Dishwasher	1300	Ceiling fan	30
Washing machine	500	Computer and printer	200
Television and videocassette recorder	200	Stereo	20
Refrigerator-conventional	500	Refrigerator high efficiency[†]	200

[†] Because refrigerators cycle on and off, they typically run about six hours per day. It is assumed that each appliance is to be 'on' during a typical day (minimum of one hour).

energy system distinguishes three conversion processes: thermal, electrical, and bio(-chemical). This section discusses solar thermal and photovoltaic systems.

Many solar radiation resource maps are published with solar radiation data. NREL, for example, provides many global and US solar radiation resource maps, Figure 5.3 shows global horizontal solar irradiance values (kWh/m²/day). According to NREL solar resource across the United States range between 1000–2500 kWh/m²/year. The southwest is at the top of this range, while only Alaska and part of Washington are at the low end. The range for the contiguous United States is about 1350–2500 kWh/m²/year. Nationwide, solar resource levels vary by about a factor of two. The weather data file can also provide solar radiation data; tabulated average global and diffuse solar radiation are provided for 50 US cities in Chapter 7.

5.3.1 SOLAR THERMAL SYSTEMS

Solar thermal systems convert the energy of the sun into thermal energy, which can be used to heat water or a building or generate electricity (thermoelectric device). Solar thermal energy can also be used for solar cooling by driving an absorption chiller, but it is a developing technology. Depending upon the specific temperature range, solar thermal systems can be distinguished as low temperature

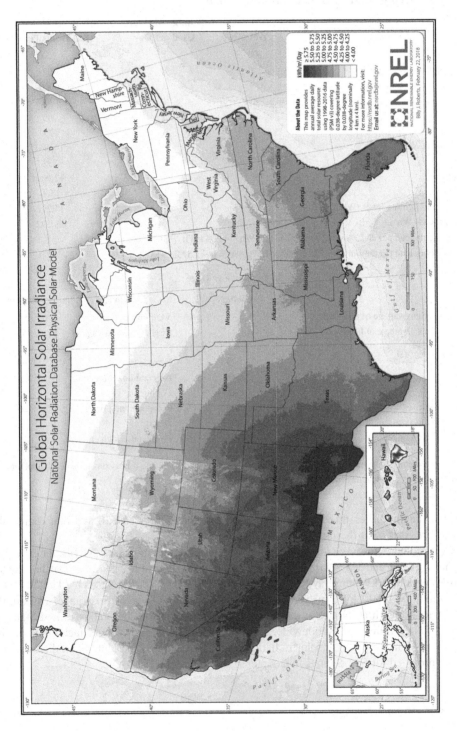

FIGURE 5.3   Global horizontal solar irradiance map of the United States, NREL.

and high temperature. The high-temperature system is used to generate electricity, while low-temperature systems are used to heat water for domestic purposes, swimming pool, and building heat. A variety of solar thermal system configurations exist today (Figure 5.4).

Solar Absorbers

Solar absorbers are simple absorber matters that are typically made of high-quality UV-resistant EPDM neoprene tubes and that have vulcanized collector and distributor piping integrated into them. Solar absorbers can be added to, or integrated into, roof surfaces, but they can also be placed 'on grade.' Solar absorbers should be temperature resistant in a range between −50° and +60°C and be capable of being cleaned out, and their main application is to heat the water of outdoor pools. The necessary solar absorber surface area for this kind of application is approximately 50–80% of the water surface of the pool that requires to be heated (Figure 5.5).

Flat Plate Collectors

The flat plate collectors are the versatile and well-developed solar thermal system, which can be used for low-temperature (<100°C) thermal purposes: water (or any fluid) heating or air heating. The solar radiation is absorbed largely due to the selective coatings on the collectors, the absorptivity of

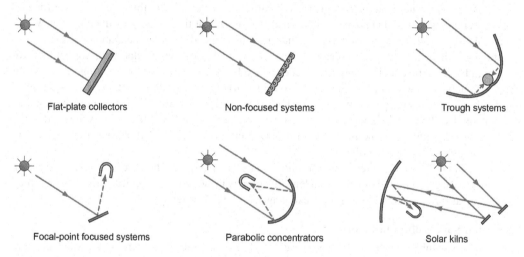

FIGURE 5.4 Configurations of solar thermal systems. Source: adapted from Klaus (2003), line work by Himanshi Jaglan

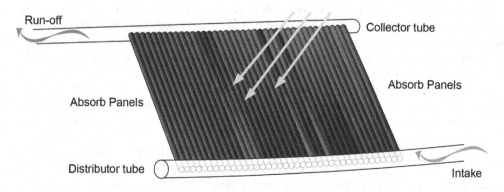

FIGURE 5.5 Solar absorber for swimming pool, 300 m² absorber surface. Source: adapted from Klaus (2003), line work by Himanshi Jaglan, color rendering by Ar Vaibhav Ahuja

which is kept high, while the emissivity is low, and the captured heat energy is transferred to water (or other fluid) or air for further use.

Flat plate collectors are insulated and weatherproofed boxes, containing these components:

- Absorber plate – black or selective coating absorbent of the incident solar energy
- One or more translucent glass or plastic (polymer) covers that transmit radiation to the absorber but prevent radiative and convective heat loss from the surface
- Water or fluid ways (tube grid or channels formed between two sheets) or air ducts to transfer the heat from the collector
- Rigid insulation covering the sides and bottom of the collector to reduce heat losses
- Support structure to protect the components and hold them in place

Solar water heating systems harness the thermal energy of the sun to heat water in homes and businesses. The systems can be installed in any climate to reduce utility bills and are composed of three main parts: the flat plate collector, insulated piping, and a hot water storage tank.

Solar water heating systems can rely on either thermosiphon (passive) to circulate water when the water tank is at a higher level than the collector or small electric pumps (hybrid) when the water tank is at a lower level. Hybrid solar water heating systems are more common in residential and commercial uses. Passive solar water heating systems are typically less expensive, but they are also less efficient. In closed-couple systems, a horizontal cylindrical tank is kept at the top edge of the collector. Water tanks need to be designed to resist corrosion and freezing, depending upon the climate location.

Although air inside flat plate collectors acts as an insulator, vacuum flat plate collectors have been found more efficient. The collectors need to be adjusted precisely to the position of the sun in the sky to achieve maximum efficiency; it should be oriented due south (northern hemisphere) or due north (southern hemisphere) with optimal azimuth angle of the sun (180° in the northern hemisphere). Ideally, the angle of solar incidence to the collector's surface is 90°, which requires that the tilt angle of solar collectors should be the same as the latitude. Figure 5.6 illustrates the orientation, tilt, and details of a flat plate collector.

An auxiliary (booster) heater may be required for cloudy days. In the case of freezing climates, anti-refrigerating compounds such as ethylene glycol may be added to the recirculating water, which transfers heat through a double wall heat exchanger to the hot water system.

Evacuated Solar-Tube Collectors

Evacuated solar-tube collectors are used to achieve higher operating temperatures and also ensure aesthetic appeal. The collectors are composed of parallel rows of exposed transparent glass tubes. Each evacuated tube consists of two glass tubes made from strong borosilicate glass. The outer tube is transparent with 75–100 mm diameter, allowing solar radiation to penetrate through with minimal reflection, and the inner tube is coated with a special selective coating to absorb solar energy but inhibits radiative heat loss. The tops of the two tubes are fused, and the air contained in the space between the two layers of glass is pumped out while exposing the tube to high temperatures. This 'evacuation' of the glasses forms a vacuum and is the critical factor in the performance of the evacuated tubes, similar to a thermos. The absorber tube contains fluid that will vaporize when heated and reaches as a vapor to the condenser of the system. Here energy is transferred to a stream of water or antifreeze solution running through manifold; the vapor then condenses and flows back to the absorber tubes, where it is reheated and vaporized. Figure 5.7 shows a diagram of the working principle of evacuated solar absorber tubes.

In both warm regions with great amounts of annual sunshine and climates with seasons of varying hot and cold temperatures, the evacuated solar tubes are capable of producing energy for both cooling and heating. However, thermal solar energy alone is typically not sufficient to supply the energy needed for large buildings. A large-scale application of evacuated tube collectors in a cold climate can be seen in the Visual Art Facility, University of Wyoming, Laramie; refer to Section 6.6 in Chapter 6.

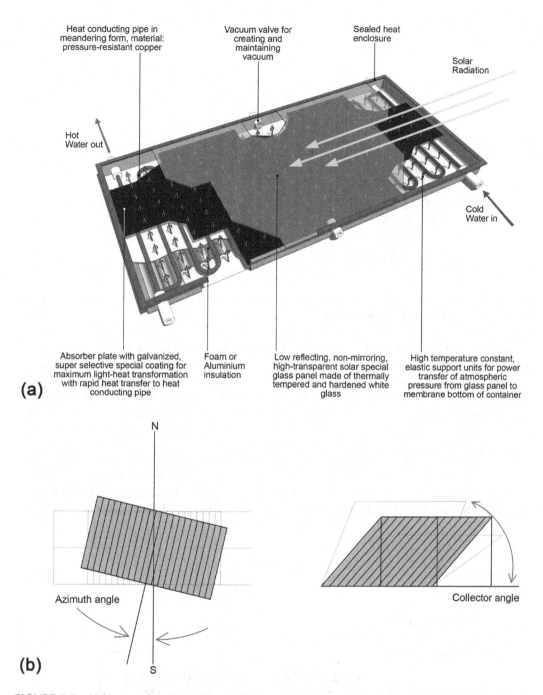

FIGURE 5.6 (a) Vacuum flat plate collector (thermosolar), (b) orientation and tilt angle. Source: adapted from Klaus (2003), line work by Ar Vaibhav Ahuja

5.3.2 Photovoltaic Systems

Photovoltaic (PV) is a term derived from 'photo' for light and 'voltaic' for a volt, a unit of electrical force. A photovoltaic or solar electric system is a technology that utilizes the semiconductor-based photoelectric cell to convert the photons of the solar light spectrum directly into electric energy. The underlying principle of this property is called the photovoltaic effect, a subcategory of the internal photovoltaic effect.

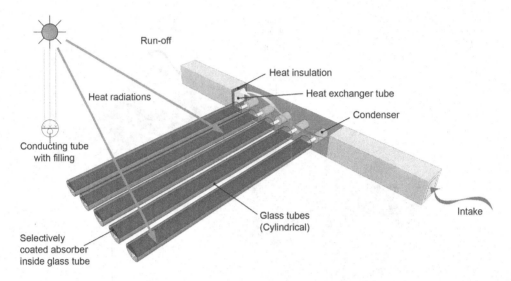

FIGURE 5.7   Evacuated solar tube collector. Source: adapted from Klaus (2003), line work by Himanshi Jaglan, color rendering by Ar Vaibhav Ahuja

The solar cell is a basic module for the PV formation, and it is typically in dark blue because of the anti-reflective coating added to the cell. The solar cell is produced from a semiconductor material, commonly silicon. The molecular structure of silicon permits electrons to be freed by the incoming energy of photons from sunlight. The solar cell consists of two-layer semiconductors, n-type (−ve) and p-type (+ve), doped with phosphorous and boron atoms, respectively; both the layers are connected by metal conductors sending n-type layers electron to p-type layer and vice-versa, completing the circuit to generate electricity (Figure 5.8).

The solar cell generates DC, and an inverter converts it into the AC. Cells come in many sizes and shapes. Panels, or modules, are assembled from many solar cells. An array is a collective name for many solar modules connected. Figure 5.9 shows an example of a single-crystal silicon technology. Each individual PV cell produces about 0.6 volts of power. Modules generally consist of

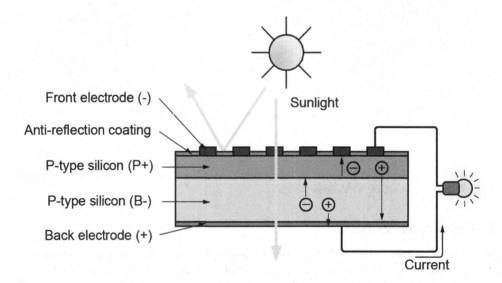

FIGURE 5.8   Configuration of PV or solar cell. Line work by Ar Vaibhav Ahuja

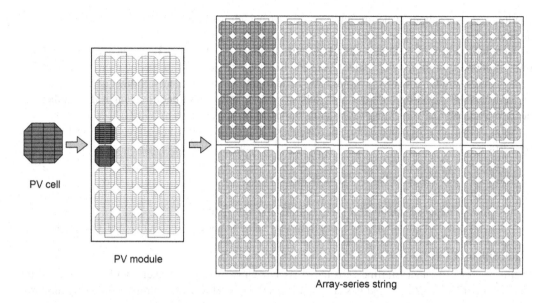

PV cell

PV module

Array-series string

FIGURE 5.9   Solar electricity – PV cell, module, and array. Line work by Ar Sakshi Singhal

100 cells to bring the module voltage up to 60 V. Ten of these modules could be connected in series to achieve a maximum voltage of 600 V per string.

A PV system can be either off-grid or on-grid connected, and the off-grid system requires energy storage (batteries) for nighttime energy use and during lean solar-producing winter days. In addition to batteries, the off-grid system requires back-up energy systems to offset lower solar energy production.

The photovoltaic systems perform based on the sunlight striking on the module, which varies according to geographical location, seasons, and other factors. Nevertheless, the photovoltaic system is a highly reliable source of renewable energy with low maintenance, quiet and pollution-free operation, and high durability.

The photovoltaic system is the most viable and widely deployed de-centralized renewable energy technology to generate electricity due to its versatility, cost-effectiveness, and capability for integration into building projects of different scales. Net or smart metering enables the important function of importing and exporting electricity from the utility grid, which makes net-zero or net-positive energy buildings more feasible. Net metering uses a single utility meter. When energy is fed back to the utility grid, the meter runs backward.

Technologies of Silicon PV

Crystal technology is the oldest silicon PV production process. The earliest crystal technology is the wafer technology, in which molten silicon is grown as a cylinder and sliced into thin round wafers. Efficiencies are typically between 10% and 18%, although the best efficiency is around 30.1% (20.4% in modules). It is the mature and the most efficient technology, but crystal devices use more material than the other types because of the wastage in cutting and assembly. An alternative method of production involves turning the silicon into thin and uninterrupted ribbons, which are then cut. This method results in less material waste and reduced energy used for production. However, module efficiency for such cells is around 13%.

Another technology, called thin film, uses a chemical vapor deposition method to deposit a thin semiconductor layer onto a surface, either coated glass or stainless steel sheets designed in the final shape of the panel. The 3 μm coatings are 100 times thinner than a typical silicon wafer, which results in an amorphous structure with a red color, which unfortunately only has an efficiency of around 6–8%.

Another production method is called crystalline silicon on glass (CSG), which deposits a < 2 µm thin semiconductor directly on a glass pane and heats it to the stage of crystallization. Efficiency for cells manufactured in this way is around 8%. In addition to glass, a variety of other carrier materials are conceivable, including stainless steel foils, polymers, and various ceramic materials, but they all need to be capable of sustaining vapor application temperatures of up to 500°C.

Solar cells can be grouped into the categories listed in Figure 5.10 based on the following properties:

Thickness (thick photovoltaic cells or thin-film cells)
Material
Structure of material (mono or polycrystalline and amorphous)
Materials of semiconductors used (organic solar cells/pentacene, dye-sensitized solar cells)

Thick Silicon Cells

In this category, monocrystalline cells (c-Si) are currently the most efficient, with conversion efficiencies up to 30.1% (20.4% modules) and voltage of 0.706 V. However, their manufacturing requires a large amount of energy, which results in a high degree of embodied or grey energy content, and there is material waste in the cutting and assembly steps. In contrast, polycrystalline cells (mc-Si) require less time and contain significantly less grey energy in crystal growing and reach an efficiency of 25.2% (16% modules). However, the grain boundaries resist current flow.

Thin-Film Silicon Cells

Thin-film silicon cells are made of amorphous silicon (a-Si) and produce the best efficiency of 21.3% (9.5% modules) and voltage of 0.664 V. This technology has many advantages like it uses less material (1% of crystalline); it has low-temperature coefficient and shading resistance due to tiny cells in series, it is lightweight, flexible, easy to fabricate multiple junctions and lowest cost, but it requires high-cost manufacturing equipment and degrades faster. Thin film technology has a good application for building-integrated PV (BIPV) involving integration of photovoltaic modules into building envelope, such as roof or facade.

Crystalline silicon – for example, monocrystalline silicon cells (c-Si) in combination with amorphous silicon – has efficiencies of 10% and also a significant share of the photovoltaic market.

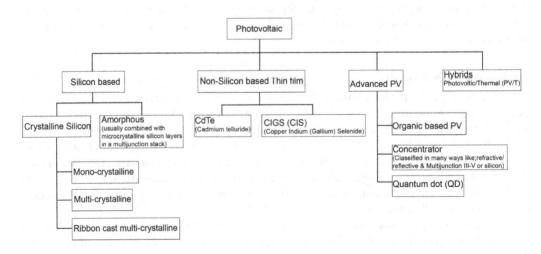

FIGURE 5.10   A taxonomy of PV technologies. Source: line work by Ar Parvesh Kumar

Non-Silicon-Based Thin Film

Besides silicon solar PV cells, many other semiconductors produce a photovoltaic effect, such as

Gallium arsenite (GaAs)
Cadmium sulfide (CdS)
Cadmium telluride (CdTe)
Germanium (Ge)
Selenium (Se)
Semiconductor solar cells

III–V-semiconductor solar cells have high performance values and are resistant to temperature and UV radiation. Because of their multiple advantages, they are often used in space applications, in which they are applied as multilayer cells with efficiencies of up to 30% (50 W/kg).

II–V-semiconductor solar cells, cadmium telluride (CdTe), give the best ratio of cost and performance because of very little material use (high solar absorptivity) and good temperature coefficient. The best efficiency of cadmium telluride PV is around 22.1% (16% modules) and voltage 0.845 V. The disadvantage is that cadmium is toxic and requires custody over life cycle.

Semiconductor solar cells for energy generation are generally joined electrically to large 'modules.' The individual cell has on its front and back thin electric conductor material. If the cells are connected in series, they achieve the desired output voltage, and if they are connected in parallel, they provide the desired amount of current source capability. Bypass diodes are included to avoid overheating of cells in case of partial shading.

CIS, CISGS Cells

Copper indium gallium selenide (CIGS) semiconductor material comprises copper, indium, gallium, and selenium. It gives the best efficiency of 24.8% (14% modules) and voltage of 0.713 V. Some of the advantages of CIGS are little material use, less exotic material use, glass or flexible substrate, and low-cost manufacturing equipment.

Advanced PV

Frensel Lens Photovoltaic Concentrator

To reduce the use of silicon, some processes employ a concentrator technology in which Fresnel lenses focus the daylight onto a smaller surface area of the solar cell, thus reducing the necessary energy-generating surface; refer to Figure 4.29. One disadvantageous thing in such a design is the fact that the solar cell needs to be always oriented perfectly toward the sun to achieve maximum efficiency.

Dye-Sensitized Solar Cells (DSC)

DSC cells are a relatively new group of low-cost solar cells. They are based on a semiconductor formed between a photo-sensitized anode and an electrolyte to create a photo-electrochemical system similar to photosynthesis in nature. These cells were invented by Michael Grätzel and Brian O'Regan at the École Polytechnique Fédérale de Lausanne in Switzerland and are also known as Grätzel cells. Their operative lifespan is limited, and they reach efficiencies of slightly more than 10%.

Titanium oxide (TiO_2) cells, using an organo-metallic dye, are less expensive and produce an output higher than the Si cells, especially at low levels of irradiance; thus, they can be used for indoor purposes.

- Multijunction solar cells use multiple junctions (compared to conventional cells that use a single n-type to p-type junction), thereby allowing greater efficiency because they can convert a wider spectrum of sunlight into energy.

Heterojunction solar cells: Amorphous silicon thin film is layered onto a crystalline silicon wafer, improving its efficiency. The hybrid approach maintains the high efficiencies of crystalline cells while also offering some of the advantages of thin film, including higher efficiencies under high temperatures and lower light levels.

Bifacial modules can be made that generate electricity from light, striking both the front and back faces of the module.

The organic solar cells currently undergoing development also have reduced efficiency and a shorter life span.

Photovoltaic/thermal

Photovoltaic and thermal hybrids (PV/T) combine solar cells and solar thermal collectors into a single module to generate both electricity and heat. The introduction of the solar thermal collector to the module captures waste heat from the photovoltaic process.

The application of a PV technology in built environment should be based on a SWOT analysis of different technologies considering site metrics. Figure 5.11 shows various types of PV cells developed from 1975 to date. As might be expected, high-efficiency single-crystal technology will take up less space where the roof area is limited but cost more. Nevertheless, low-efficiency thin-film technologies may take up more roof spaces and sometimes give more output per dollar (i.e. capital cost of PV). The roof area of most residences is more than adequate to use low-efficiency cells and still generate all the power needed, especially if the PV system is oriented due south (northern hemisphere) or north (southern hemisphere) and tilted at the angle same as the latitude. In addition to the technology type, the manufacturer and market conditions will impact PV technology selection for a given project.

Example

You are interns in a design firm that is designing an elegant (green/efficient) dream home and studio for a recently retired executive couple. The couple has a great interest in a PV system that is to be thoughtfully integrated into the design of the studio from the beginning (not just 'slapped-on' as an afterthought). As an initial step, you have been given the task of determining the basic power/electricity needs and find the size of the PV array for a stand-alone building by the short calculation method.

The size of the PV array depends on the following factors:

1. Amount of electrical energy needed per day (kWh/day)
2. System efficiency – losses in inverters, controllers, etc. (50% is typical)
3. Amount of solar radiation available per day (kWh/m²/day)
4. The power produced by the PV module (kW/m²)

PV array sizing is explained taking the example given in Table 5.1 for 43,038 Wh/day electrical adjusted plug load. The solar energy available per day to PV system tilted towards the equator at an angle equal to the latitude on the site is calculated from the solar resources available at https://maps.nrel.gov/re-atlas. Each kWh/m²/day of solar energy is considered to be one peak 'sun-hours;' for the United States the sun-hours value range is 3.5–6 hours. For instance average solar radiation available to PV system in Austin, Texas is about 5300 kWh/m²/day, therefore, peak solar hour is 5.3. The required peak-watts (Wp) is calculated by dividing the adjusted load by the sun-hours.

$$Wp = \frac{adjusted\left(\dfrac{Wh}{day}\right)}{sun-hours} = \left(\frac{43038}{5.3}\right) = 8120.4$$

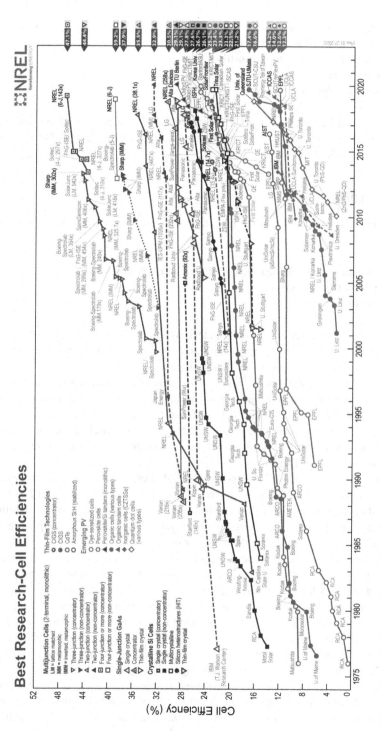

FIGURE 5.11 PV cells' efficiency from 1975 to date, NREL.

The array size is calculated by dividing peak-watts (*Wp*) by conversion efficiency of the silicon cells.

$$A = \frac{W_p}{W/ft^2} = 8120.4/12 = 676.7 \text{ ft}^2 \sim 60.9 \text{ m}^2 \text{ (for m}^2\text{, multiply ft}^2 \text{ by 0.09)}$$

12 W/ft^2 for single-crystal silicon cells
Or 10 W/ft^2 for polycrystalline silicon cells
Or 5 W/ft^2 for amorphous silicon or thin-film cells
Or 5 W/ft^2 for PV standing seam roof
Or 2.5 W/ft^2 for PV shingles

A study by the National Renewable Energy Laboratory found 2.58 hectares (6.38 acres) per megawatt to be the average capacity–based area requirement or the number of acres of PV panels required to generate 1 megawatt (MW) of electricity.

5.4 WIND ENERGY

The kinetic energy in the wind is a prominent source of renewable energy, caused by a combination of three concurrent events:

i)   The solar radiation unevenly heating the atmosphere day and night
ii) The irregular earth's surface
iii) The rotation of the earth

Although the sun and the wind energy are characterized by intermittent energy sources, their availability timings are different. Wind energy, therefore, can be a complementary or supplementary solution to solar energy in a sustainable built environment.

The kinetic energy of the wind is harnessed and converted to generate useful energy through windmills of yesteryears and contemporary wind turbines. Windmills are a high solidity system and have been prevalently used for pumping water from the wells (Figure 5.12).

The increased awareness of energy shortage led to intense research, development, and implementation of efficient wind turbine designs. The wind turbine is used for converting the kinetic energy of wind into electricity. The blades and rotor assembly of a wind turbine turn in the wind and spin a generator, which produces AC electricity. A wind energy system can be grid-connected or off-grid. A grid-connected wind turbine would need a suitable inverter and power conditioner. The net-metering grid-connected application is a necessary function because of the variability of wind and corresponding energy generation. A stand-alone system would need back-up storage batteries to ensure the availability of power when there is no wind.

All wind turbine technologies and scale require ongoing maintenance because energy generation is through moving parts that function year-round, often in high winds and poor weather conditions. Wind turbines have power ratings in terms of kW of power and also annual energy output (AEO). Wind turbines perform best under a specific range of average wind speeds. There are two main types of turbines: horizontal axis wind turbine and vertical axis wind turbine. A utility-scale wind turbine is an extremely successful and well-developed renewable energy application. Small- and medium-scale wind turbines can provide all or part of the energy requirements for a sustainable built environment. Average wind speed is the critical metric used in the estimated annual energy generation. Many wind resource maps are published with average wind speeds data. NREL, for example, provides many global and US wind resource maps, such as Figure 5.13 shows annual average wind speed at 80m height. The weather data file can also provide wind data; tabulated average wind speed and wind roses are provided for 50 US cities in Chapter 7.

FIGURE 5.12 Windmill of yesteryears, Amsterdam, The Netherlands. © Tapan Kumar Ghoshal

5.4.1 Horizontal Axis Wind Turbine (HAWT)

HAWT is the most common and mature technology featuring a three-blade rotor mounted on a horizontal axis, geared or gearless. These turbines operate 'upwind,' with the turbine pivoting at the top of the tower at 30 m (100 feet) or more above the ground such that blades face into the wind and can track with the wind direction to capture the most energy while avoiding turbulent wind near the ground. The energy generated by a wind turbine is a function of the energy collected by the blades, meaning larger blades and higher wind speeds both contribute to greater energy generation. The energy generated is a function of the rotor diameter and the wind speed. Kinetic energy as per Equation 5.2 is

$$KE = \tfrac{1}{2} * MV^2$$

If the density of air is taken as 1.2 kg/m³, the mass flow rate is

$$M = A \times 1.2\ V$$

Then the power of wind (*Pk*) over a swept area *A* is ideally given by Equation 5.7.

$$Pk = \tfrac{1}{2} \times A \times 1.2\ V \times V^2 = \tfrac{1}{2}\ A\ 1.2\ V^3 \tag{5.7}$$

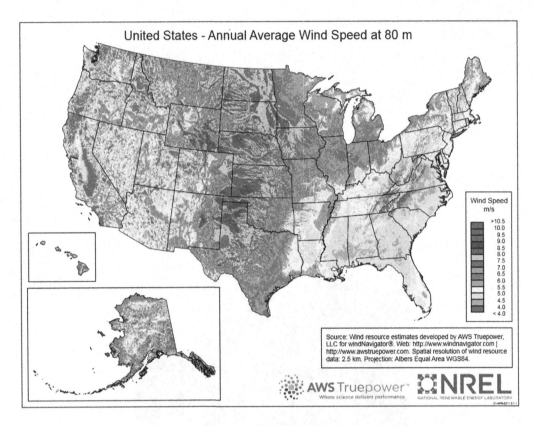

FIGURE 5.13   Wind speed map of the United States, NREL.

However, a German physicist Albert Betz concluded in 1919 that no wind turbine can convert more than 16/27 (59.3%) of the kinetic energy of the wind into mechanical energy turning a rotor. The theoretical maximum power coefficient of a wind turbine is 0.59, Betz limit or Betz law. Practically, the maximum power coefficient (C_p) may range from 0.25 to 0.45, and the power from a wind turbine can be calculated by Equation 5.8.

$$Pk = C_p \tfrac{1}{2}\, A\, 1.2\, V^3 \qquad\qquad (5.8)$$

AEO can be estimated using Equation 5.9, given by the US Department of Energy, Energy Efficiency and Renewable Energy, 'Small Wind Electric Systems: A US Consumer's Guide,' 2007, p. 10.

$$AEO = 0.01328 \times D^2 \times V^3 \qquad\qquad (5.9)$$

where
   AEO = annual energy output in kWh
   D = turbine diameter in feet
   V = wind speed in miles per hour

It is, therefore, said that the power of the wind turbine is proportionate to velocity cubed. The wind velocity increases with the higher elevation above the ground. The enormous towers can support larger blade spans or swept area and also a position high above the ground to gain access to greater wind speed. As HAWT is mounted high off the ground, to optimize energy generation, turbine access for maintenance is a greater challenge.

Figure 5.14 illustrates various components of HAWT. Figure 5.15 provides a detailed view of the inside of a wind turbine and its components, and their functionality is explained as under:

Blades: lift and rotate when the wind is blown over them, causing the rotor to spin. Most turbines have either two or three blades.

Rotor hub: holds the blades in proper position as well as rotates to drive the generator and connects them to the main tower of the wind turbine.

Pitch system: turns (or pitches) blades out of the wind to control the rotor speed, and to keep the rotor from turning in winds that are too high to produce electricity safely.

Power conditioning electronics: similar to the inverter of a PV system, takes the variable-frequency power from the generator and converts it into 60 Hz power in phase with the larger utility system. This inverter mounted at the base of the tower or is hung from the bottom of the nacelle in off-shore turbines.

Low-speed shaft: turns the low-speed shaft at about 30–60 rpm.

Gearbox: connects the low-speed shaft to the high-speed shaft and increases the rotational speeds from about 30–60 rpm to about 1,000–1,800 rpm: this is the rotational speed required by most generators to produce electricity.

The gearbox is a costly (and heavy) part of the wind turbine, and engineers are exploring 'direct-drive' generators that operate at lower rotational speeds and don't need gearboxes.

Brake: stops the rotor mechanically, electrically, or hydraulically, in emergencies.

Controller: starts up the machine at wind speeds of about 3.57–7.15 (8–16 mph) and shuts off the machine at about 24.6 m/s (55 mph). Turbines do not operate at wind speeds above about 24.6 m/s (55 mph) because they may be damaged by the high winds.

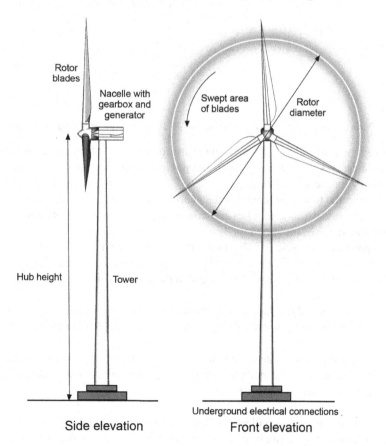

FIGURE 5.14 Horizontal axis wind turbine. Line work by Ar Vaibhav Ahuja

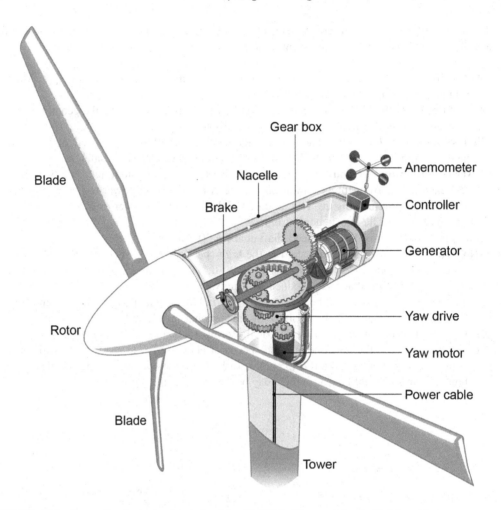

FIGURE 5.15   Parts of horizontal axis wind turbine (HAWT). Line work by Ar Vaibhav Ahuja

Anemometer: measures the wind speed and transmits wind speed data to the controller.

Wind vane: measures wind direction and communicate with the yaw drive to orient the turbine properly for the wind direction.

Nacelle: sits atop the tower and contains the gearbox, low- and high-speed shafts, generator, controller, and brake. Some nacelles are large enough for people to work inside.

High-speed shaft: drives the generator.

Generator: produces 60-cycle AC electricity. It is used to be an off-the-shelf induction generator, but newer technology uses variable speed generators that then use power conditioning electronics to create power compatible with the utility grid.

Yaw motor: powers the yaw drive.

Yaw drive: orients upwind turbines to keep them facing the wind when the direction changes. Some small wind turbines use a tail fin rather than a yaw drive to keep the turbine facing into the wind.

Tower: Is made from tubular steel, concrete, or steel lattice. Supports the structure of the turbine. Because wind speed increases with height, taller towers enable turbines to capture more energy and generate more electricity.

HAWT ranges in size from very small to very large utility-scale turbines and can be sized to meet a variety of applications. Single small turbines – below 100 kW – are typically used for residential, agricultural, and small commercial and industrial applications. Figure 5.16 presents the impact of increased turbine height, together with increased wind speed and blade diameter, on potential energy generation based on Equation 5.9. Small turbines can be used in hybrid energy systems with other distributed energy resources, such as microgrids powered by diesel generators, batteries, and photovoltaics. These systems are called hybrid wind systems and are typically used in remote, off-grid locations (where a connection to the utility grid is not available) and becoming more common in grid-connected applications for resiliency.

Wind turbines perform best under a specific range of average wind speeds. If the average wind speed is too slow, less than 3.129 or 3.576 m/s (7 or 8 mph), turbines produce very little energy. Small wind turbines have a minimum start-up speed and cut-in speed near this lower end of the range. Most wind turbines perform typically in the range of 3.5–8.5 m/s (8–19 mph). Too-high wind speeds can damage a wind turbine, so they can be designed with a cut-out speed (or furling speed) and a governing system to protect the turbine. A good rule of thumb, the rotor height should be at least 9 m (30 feet) above any obstructions within a 91.44 m (300 feet) radius of the tower.

5.4.2 VERTICAL AXIS WIND TURBINE (VAWT)

VAWT has a rotor mounted on vertical axis (Figure 5.17) and a range of rotor configurations (Figure 5.18), including the eggbeater-style Darrieus rotor, named after its French inventor. These turbines are omnidirectional, meaning they don't need to be adjusted to point into the wind to operate; it can capture wind from any direction. Vertical axis turbines are generally very small scale and, therefore, have proportionately limited energy generation capability.

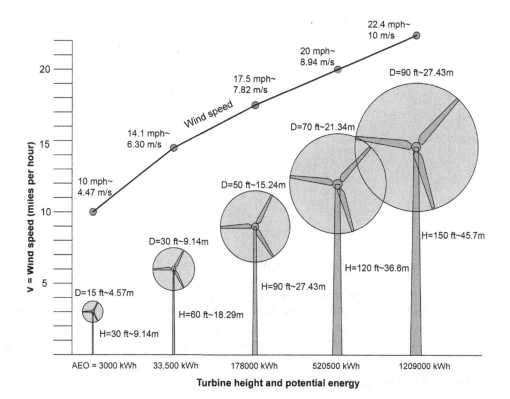

FIGURE 5.16 Impact of increased turbine height with increased wind speed and blade diameter on potential energy generation. Source: adapted from Hootman (2012). Note: AEO estimated using AEO = $0.01328 \times D^2 \times V^3$ from the U.S. Department of Energy, Energy Efficiency and Renewable Energy, 'Small Wind Electric Systems: A.U.S. Consumer's Guide,' 2007, p. 10; wind speed based on height estimated from the graph on p. 11.

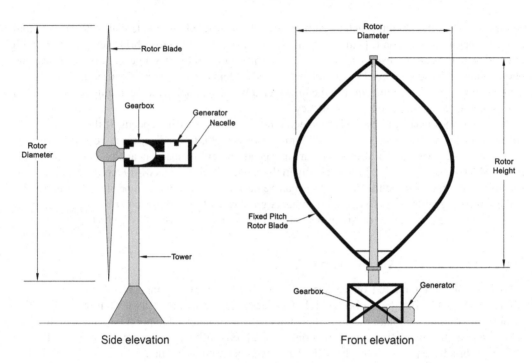

FIGURE 5.17 Parts of vertical axis wind turbine (VAWT). Line work by Ar Vaibhav Ahuja

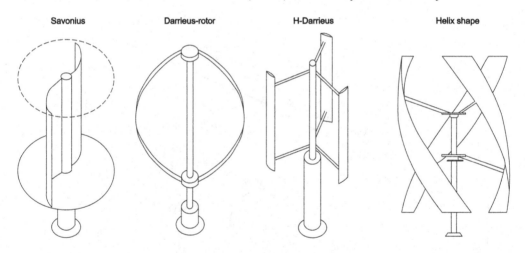

FIGURE 5.18 Range of rotor configurations of VAWT. Line work by Ar Rohan Bhatnagar

VAWT is suitable for roof mounting, a low solidity device, serving the individual consumer. It is available in 1.2–2 kW outputs, with 0.8–1.5 m diameter and with suitable current conditioning. It may be used as a grid-connected device with suitable inverter and power conditioning. A stand-alone system would need back-up storage batteries to ensure the availability of power when there is no wind.

5.5 OTHER RENEWABLES

5.5.1 BIOMASS

Historically, biomass has been in use since people first began burning wood for domestic cooking and space heating. The modern use of biomass in more focused applications to produce biopower (heat and electricity), biofuels, and bioproducts began around the late 20th century, ushering a true renaissance of this traditional renewable energy source. For many countries, the use of biomass will be an important pillar of their energy portfolio in the future. Biomass is the largest source of renewable energy in the United States, comprising 43.5% of all renewable energy generation in the year 2019. Biopower, which uses biomass to generate heat and/ electricity, is the application of interest for the sustainable built environment, and so is the thrust of this section.

Today the term 'biomass' connotes a wide spectrum of natural resources derived from biological or organic matter, and, in a broader sense, also those materials that are the result of a transformation of organic matter or its final products, such as paper, organic household waste, vegetable oils, and biogas. Primary biomass products are energetically processed energy sources such as wood chaff, wood pellets, vegetable oils, and substances such as liquid manure and sludge that are derived as a result of organic transformation processes in sewage treatment plants. The most important groups are:

- Residual products from harvesting and forestry
- Grasses with a high degree of cellulose content, and bush and tree species with low maintenance requirements
- Sugar and starch crops
- 'Energy plants (crops)' such as the oil-rich species of sunflowers, canola, and corn
- Animal fats
- Organic byproducts as a result of production processes in woodworking, liquid manure from large cattle operations, and biogas
- Organic waste such as sludge or methane mine gas

Typically plant-based biomass material utilizes the stored energy from the sun through the photosynthesis process that converts solar energy into chemical energy in the form of glucose or sugar. The glucose in plant-based material is an important biological energy source and part of the earth's carbon cycle. Plant-based biomass is capable of delivering carbon-neutral energy because when biomass is combusted to generate heat or electricity, that amount of CO_2 is released to the atmosphere that was sequestered (absorbed and stored) by the plant during its growth period as part of photosynthesis. However, negative aspects of biomass fuel burning are important to recognize and alleviate. Fine particulate matter (PM) or, simply, fine dust and pollutants such as carbon monoxide and nitrogen oxides cause health-related risks, so its release into the air needs to be eliminated in the process, and it must be fully addressed through state-of-the-art pollution controls and strict regulations.

The biofuel application involves making liquid (e.g., ethanol, biodiesel) or gas fuels from biomass. Ethanol (alcohol) can be produced from the sugar of starch-rich plants such as sugar cane, corn, sugar beet, and potatoes, using fermentation and distillation, a commercial process but inefficient in terms of land use. Biodiesel can be commercially produced from vegetable oils, especially soybean and canola oil, using simple chemical process of trans-esterification of their short-chain alkyl (methyl or ethyl) esters, but it is too limited by land-use constraints. Jatropha, an inedible plant that can be cultivated on non-arable land and thus is not competing with food crops, has seeds that are capable of producing three times more oil than soybeans. Large-scale plants are also capable of producing gas from biomass such as sludge, organic waste, and liquid manure using fermentation processes in the absence of air. New processes such as biomass-to-liquid (BML, or BMTL, a multistage process to convert biomass into biofuels) are capable of transforming various biomass sources into universal oil, comparable to fossil oils. All liquid biofuels have quite high-energy densities (but ethanol and methanol are significantly below that of petrol), and for ease of handling, these liquid biofuels can be combusted conveniently in gas turbines and gas engines. Thus all liquid biofuels are primarily being used in the transportation sector to replace and/or supplement a variety of fuels as well as for power-heat coupling and cogeneration process.

Several combustion technologies are available for the burning of biogenic energy sources. Biogas and bio-oil, on the other hand, are already used in combustion motors and gas-fired applications such as furnaces.

Bioproducts are products that formerly were made with petroleum but now produced with one of the many forms of biomass; bio-based plastics are one example.

Figure 5.19 indicates the various sources of biomass, the conversion processes, and the final products of biomass conversion.

Biopower: Energy for Heat and Electricity

Biomass is one the largest carriers of renewable energy, and it is distinctly important because it can be readily replenished and is available at any given time – in contrast to intermittent renewable energy of solar and wind. Biopower energy systems can be used at a utility or a distributed scale for individual buildings or communities. In 2018, an installed global total power capacity of 130 GW exists for this technology. Biopower technologies harvest the energy stored in renewable biomass fuels to produce heat and/or electricity using three processes: burning (direct firing or co-firing), bacterial decay (anaerobic), and conversion to a gaseous or liquid fuel by gasification and pyrolysis, respectively.

Most of the utility-scale power plants are in the 20–50 MW range; biomass is currently being burned in furnace units to gain the thermal energy necessary to drive turbines and feed utility grid. These power plants typically use wood and agricultural residues as the source, but in many cases biomass power plants also burn household waste, which further increases energy generation. Biomass is also used as a substitute for a portion of coal in existing coal-fired power plant furnace that has been converted to co-firing technologies that allow two different materials to be combusted for electricity generation. Thus biopower can offset the need for carbon fuels and also lower the carbon intensity of electricity generation. In addition to solid biomass, bio-oils and biogas are capable of being used.

Biopower technologies come in many forms that have applications at the building scale; this is often called a modular biopower system that generates electricity at a capacity of 5 MW or less. Distributed energy resources refer to a variety of small, modular power-generating technologies that can be combined to improve the operation of the electricity delivery system. In a modular biomass system, solid biomass is directly burned in a boiler to create high-pressure steam that flows over a series of turbine blades, causing them to rotate. The rotation of the turbine drives a generator producing electricity. This process can also be used to heat a building or water or for some process

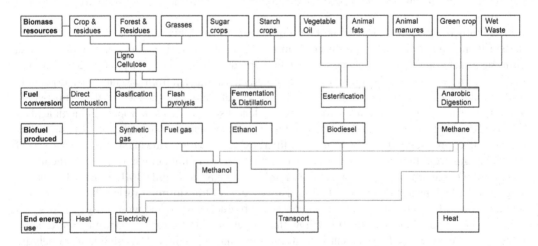

FIGURE 5.19 Biomass conversion processes. Source: after Stuckley et al. (2004), Diesendorf (2007), Szokolay (2008), line work by Manoj Sehgal

heating use. An efficient application of biomass is to generate both electricity and heat or cogeneration (also called combined heat and power). The process to generate electricity through burning is not, however, very efficient (20–40%), and this inefficiency is regarded as waste heat. But the waste heat can be captured in a cogeneration application and used for a variety of heating purposes. Trigeneration can be achieved where cooling is also provided. Waste heat can be used to drive an absorption chiller, in addition to providing heating energy. In cogeneration or trigeneration systems, high efficiencies of 70–90% can be achieved.

The integrated gasification technology converts the solid biomass into gas; the solid biomass is exposed to high temperatures with little oxygen present, to produce a synthetic gas (or syngas), which is a mixture of carbon monoxide and hydrogen. The gas can then be combusted in a conventional boiler to generate electricity, or it can also be used to replace natural gas in a combined-cycle gas turbine. The use of syngas as a fuel makes it possible to use more efficient electricity generation technologies, including combined steam and gas turbines, Stirling engines, microturbines, and fuel cells. This technology can segregate biomass, such as wood, agricultural residues, and energy crops. It is also a scalable technology, often referred to as modular biopower systems.

Another method of utilizing a gas-fired biopower system is to create biogas using oxygen-free tanks called anaerobic digesters (Figure 5.20). The wet organic waste material, such as animal dung or human sewage, is decomposed by anaerobic bacteria that produce biogas a renewable natural gas comprising 50–80% methane, which can then be purified and used to generate electricity. The process is considered anaerobic because it is completed without the presence of oxygen.

The term 'pyrolysis' is derived from the Greek words 'pyro,' means 'fire' and 'lysis' means 'separating.' Pyrolysis is the thermal decomposition of biomass in the complete absence of oxygen to produce a crude bio-oil. This bio-oil is then substituted for fuel oil or diesel in furnaces, turbines, and engines for electricity production.

One of the most important factors in analyzing the applicability of a biopower system for a sustainable built environment is the availability of appropriate biomass. Biogenic fuel availability is limited, fluctuates significantly from region to region, and will be capable of supplying only a small

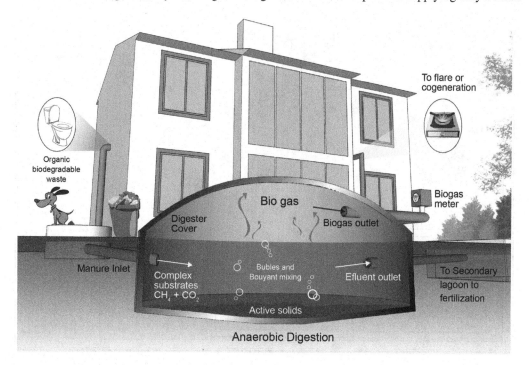

FIGURE 5.20   Application of anaerobic digester at the building level. Credit: Ar Vaibhav Ahuja

percentage of the energy requirement of a building. This makes biomass a regionally specific solution, as the biomass should be locally available and have a reliable distribution system in place. Readily available solid biomass derived from wood or straw need to be prepared for energy sources use by shredding (cut wood) or cutting or sawing (wood pellets), whereas a specific type of biomass for bioenergy application will entail cultivation and harvesting. The map in Figure 5.21 from NREL shows both the availability of biomass resources in the United States and the composite availability of a wide range of biomass resources.

In addition to the greenhouse gas emissions of using biomass energy applications, other environmental, social, and economic issues must be addressed:

- Land-use changes decreasing biodiversity and compromising natural habitat
- High water demand
- Agricultural practices imposing pressure on food crop production
- Fertilizers causing eutrophication of water bodies affecting aquatic habitat and water quality

5.5.2 Geothermal Energy

The term 'geothermal' means 'earth heat' or 'heat of the earth,' based on Greek words 'geo' (earth) and 'therme' (heat). Geothermal energy is a renewable energy source because incredible amounts of heat or thermal energy can be found beneath the earth's surface. The earth consists of four primary layers, starting with an inner core at the planet's center, enveloped by the outer core, mantle, and crust. The temperature at the center of the earth, about 6500 km (4000 miles) deep, is about the

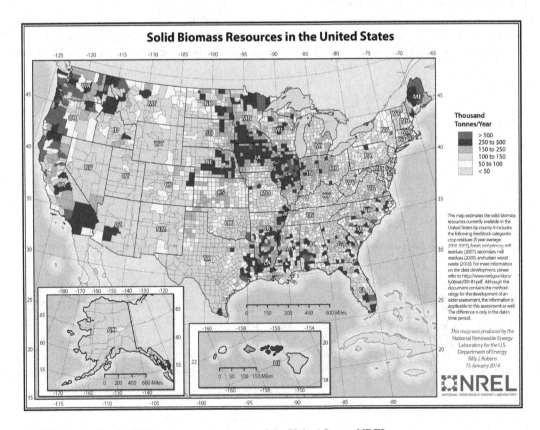

FIGURE 5.21   Total solid biomass resources map of the United States, NREL.

same as the surface of the sun, nearly 6000°C (>10,800°F). This heat is continually replenished by the decay of naturally occurring radioactive elements in the earth's interior. The heat energy of the crust is the result of the heat flowing from the earth's interior and the solar radiation falling on the upper surface of the earth. The average geothermal gradient is approximately 3°C per 100 m, but it is much higher at the hot and active geothermal region. This energy can be integrated with the built environment and used for water and space heating, electricity production, and industrial use. Its utilization – in contrast to other renewable energy sources – is not surface intensive because it can be extracted by pointed exploration.

It is understood that the energy stored in the upper 3 km (~ 2 miles) of the earth's crust is estimated to be 3 million quads, sufficient to supply the total energy demand of mankind for the next 100,000 years or the energy demand of the United States for the next 30,000 years. The heat flowing from the earth's interior is estimated to be equivalent to 44.2 terawatts-thermal (TWth) of power (Klaus 2008).

Depending upon the depth from which geothermal energy is harnessed, the following distinctions are made: shallow, near-surface geothermal energy extraction takes place up to a depth of 500 m below the earth's surface, while deep geothermal applications reach beyond 3000 m (Figure 5.22). The probes are normally inserted into the ground to make use of shallow geothermal energy, while the test boreholes are drilled to evaluate the potential of the soil for energy extraction before large-scale geothermal well fields are installed. Many technologies have been developed to harness geothermal energy and are categorized as given below.

Shallow Geothermal Systems

Direct use and district heating systems use low-temperature geothermal water (20–150°C ~ 68–302°F) from the springs or reservoirs of hot water and steam located near the surface of the earth. Traditionally, hot springs have been used for bathing, and for natural healing powers, they can be used directly to heat buildings or can also be used in district central heating systems.

Geothermal heat pumps use the constant temperatures near the surface of the earth to heat and cool buildings through a ground heat exchanger. The shallow ground, or the upper surface at a depth of 3 m (10 feet) of the earth, maintains a relatively constant temperature between 10°C and 15.5°C (50–60°F) all-year-round; this temperature is warmer than the air above it in the winter and cooler in the summer.

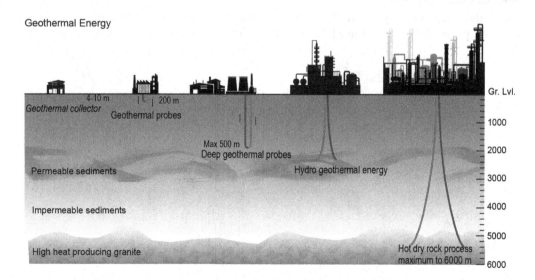

FIGURE 5.22 A taxonomy of geothermal energy systems (shallow to deep). Source: adapted from archplus 184, October 2007, line work by Himanshi Jaglan, color rendering by Ar Vaibhav Ahuja

Heat pumps, which work according to the concept of a chiller, reject cooling energy (evaporator power) into the soil during winter operations, simultaneously extracting heat energy from the heat exchanger. The cold water supplied to the ground then heats up by approximately 3–4 K. The gained heat energy is conducted through the compressor of the heat pump and can be stepped up to the temperature level – typically between 40°C and a maximum of 50°C – needed in building for space heating. During the summer, the heat pump will work as a cooling machine, now supplying warm water to the ground through the heat exchanger. This shallow thermal resource is used in geothermal heating and cooling, known as geo-exchange.

According to the US Environmental Protection Agency (EPA), geothermal heat pumps are the most energy-efficient, environmentally clean, and cost-effective systems for heating and cooling buildings. All types of buildings, including homes, office buildings, schools, and hospitals, can use geothermal heat pumps. The ground heat exchanger is a system of pipes called a loop (usually made of plastic), which is buried in the shallow ground or submerged in water near the building. A fluid (usually water or a mixture of water and antifreeze) circulates through the pipes to absorb or relinquish heat within the ground. The loop can be in a horizontal, vertical, or pond/lake configuration. The horizontal looping system is generally the most cost-effective for residential installations, particularly for new constructions where sufficient land is available. The loop layouts either use the conventional method of two pipes placed side-by-side at 1.5 m (5 ft) in the ground in 0.6 m (2 ft) wide trench or the Slinky method (in the form of a coil) in a shorter trench. Vertical looping system uses a similar method but saves earthworks and minimizes the disturbance to existing landscaping by drilling a large number of boreholes (approximately 100 mm–4 inches diameter) about 6 m (20 ft) apart and 30.5–121.92 m (100–400 ft) deep. Into these holes go two pipes that are connected at the bottom with a U-bend to form a loop. The vertical loops are connected with a horizontal pipe (i.e., manifold), placed in trenches, and connected to the heat pump in the building. Figure 5.23 illustrates the Slinky method and vertical looping systems.

If the site has an adequate water body, this may be the lowest-cost option. A supply line pipe is run underground from the building to the water and coiled into circles at least 2.44 m (8 ft) under the surface to prevent freezing. The coils should be placed only in a water source that meets minimum volume, depth, and quality criteria. If the local conditions of the site do not provide a suitable natural aquifer for heat storage, under certain circumstances an artificial aquifer can be designed.

Deep Geothermal Systems

The geothermal power supplies are about 2% of the energy consumption in the United States; it has enormous potential to supply a much larger portion. The geothermal resources are considerable,

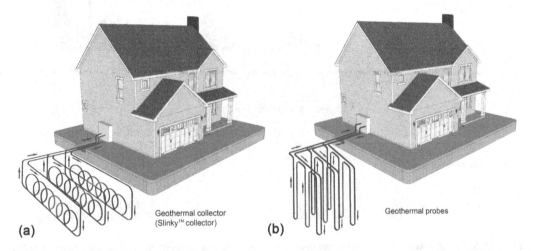

FIGURE 5.23   Application of geothermal at the individual building level. Credit: Ar Shiva Bagga

especially the high-temperature resources found throughout the western United States illustrated on the map in Figure 5.24.

Geothermal power plants use high temperature 148.89–371.11°C (300–700°F) hydrothermal resources, from either dry steam wells or hot water wells, deep in the earth that is accessed by drilling. Then piping steam or hot water is sprayed to the surface to drive a turbine that activates a generator to produce electricity at utility scale. Some geothermal wells are as much as 3.22 km (2 miles) deep. There are three basic types of geothermal power plants: dry steam, flash steam, and binary cycle.

Dry steam power plants make use of steam directly from a geothermal reservoir to power a turbine/generator unit. This is the oldest and mature technology of geothermal plants, but access to steam reservoirs limits its application.

Flash steam power plants are the most common, and they take high-pressure hot water from deep inside the earth and convert it to steam to drive a turbine/generator. Any leftover water and condensed steam are injected back into the reservoir, making this a sustainable resource. However, the water temperature needed to run a flash steam plant is 182.2°C (360°F), and it depends on regions with high geothermal temperatures raised by volcanic activity along tectonic plate boundaries.

Binary cycle power plants transfer the heat from geothermal hot water to another working liquid, usually an organic compound, through the heat exchanger. The heat causes the working liquid to vaporize, which is used to drive a turbine/generator unit. The water is then injected back into the ground to be reheated. The water and the working fluid are kept separated during the whole process, so there are little or no air emissions. Binary cycle technology operates at low temperatures of about 107–182°C (225–360°F). Currently, two types of geothermal resources can be used in binary cycle power plants to generate electricity: enhanced geothermal systems (EGSs) and low-temperature or co-produced resources.

An ideal deep geothermal resource site has heat, water, and permeable geology that allows water to move through the heated rock. However, many sites have heated rock layers but lack the water and permeable geology. EGSs are engineered deep reservoirs created from geothermal resources that lack water or permeable rock. The US Geological Survey estimates that potentially 500,000 megawatts of EGS resource is available in the western United States or about half of the current installed electric power-generating capacity in the United States,

Low-temperature and co-produced geothermal resources are typically found at temperatures of 300°F (150°C) or less. Some low-temperature resources can be harnessed to generate electricity using binary cycle technology. Co-produced hot water is a byproduct of oil and gas wells in the United States. This hot water is being examined for its potential to produce electricity, helping to lower greenhouse gas emissions and extend the life of oil and gas fields.

5.5.3 Hydrogen and Fuel Cell

Like electricity, hydrogen is considered an alternative fuel under the Energy Policy Act of 1992. Hydrogen is a secondary source of energy. It stores and transports energy produced from other resources (fossil fuels, water, and biomass). Hydrogen is abundant in our environment; it is stored in water (H_2O), hydrocarbons (such as methane, CH_4), and other organic matter. One of the challenges of using hydrogen as fuel comes from being able to efficiently extract it from these compounds. Hydrogen (H_2) can be produced from diverse resources such as water electrolysis, biomass, natural gas, propane, and methanol, with the potential for near-zero greenhouse gas emissions. Most hydrogen production today is by steam reforming natural gas, a fossil fuel, so the carbon dioxide released in the reformation process adds to the greenhouse effect. There is ongoing research to derive more and more hydrogen from renewable energy sources. Hydrogen generated by water electrolysis, which is the splitting of water molecules into hydrogen and oxygen using electricity, is energy-intensive but can take advantage of variable renewable energy sources such as solar and wind; thus the hydrogen stores the unused energy for later use and is classified as a renewable fuel and energy

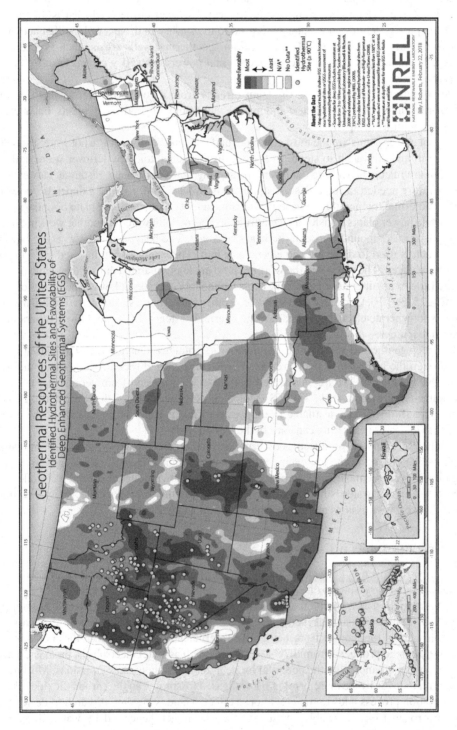

FIGURE 5.24   Geothermal resources map of the United States, NREL.

resource. Other hydrogen-producing technologies, solar energy, and the photo-biological process to split water into hydrogen and oxygen, are in the development stage. Fermentation of pretreated biomass such as sugar-rich feedstock and agricultural residues like corn stover, wheat straw, and rice straw can be used to generate hydrogen. The wind to hydrogen project of NREL involves finding ways to use wind and solar energy to harvest hydrogen.

Hydrogen's energy content by volume is low. This makes storing hydrogen a challenge because it requires high pressures, low temperatures, or chemical processes to be stored compactly. Once produced, hydrogen generates electrical power in a fuel cell, emitting only water vapor and warm air.

A fuel cell is a device that uses the chemical energy of hydrogen (or hydrogen-rich fuel) and oxygen to generate electricity with efficiencies of up to 60%. Fuel cells are more energy efficient than combustion engines, and hydrogen used to power them can come from a variety of sources. If pure hydrogen is used as a fuel, fuel cells emit only heat and water as byproducts, eliminating concerns about air pollutants or greenhouse gases. Fuel cell technology is still in its early development, needing improvements in efficiency and durability.

Most fuel cells have similar designs and working principles, 'flameless' combustion of natural gas or hydrogen in the presence of an electrolyte. They are typically either distinguished by the employed electrolyte or according to their operating temperature (high- and low-temperature fuel cells). The main types of fuel cells, classified by their electrodes, are (1) polymer electrolyte membrane (PEM) (2) alkaline (AFC) (3) phosphoric acid (PAFC), (4) molten carbonate (MCFC), and (5) solid oxide (SOFC). The first four types of fuel cells use liquid electrolytes while the SOFC is a ceramic – or, as the name implies, a solid oxide. Large phosphoric acid fuel cells are commercially available. Table 5.2 presents a comparison of current fuel cell technologies.

In the cell, a hydrogen-rich fuel passes over the anode, while an oxygen-rich gas (air) passes over the cathode. Catalysts help split the hydrogen into hydrogen ions and electrons. The hydrogen ions move through an external circuit, thus providing a direct current at a fixed voltage potential.

One of the more common types of fuel cells is the polymer electrolyte membrane (PEM) fuel cell. The PEM is a thin, solid organic compound, typically the consistency of plastic wrap and about as thick as 2–7 sheets of paper. This membrane functions as an electrolyte: a substance that conducts charge ions (in this case protons) but does not conduct electrons. This allows the solution to conduct electricity. This membrane must be kept moist to conduct particles through it. This membrane is sandwiched between an anode (negative electrode) and a cathode (positive electrode) (Figure 5.25).

The anode is the electrode at which oxidation (loss of electrons) takes place. In a fuel cell, the anode is electrically negative. The anode is composed of platinum particles uniformly supported on carbon particles. The platinum acts as a catalyst, increasing the rate of the oxidation process. The anode is porous so that hydrogen can pass through it.

The cathode is the electrode at which reduction (gaining of electrons) takes place. In a fuel cell, the cathode is electrically positive. The cathode is composed of platinum particles uniformly supported on carbon particles. The platinum acts as a catalyst, increasing the rate of the reduction process. The cathode is porous so that oxygen can pass through it.

Flow plates perform several important functions: (i) they channel hydrogen and oxygen to the electrodes, (ii) they channel water and heat away from the fuel cell, and (iii) they conduct electrons from the anode to the electrical circuit and from the circuit back to the cathode.

The amount of power produced by a fuel cell depends on several factors, including fuel cell type, cell size, the temperature at which it operates, and the pressure at which the gases are supplied to the cell. A single fuel cell produces less than 1.16 volts – barely enough electricity for even the smallest applications. To increase the amount of electricity generated, individual fuel cells are combined in series, into a fuel cell 'stack.' A typical fuel cell stack may consist of hundreds of fuel cells. The electricity produced by the fuel cell stack is DC. A power electronics module is typically incorporated into the fuel cell system design to manage the power output and quality from the fuel cell system. If the end-user requires an AC output, the conversion from DC to AC power is accomplished in the power electronics module. A control system is used to operate the entire fuel cell system and

TABLE 5.2
Comparison of Fuel Cell Technologies

Fuel Cell Type	Polymer Electrolyte Membrane (PEM)	Alkaline (AFC)	Phosphoric Acid (PAFC)	Molten Carbonate (MCFC)	Solid Oxide (SOFC)
Common electrolyte	Perfluorosulfonic acid	Aqueous potassium hydroxide soaked in a porous matrix or alkaline polymer membrane	Phosphoric acid soaked in a porous matrix or imbibed in a polymer	Molten lithium, sodium, and/or potassium carbonates, soaked in a porous matrix	Yttria stabilized zirconia
Typical stack size	<1–100 kW	1–100 kW	5–400 kW, 100 kW module (liquid PAFC), <10 kW (polymer membrane)	300 kW–3 MW, 300 kW module	1 kW–2 MW
Efficiency (LHV)	60% direct H_2; 40% reformed fuel	60%	40%	50%	60%
Operating temperature	<120°C	<100°C	150–200°C	600–700°C	500–1000°C
Applications	Back-up power Portable power Distributed generation Transportation Specialty vehicles	Military Space Back-up power Transportation	Distributed generation	Electric utility Distributed generation	Auxiliary power Electric utility Distributed generation
Advantages	Solid electrolyte reduces corrosion and electrolyte management problems Low-temperature Quick start-up and load following	A wider range of stable materials allow lower-cost components. Low-temperature Quick start-up	Suitable for CHP Increased tolerance to fuel impurities	High efficiency Fuel flexibility Suitable for CHP Hybrid/gas turbine cycle	High-efficiency Fuel flexibility Solid electrolyte Suitable for CHP Hybrid/gas turbine cycle

Source: Adapted from DOE (2016).

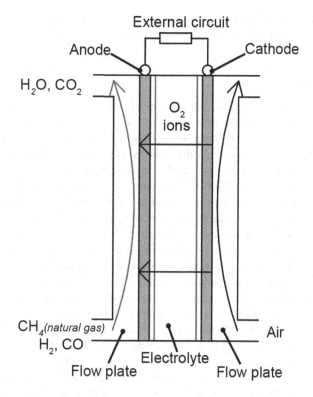

FIGURE 5.25 PEM fuel cell configuration.

typically interfaces with the electrical system at the customer site. Fuel cells are a flexible technology and have a broad range of potential applications; they can provide power for systems as large as a utility power station and as small as a laptop computer.

Platinum represents one of the largest cost components of a fuel cell, so much of the R&D focus on approaches that will increase activity and utilization of current platinum group metal (PGM) and PGM-alloy catalysts, as well as non-PGM catalyst approaches for long-term applications. DOE durability targets for stationary and transportation fuel cells are 40,000 and 5000 h, respectively, under realistic operating conditions.

A typical packaged fuel cell power plant consists of a fuel reformer (processor), which generates hydrogen-rich gas from fuel, a power section (stack) where the electrochemical process occurs, and a power conditioner (inverter), which converts the DC power generated in the fuel cell into AC power. Most fuel cell applications involve interconnectivity with the electric grid; thus, the power conditioner must synchronize the fuel cell's electrical output with the grid (ASHRAE 2001; *ASME Standard PTC 50*). A growing number of fuel cell applications are grid independent to reliably power remote or critical systems.

Fuel cells for the generation of electric and thermal energy are in use in many applications and have an electric power generation capability of around 200 kW. Additional systems, with power outputs of around 5 kW, are being offered for applications in residential units. For applications in buildings, high-temperature fuel cells using gas and hydrogen are being utilized. They provide the efficiency of electricity generation of 50% and produce exhaust gas temperatures that are between 80°C and 180°C. Fuel cell technology is seen as an environmentally friendly alternative to cogeneration power plants because no carbon oxide is released into the atmosphere. Emissions from fuel cells are very low; NO_x emissions are less than 20 ppm.

Nevertheless, the use of fuel cell is very interesting in cases in which great and long duration solar energy gains exists. In such cases, the production of hydrogen with electricity gained from photovoltaic systems may be economical. It can be assumed that large-scale fuel cell applications, which for the most part still need to be developed, will take place in sunny regions of the earth in the future.

5.5.4 HYDROPOWER

Human civilization in antiquity started on the banks of rivers and streams as water is the lifeline of sustenance as well as it can be harnessed to perform work. Water is an extraordinary source of renewable energy, especially moving water via gravity, waves or tides is a source of kinetic energy. Historically the force of water flowing in the streams and rivers was used to produce mechanical energy for grinding grains into flour, saw wood, and other works. It is also a tremendous source of thermal energy stored from the sun. The earth's oceans are sources, too, of tidal energy, wave energy, and thermal energy, all of which can be harnessed to generate electricity. A hydroelectric power or hydropower is the electricity created when the kinetic energy is harnessed from flowing water. Moving water turns a turbine and shaft, which spins a generator that produces electricity. Hydroelectric power produced 21.76% of the total renewable electricity in the United States in the year 2019, and 2.5% of the total US electricity. The water cycle makes the hydropower a sustainable option since solar energy evaporates surface water into clouds and recycles back to the earth as precipitation.

Hydropower systems come in all sizes and can be distinguished in three categories as per DOE: large-scale hydropower plants with a capacity of more than 30 MW, small-scale hydropower system that generates 10 MW or less of power, and micro hydropower systems having a capacity less than 100 kW. The large hydropower plants are three types: impoundment, diversion, and pumped storage. An impoundment facility, the most common type of hydroelectric power plant, involves the construction of a large dam to store river water in a reservoir. Water released from the reservoir flows through a turbine, spinning it, which in turn activates a generator to produce electricity (Figure 5.26). This type of hydropower plant has environmental concerns

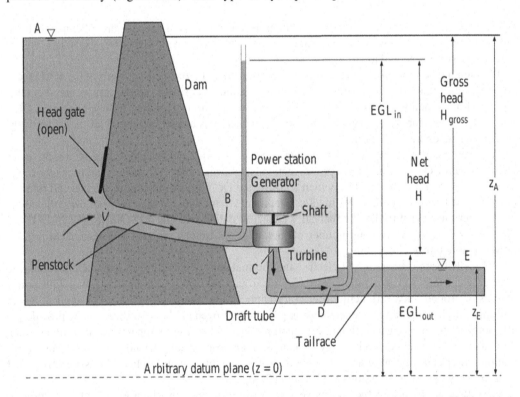

FIGURE 5.26　Hydropower plant. Source: line work by Ar Sakshi Singhal, color rendering by Ar Vaibhav Ahuja

because of displacement and/or destruction of natural habitats, terrestrial, and aquatic ecosystems resulting from the construction of the dam and hydropower plants. Hydropower is also becoming even more environmentally friendly as research and development works are directed on new technologies that are more environmentally sensitive than the extant technologies. The 'dam-less,' hydropower system, known as diversion or 'run-of-the-river,' facility channels a part of a river or stream through a canal or penstock to a powerhouse before the water rejoins the main river downstream through a trail race.

Small hydropower systems are applicable at the city municipal level, while micro hydropower can be implemented at individual property or residence. Small or micro hydropower systems can be a grid-connected system or off-grid system; battery storage can be added in both cases. A grid-connected system should include a net or a smart meter.

Application of hydropower at individual building level entails having access to an appropriate waterway with a quality water source, compliance with the normative framework of the city, and security of water rights. The energy available through hydropower is a function of the water's head (elevation drop to the turbine) and flow. The taller the head height and the higher the flow rate, the greater the energy potential for the system. Hydropower is the most consistent renewable energy source because the stream or river flows day and night. The flow may be subject to seasonal variations, but the elevation change in a hydro system is fixed based on the system design. Hydropower is one of the lowest-cost renewable energy systems available, and installations can generate energy for many decades.

Micro and small hydropower systems may use two types of hydro turbines: impulse and reaction turbines. Each one has its applications based on the height of standing water – 'head' and the flow or volume of waterways to be used for hydropower.

The impulse turbine generally uses the velocity of the water to move the runner (wheel) that is not immersed in water. Rather, the water stream hits each nozzle/bucket on the runner at high velocity, and the water flows out the bottom of the turbine after hitting the runner. Impulse turbines are generally applicable for high-head (high-pressure) and low-flow conditions. Three types of impulse turbines are the Pelton, Cross-flow, and Turgo.

The reaction turbine develops power from the combined action of pressure and moving water. The runner/blades are fully immersed in the flowing water stream and are applicable for sites with lower to medium head and higher flows than compared with the impulse turbines. There are several variants of reaction turbines, including the Propeller, Francis, and Kalpan. Each has a unique blade design to optimize high flow water energy capture. The water flow acts on all blades simultaneously.

Hydro turbines can attain efficiencies as high as 90% and most will operate in the 80–90% range if incorporated into a well-designed system. However, the final water-to-wire efficiency will be less than the turbine efficiency because of other losses in the system. Total system efficiencies run at 50% for micro-hydropower and 60–70% for small hydropower systems.

Ocean Energy

The ocean's kinetic energy (tidal and wave energy) can be harnessed and converted into electricity by marine and hydrokinetic technologies. A buoy can harness energy from the vertical rise and fall of ocean waves, as well as the back-and-forth and side-to-side movements, while currents and tides can generate electricity by spinning a turbine. With oceans covering 75% of the planet and many water resources located near the most populated areas, ocean energy has great potential as a plentiful renewable resource.

Tides are the result of gravitational forces of the sun and the moon onto the earth's oceans. The tidal energy has the advantage that low and high tides can be precisely calculated and predetermined, in contrast to solar and wind power, tides are entirely independent of outside influences. The technology to harness the kinetic energy of the water flow and convert into electricity includes an axial flow turbine, a cross-flow turbine, and a reciprocating device. Axial flow turbines look similar to HAWT, and cross-flow turbines are like VAWT. They can be placed on the seafloor where there is strong tidal flow. Because water is about 800 times denser than air, tidal turbines have to be much sturdier and heavier than wind turbines. Tidal turbines are more expensive to build than wind turbines but capture more energy with the same-size blades. Figure 5.27 illustrates an axial flow turbine.

Waves form as wind blows over the surface of open water in oceans and lakes. Ocean waves contain tremendous energy. The theoretical annual energy potential of waves off the coasts of the United States is estimated to be as much as 2.64 trillion kilowatt-hours or the equivalent of about 64% of US electricity generation in 2018.

A wave energy converter (WEC) is defined as a device that converts the kinetic and potential energy associated with a moving wave into useful mechanical or electrical energy. WEC can be classified into six categories: point absorbers/submerged pressure differential, oscillating water column, attenuators, overtopping or terminator devices, oscillating wave surge converter, and rotating mass devices.

Devices use to capture hydrokinetic power must be able to withstand turbulent and harsh conditions and be designed to preserve the integrity of the marine environment.

Attenuator is a wave energy capture device with the principal axis oriented parallel to the direction of the incoming wave to bend or focus waves into a narrow channel to increase their size and power and to spin the turbines that generate electricity (Figure 5.28). Waves can also be channeled into a catch basin or reservoir where the water flows to a turbine at a lower elevation, similar to the way a hydropower dam operates.

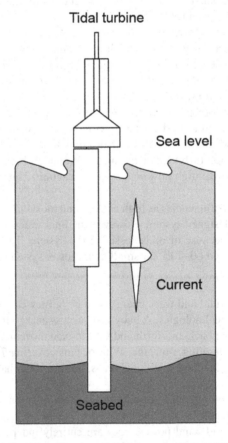

FIGURE 5.27 Axial flow tidal turbine. Source: adapted from the National Energy Education Development Project (public domain), line work by Ar Sakshi Singhal

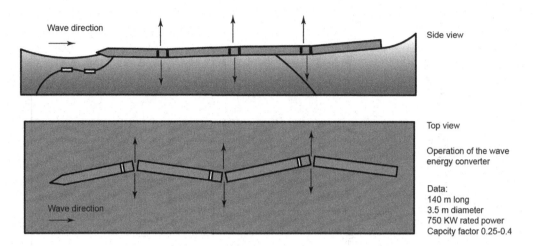

FIGURE 5.28   Principle of wave energy converter-attenuator. Source: adapted from Klaus (2008), line work by Ar Sakshi Singhal, color rendering by Ar Vaibhav Ahuja

5.6   ENERGY STORAGE AND SMART GRID

The subject of energy storage, especially storage of renewable energy sources at the utility-scale and end-user scale, is not only important but also, in some respects, highly complex. It needs to be an integral and significant aspect of future energy supply. The first major issues are the renewable energy time shift because of mismatch in the timing of supply availability and demand. The other two issues are renewable capacity firming and renewable energy grid integration. The storage requirements may be short term (e.g., the 24-hour cycle) or long term (inter-seasonal). Much work has been devoted to the latter, especially in cold winter climates to store heat collected in the summer to be available in the winter. Such storage must be inexpensive. Figure 5.29 and Table 5.3 present an overview of the particular operation of various storage technologies that are currently available. This section particularly addresses energy storage technologies pertaining to renewable energy and details of electrochemical, mechanical, chemical storage and phase change materials. The smart grid is also discussed as futuristic technology integrating utility grid with renewable energy systems at building level.

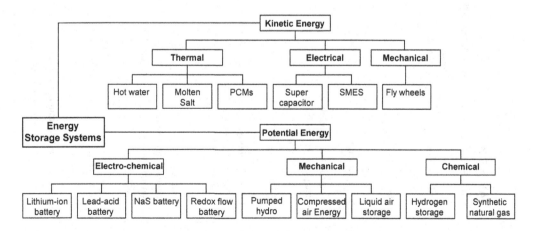

FIGURE 5.29   A taxonomy of energy storage systems. Source: World Energy Council (2016)

TABLE 5.3
Characteristics of Major Energy Storage Technology

Technologies	Power Saving (MW)	Storage Duration (h)	Cycling or Lifetime	Self-Discharge (%)	Energy Density (Wh/l)	Power Density (W/l)	Efficiency (%)	Response Time
Thermal technologies								
Molten salt	1–150	hours	30 years	NA	70–210	N/S	80–90	min
Electrical technologies								
Super capacitor	0.01–1	ms–min.	10,000–100,000	20–40	10–20	40,000–120,000	80–98	10–20 ms
Superconducting magnetic energy system (SMES)	0.1–1	ms–min.	100,000	10–15	~6	1000–4000	80–95	<100 ms
Mechanical technologies								
Flywheels	0.001–1	sec–hours	20,000–100,000	1.3–100	20–80	5,000	70–95	10–20 ms
Pumped hydro system	100–1,000	4–12 hours	30–60 years	~0	0.2–2	0.1–0.2	70–85	sec–min
Compressed air energy system (CAES)	10–1000	2–30 hours	20–40 years	~0	2–6	0.2–0.6	40–75	sec–min
Electrochemical technologies (batteries)								
Lead acid	0.001–100	1 min–8 hours	6–40 years	0.1–0.3	50–80	90–700	80–90	<sec
Sodium sulfur (NaS)	10–100	1 min–8 hours	2,500–4,400	0.05–20	150–300	120–160	70–90	10–20 ms
Lithium-ion	0.1–100	1 min–8 hours	1,000–10,000	0.1–0.3	200–400	1,300–10,000	85–98	10–20 ms
Redox flow	1–100	1–10 hours	12,000–14,000	0.2	20–70	0.5–2	60–85	10–20 ms
Chemical technologies								
Hydrogen	0.01–1.000	min–weeks	5–30 years	0–4	600 (200 bars)	0.2–20	25–45	sec–min
SNG	50–1,000	hours–weeks	30 years	Negligible	1800 (200 bars)	0.2–2	25–50	sec–min

Source: Adapted from Deloitte (2015).

5.6.1 ELECTROCHEMICAL STORAGE

Rechargeable batteries as energy storage devices can be used to store electricity generated by renewable energy sources such as PV or wind turbines, but these have a limited capacity and are expensive and their useful life is limited (maximum 10 years). Nevertheless, several companies and organizations like the National Renewable Energy Laboratory (NREL) are developing very large-scale battery systems to the size of several shipping containers to store up to 4 MWh of energy. Such extremely large batteries will be necessary for the future in some applications because of the electrical energy produced by wind-power farms, tidal power plants, and, in part, solar generations will need to be stored.

5.6.2 MECHANICAL STORAGE

Pumped Hydro System

Energy can be stored by pumping water up to a reservoir at a higher altitude during a power surplus condition, and then when power demand increases, it is run through generators back to the lower reservoir in hydroelectric power plants. Conditions such as these may exist, for instance, when no alternative renewable energy sources such as wind or solar applications exist at the time of energy demand. Unfortunately, the use of large-scale water reservoirs will play only a minor role in a global future of renewable power supplies because in most regions neither large water bodies nor sufficient differences in topographic altitude exist.

Compressed Air Energy System (CAES)

A compressed air energy system, an alternative to the pumped hydro system, operates on a similar principle of a pumped hydro energy storage system. During the lower demand period, energy can be stored by generating compressed air at pressures 5000–7000 kpa (~50–70 to 725–1015 psi), which is stored in appropriate airtight subterranean geological formations in the earth's crust or large surface tank farms. Then when power demand increases it can be released to operate a turbine or a reciprocating engine for recovery of energy. In the future, these facilities will also play only a lesser role, similar to that of hydroelectric power generation.

Flywheel Energy Storage System (FES)

Under certain circumstances, electrical energy can be stored when surplus power is used as input for the acceleration of a flywheel with high mass to around 15,000 RPM. When used for uninterruptible power supply operations, the kinetic energy of the flywheel is given off driving a generator as recovery when the regular power supply is interrupted. Flywheel storage represents a limited possibility for energy storage and is typically used in local building applications.

As an alternative, supercapacitor technology can be used for short-term storage. These storage elements have 20 times the power density of regular accumulators, and they are incapable of storing energy over longer periods.

5.6.3 CHEMICAL STORAGE

Chemical storage is one of the most promising technology for the long-term storage of high-grade energy. Chemical storage can use electricity from the renewable energy source to produce a chemical, which later can be used as a fuel to serve a thermal load or for electricity generation. In principle, all gaseous fuels derived from renewable sources are excellent candidates for chemical energy storage because they can be stored and used whenever a power demand arises and, as such, they are independent of the cycle of their generation. This section discusses two potential chemical energy storage hydrogen (H_2) and ammonia (NH_3). However, both hydrogen and ammonia are hazardous, so extensive safety controls are paramount.

Electricity generated from renewable energy sources (e.g., sun or wind) can be used for electrolysis to split water into H_2 and O. After its separation, hydrogen can be stored in storage tanks with an operating pressure of up to 30 MPa (~300 bar–4351.13 psi) in the form of H_2 and O for later use. Stored hydrogen can be recombined with oxygen in a fuel cell or combustion turbine to generate electricity, which would make the resulting storage cycle more similar to an electricity-in/electricity-out system. Hydrogen can be transported by pipe to different locations in a city or other cities. Alternatively, a hydrogen storage system can be co-located with a source of charging energy (e.g., a large wind or solar farm) to avoid transportation and power system-interconnection costs.

The ammonia molecule is composed of three hydrogen atoms that are bonded to one nitrogen atom. With high-temperature input it can be split into nitrogen and hydrogen (thermal dissociation); the two gases can be separately stored in a conventional tank and transported by conventional pipelines. Later two gases may be made to re-combine (in the presence of some catalyst), which is a highly exothermic process. Ammonia can be used as a fuel in conventional power plants, but its resulting nitrogen oxide emissions are an important consideration.

5.6.4 Phase Change Materials

Phase change materials (PCMs) are substances with a high 'latent heat of fusion' that, through melting and solidifying at specific temperatures, are capable of storing or releasing large amounts of energy. PCMs use chemical bonds to store and release heat. The latent heat of fusion that is employed to break the chemical bond for an upward change (solid to liquid) is released during the corresponding downward change (liquid to solid) (Figure 5.30).

- Phase change materials can be classified into organic and inorganic compounds and their eutectic mixtures, as shown in Figure 5.31. The inorganic materials are characterized by the high latent heat of fusion per unit volume. The organic compounds include paraffin and non-paraffin organics.
- Phase change materials represent a potential technology to reduce peak cooling/heating loads and energy consumption of HVAC systems in buildings. PCMs, with melting temperatures between 20°C and 50°C, can be applied to enhance thermal storage in conjunction

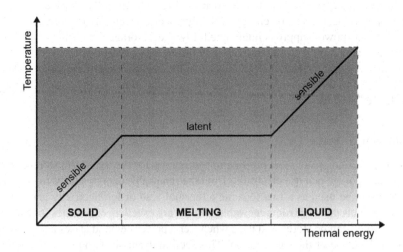

FIGURE 5.30 Specific thermal energy input vs. temperature. Source: adapted from Aa et al. (2011), line work by Ar Vaibhav Ahuja

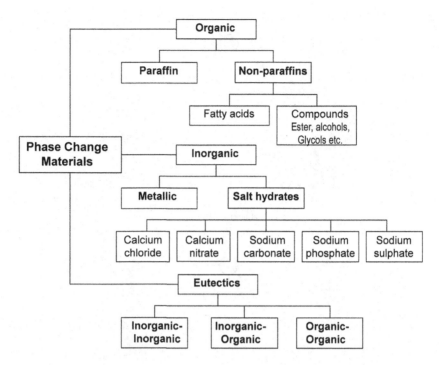

FIGURE 5.31   A taxonomy of phase change materials (PCMs).

with both passive and active storage for heating and cooling in buildings in three different ways:

i)   PCMs in building facade (walls or smart glazing)
ii) PCMs in other building components (for example, subfloor or ceiling systems)
iii) PCMs in separate heat or cold storage systems

- The first two are (or can be) completely passive systems, where the stored energy ('heat' or 'cold') is automatically released when indoor or outdoor temperatures rise or fall beyond the temperature of melting. Figure 5.32 illustrates the application of PCMs as an internal latent storage layer, external latent storage layer, and solar wall. The third one is an active system, where the stored energy ('heat' or 'cold') is kept thermally separated from the building structure and the environment using a suitably insulated device (storage tank, heat exchanger, …). In this case, the energy is used only on demand and not automatically.
- Stanford Central Energy Facility, Stanford University, Chapter 6, Section 6.3, has used vegetable-based bio-phase change material for temperature control.

5.6.5   SMART GRID

Smart grid is a concept and vision that captures a range of advanced information, sensing, communications, control, and energy technologies. Taken together, these result in an electric power system that can intelligently integrate the actions of all connected users – from power generators to electricity consumers to those that both produce and consume electricity ('prosumers') – to efficiently deliver sustainable, economic, and secure electricity supplies. (definition adapted from the European Technology Platform Smart Grid – ETPSG).

The power supply system in a built environment is composed of the connection to the electric utility grid, on the one hand, and the building integrated renewable energy system(s), on the other. These systems with integrated renewable energy generators operate in dual mode – when

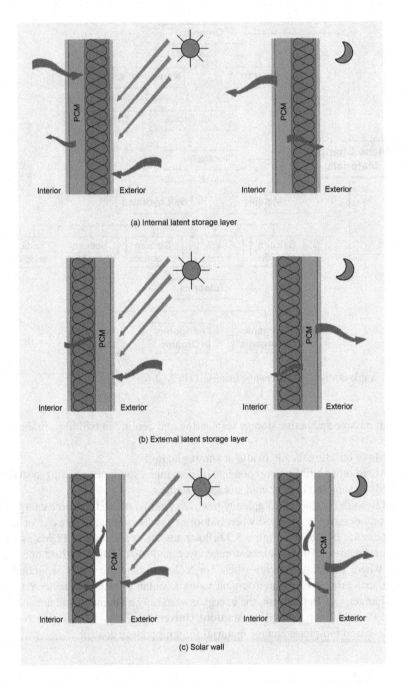

FIGURE 5.32   Application of PCM in-wall and its functioning during day and night. Source: adapted from Aa et al. (2011), line work by Ar Vaibhav Ahuja

renewable generation is surplus, it needs to be supplied to the utility grid, and conversely, when renewable energy generation is in deficit, energy needs to be drawn from the utility grid. However, renewable energy sources are both more uncertain and more variable than conventional genera-tors. A wind turbine may have a reliable estimate of AEO (annual energy output) and efficiency, but it is not easy to predict far in advance when generation will occur. Similarly, a rooftop solar installation is subject to transient events such as cloud cover can reduce output quite instantly. Generally, solar PV output is more variable than wind (changing faster on a minute-to-minute basis), but it is less uncertain. Consequently, the supply of wind and solar may not coincide closely with demand, introducing challenges at the power system at the utility and distribution level. System operators' need to balance supply and demand in situations of high renewable energy production and low demand or low renewable production and high demand.

- **Smart grid** technologies offer new solutions to integrate variable renewable energy and to manage the continuous balancing of the system.
- **Better forecasting**: Widespread instrumentation and advanced computer models allow system operators to better predict and manage renewable energy variability and uncertainty.
- **Smart inverters**: Inverters and other power electronics can provide control to system operators, as well as to automatically provide some level of grid support.
- **Demand response**: Smart meters, coupled with intelligent appliances and even industrial-scale loads, can allow demand-side contributions to balancing.
- **Integrated storage**: Storage can help to smooth short-term variations in renewable energy output, as well as to manage mismatches in supply and demand.
- **Real-time system awareness and management**: Instrumentation and control equipment across transmission and distribution networks allows system operators to have real-time awareness of system conditions, and increasingly, the ability to actively manage grid behavior.

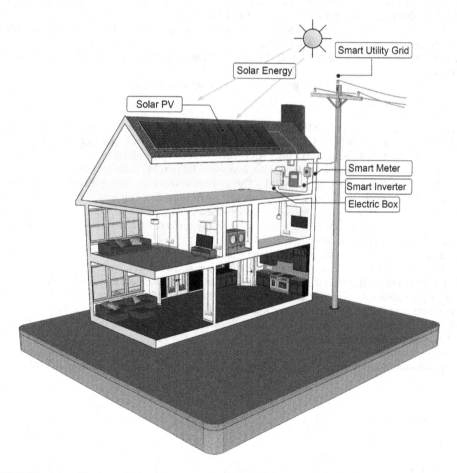

FIGURE 5.33   Visualization of net or smart meter and smart grid interface at the building level. Credit Ar Shiva Bagga

Besides smart grid technology, the other important area of focus – innovative policy, regulation, and business models, are needed to promote and implement next-generation grid architectures. Figure 5.33 is a visualization of net or smart meter and smart grid interface at the individual building scale. Finally, it is the smart building users, who are going to contribute to energy efficiency in the built environment.

> Buildings don't need energy, it is function, comfort, and occupancy which require energy.

(Janda 2011)

REFERENCES

Aa, AvD, Heiselberg CP, and Perino M (2011) Designing with Responsive Building Components. IEA-ECBCS, Annex 44 Integrating Environmentally Responsive Elements in Buildings, Aalborg University, Denmark.

Deloitte (2015) Energy Storage: Tracking the Technologies that Will Transform the Power Sector. Available at https://www2.deloitte.com/content/dam/Deloitte/us/Documents/energy-resources/us-er-electric-sto rage-paper.pdf, accessed on 14 April 2020.

Diesendorf M (2014) *Sustainable Energy Solutions for Climate Change*. Routledge, Abingdon, England.

DOE (2016) Comparison of Fuel Cell Technologies. U.S. Department of Energy. Available at https://www.energy. gov/sites/prod/files/2016/06/f32/fcto_fuel_cells_comparison_chart_apr2016.pdf, accessed on 10 April 2020.

Energy Independence and Security Act (2007) Available at https://www.govinfo.gov/content/pkg/PLAW-110p ubl140/pdf/PLAW-110publ140.pdf, accessed on 10 April 2020.

Energy Policy Act (2005) Available at https://www.govinfo.gov/content/pkg/BILLS-109hr6enr/pdf/BILLS-109hr6enr.pdf, accessed on 10 April 2020.

Hootman T (2013) *Net Zero Energy Design, A Guide for Commercial Architecture*. John Wiley & Sons, Inc, Hoboken, NJ.

https://rredc.nrel.gov/solar/old_data/nsrdb/1961-1990/redbook/atlas/colorpdfs/208.PDF

https://cleanet.org/index.html

https://www.energy.gov/eere/

Janda K (2011) Buildings Don't Use Energy: People Do. *Architectural Science Review*, vol. 54 (1), pp 15–22.

Klaus D (2003) *Advanced Building Systems: A Technical Guide for Architects and Engineers*. English translation by Elizabeth Schwaiger, Birkhäuser Architecture, Basel, Switzerland.

Klaus D, Hammann RE (2008) *Energy Design for Tomorrow*. Edition Axel Menges, Munich.

National Renewable Energy Laboratory (NREL) Energy Basics. Office of Energy Efficiency and Renewable Energy, US Department of Energy, https://www.nrel.gov/research/learning.html, accessed on 12 April 2020.

Office of Energy Efficiency and Renewable Energy. Distributed Energy Technologies for Federal projects. US Department of Energy. Available at https://www.energy.gov/eere/femp/distributed-energy-technologies-federal-projects, accessed on 15 April 2020

Sengupta M, Xie Y, Lopez A, Habte A, Maclaurin G, Shelby J (2018) The National Solar Radiation Data Base (NSRDB). *Renewable and Sustainable Energy Reviews*, vol. 89 (June), pp 51–60.

Stuckley, CR, Schuck, SM, Sims, REH, et al. (2004) *Biomass Energy Production in Australia: Status, Costs and Opportunities for Major Technologies*. A report for the Joint Venture Agroforestry Program, RIRDC Publication no 04/031, Rural Industries Research & Development Corporation, Canberra.

Szokolay SV (2008) *Introduction to Architectural Science, the Basis of Sustainable Design*. Architectural Press/Elsevier Science, Oxford.

World Energy Council (2019) Energy Storage Monitor, Latest Trends in Energy Storage. https://www.worldene rgy.org/assets/downloads/ESM_Final_Report_05-Nov-2019.pdf, accessed on 14 April 2020.

6 Design Case Studies

6.1 INTRODUCTION: BACKGROUND AND DRIVING FORCES

Architecture needs to be understood as a holistic task, where the aim for any building design project should be to respond to the design, structural, ecological, economic, cultural, and social aspirations. It is self-evident that fulfillment of the complex objectives at hand is only reached when all cardinal aspects of a project are brought to a conclusive balance and synergy. To illustrate a point, it's important to give case studies, making a complicated subject easier to comprehend. The hope is that by providing these contemporary case studies, architects striving for sustainability will get insight into the process of synergistic design.

This chapter aims to illustrate a case study from each of the five major climate zones in the United States. In 1997, the Committee on the Environment (COTE) of the American Institute of Architects (AIA) launched its Earth Day Top Ten where it felicitates ten buildings as exemplary case studies of sustainable design every year. Over the past 22 years, the program has generated a repertoire of 220 most environmentally friendly buildings in the world. Similarly, the American Society of Heating, Refrigerating Air-Conditioning Engineers (ASHRAE) has developed a database of about 100 high-performance buildings.

The databases of COTE and ASHRAE were segregated according to the climate zones, and the selection of the design case studies is done based on three criteria:

i) Paradigms of sustainable building design
ii) Representative design for each of the five major climate zones in the United States
iii) Accessibility of data

This chapter presents design case studies to illustrate the principles and practices of synergistic design of sustainable built environments in each of the five major climate zones; Table 6.1 enlist these case studies. It is noteworthy that the design strategies illustrated for the cool (zone 5) and the cold (zone 6) climates could be expanded for the very cold (zone 7) and subarctic/arctic (zone 8) climates. Similarly, the design principles illustrated for the very hot (zone 1) climate could be adapted for the hot (zone 2) climate.

The projects presented in the following sections are environmentally sensitive designs by the eminent architects. The NOAA Daniel K. Inouye Regional Center, Honolulu, and the Edith Green–Wendell Wyatt (EGWW) Federal Building, Portland, are renovation and reconstruction projects and underline the principles of resource optimization. Stanford University Central Energy Facility, Stanford and Research Support Facility, NREL, Golden are net-zero energy projects and showcase the principles of energy optimization. University of Wyoming- Visual Arts Facility, Laramie, a LEED platinum-certified project, demonstrates how environmental sustainability can be pursued on a long-term basis.

Each case study is explained with illustrations under five headings: (i) design intentions, (ii) climate and site, (iii) daylight and thermal design, (iv) energy systems, and (v) sustainable thinking. At the end of each case study, there is a design profile table giving an overview of the salient features.

TABLE 6.1
Design Case Studies

S. No.	Climate Zone	Design Case Study	Green Certification	Distinctions
1.	1A Very hot humid	National Oceanic & Atmospheric Administration (NOAA) Daniel K. Inouye Regional Center, Honolulu, Hawaii	LEED Gold	COTE 2017
2.	3C Warm marine	Stanford University Central Energy Facility, Stanford, California	ZNE	COTE 2017
3.	4C Mixed marine	Edith Green–Wendell Wyatt (EGWW) Federal Building, Portland	LEED Platinum	COTE 2016
4.	5B Cool dry	Research Support Facility, NREL, Golden, Colorado	LEED-NC Platinum	AIA COTE 2011, HPB ASHRAE
5.	6B Cold dry	University of Wyoming- Visual Arts Facility, Laramie, Wyoming	LEED-NC Platinum	AIA COTE 2016

6.2 NATIONAL OCEANIC AND ATMOSPHERIC ADMINISTRATION DANIEL K. INOUYE REGIONAL CENTER, HONOLULU, HAWAII (ZONE 1A VERY HOT HUMID, COTE 2017)

6.2.1 DESIGN INTENTIONS

The National Oceanic and Atmospheric Administration (NOAA) is the US government's oldest scientific agency, with a mission to understand and predict changes in the earth's environment (climate, weather, oceans, and coasts), to share that knowledge and information with others and to conserve and manage coastal and marine ecosystems and resources.

Working toward 'One NOAA' concept, the agency decided to consolidate more than 800 staff from 15 NOAA offices that had been spread across Oahu into a single high-performance research and office facility on Ford (Island, at the heart of Pearl Harbour National Historic Landmark District on Oahu (Figure 6.1). The facility is named for the late former Senator Daniel K. Inouye, in recognition of his significant contribution to ocean and environmental issues and his steadfast support for the construction of the campus (The Rafu Shimpo 2013, Manahane 2013).

The NOAA Daniel K. Inouye Regional Center (IRC) is 12.14 hectares (30 acres) parcel on federally owned property and combines new facilities with the historic preservation of four buildings culminating into a campus. The HOK, in collaboration with Hawaii-based architect Ferraro Choi and associates, worked on this project that is environmentally sustainable, state-of-the-art, Gold-certified in Leadership in Environmental and Energy Design (LEED), and is the AIA COTE Top Ten winner for sustainable design excellence in 2017; the jury commented:

> This project transformed a historic Albert Kahn industrial building into a new research building with a laboratory-focused program. The building achieves significant energy reduction from an innovative passive downdraft system, providing 100 percent natural ventilation in office and public spaces. The project rejuvenated the site through the creation of a new waterfront public space. The building is designed to resist a 500-year storm event, and the flexible space can be used for critical response in the event of a natural disaster.

> (AIA 2017a)

The main facility of NOAA IRC features the adaptive reuse of the historic twin airplane hangars (Buildings 175 and 176) and a new steel and glass central entry structure (Building A) – that links

FIGURE 6.1   A new glass and steel structure links a pair of 1941 aircraft hangars on either side. Image courtesy HOK

the hangers into a unified composition, providing 34,374.12 m² (370,000 ft²) of space to bring together diverse entities – from the National Weather Service Pacific Region Headquarters (Tsunami Warning Center) to the National Environmental Satellite Service (Global Environment data and Earth system monitoring), the National Maritime Fisheries Service, the National Ocean Service and NOAA Research (Climate observation and Marine & atmospheric research), and others. The twin hangars, designed in 1939 by Detroit-based Albert Kahn, were originally meant to repair World War II fighter aircrafts, survived the Pearl Harbour attack in 1941, and had been derelict – with smashed windows and a leaky roof. The exterior of the hangars has been restored and repaired to meet the functional requirements of NOAA as well as the requirements of historic preservation partners to maintain the visual integrity of the exterior; Figure 6.2 shows the historical review of the NOAA IRC. Viewed through lenses of history and biomimicry, the main facility features a range of key sustainable and innovative features, harnessing the power of water, wind, and sun (Figure 6.3).

6.2.2   CLIMATE AND SITE

The very hot humid climate (zone 1A) of Honolulu is characterized by high temperature, dampness in the atmosphere nearly throughout the year, and moderate rainfall. The average annual rainfall is 434.3 mm in Honolulu; about 70% of annual rainfall in Honolulu is received during the six months from October to March. August is generally the hottest month when the mean daily maximum temperature is 31.1°C, and the mean daily minimum temperature is 23.9°C at Honolulu. January is the coldest month when the mean daily maximum temperature is 26.1°C, and the mean daily minimum temperature is 19.6°C at Honolulu. Throughout the year the morning maximum relative humidity is above 70%, and the afternoon minimum relative humidity ranges between 49% and 63%.

Throughout the year cloud cover ranges between 44% and 57.4%. Sunshine hours of the year are 3035.9 h. Wind speed ranges between 3.7 and 6 m/s, and the prevailing wind direction is the

NOAA consolidated resources and repurposed and preserved a 75 year old world war era II building, saving cost on both construction and operations.

1941

2014

1939

2005

FIGURE 6.2   Historical review – adaptive reuse of aircraft hangers as research facility. Image courtesy HOK

ADDRESSED HAZARDOUS MATERIALS

REUSED STRUCTURE

PASSIVE DOWNDRAFT SYSTEMS

PREVAILING SEA BREEZES   PREVAILING SEA BREEZES

NEW ADDITION

SKYLIGHTS AND ATRIA

RESTORED   ACTIVE CHILLED BEAM IN LABS
HANGARS   RAINWATER HARVESTING   HIGH PERFORMANCE ENVELOPE
           GREYWATER HARVESTING   RECLAIMED WATER BUILDING

                    BIOFILTERATION LANDSCAPE   GEOEXCHANGE SYSTEM
           SOLAR THERMAL PANELS

                                                          BIOSWALE

                                    REINFORCED SHORELINE, COASTAL BARRIER
                                              EROSION CONTROL

FIGURE 6.3   Key sustainable strategies to harness the sun, air, and water. Image courtesy HOK

northeast. The global solar radiation goes as high as 6 kWh/m². The climatic data of Honolulu are given in Table 7.23, and the sun-path diagram is in Figure 7.22 (online resource).

The main facility is raised 600 mm (2 ft) from the previous existing grade to safeguard it from a rise in the sea level and flooding from a 500-year storm event. The NOAA IRC is designed to resist a Category 3 hurricane, exceed seismic requirements, and meet AT/FP requirements.

The circulation path unites the historic tarmac, the historic hangars, and the Makai waterfront into a unified NOAA IRC campus. Along the waterfront there are places for large gatherings, Polynesian festivities, informal dining, and exhibits highlighting the islands' voyaging history; the site plan is shown in Figure 6.4.

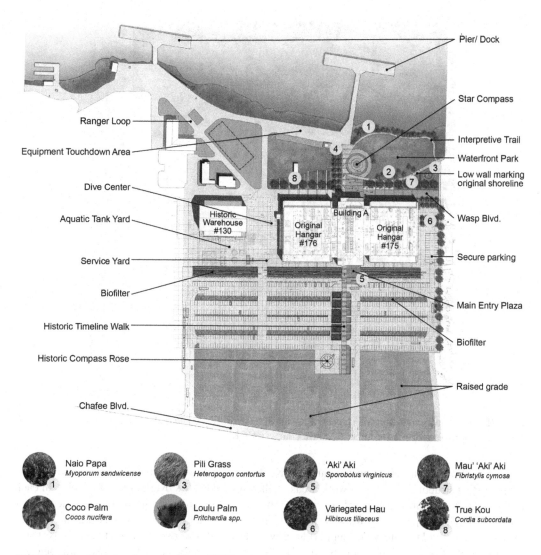

FIGURE 6.4 Site plan. Image courtesy HOK

The historic site's paved airfield had been impervious; most of the stormwater on the 12.14-hectare (30-acre) site had previously flowed directly into the harbor without any treatment. Large sections of the historic paving were removed to add 29% native Hawaiin vegetation including drought-tolerant aki-aki grass and kou trees that thrive without potable water. Grassed stripped and grassed bio-filter reduce, slow, and cleanse the stormwater runoff from the building and parking area. Native Hawaiian planting along the water's edge at the harbor helps filter site runoff before it enters the harbor, while educating building occupants and visitors about Hawaiian flora.

Entering the main facility on the ground floor of Building A, one is greeted by a three-story high central atrium, which serves as a visitor center for education and outreach (Figure 6.5). The spaces at the ground floor contain an extensive exhibit space, reception and pre-function space, a reference library, a 200-seat auditorium, training and meeting rooms, an occupational health/fitness center, and a dining facility with a panoramic view to the water and the mountain range in the distance. Located in Building A, the National Weather Service continuously monitors and anticipates significant upcoming weather events through the Pacific Regional Operations Center and provides reports, status updates, and support as necessary for hazardous weather and tsunami events.

FIGURE 6.5   The ground floor lobby, central atrium showcasing dynamic and diffuse light along with complexity and order of space in Building A. Image courtesy HOK

Within the flanking airplane hangars, HOK inserted two additional floor levels, reusing the original steel structure, with open work areas, walkways, and stairs overlooking the lobby. In each of the hangar structures, courtyards open up from the roof down to the ground level. These areas serve as gathering spaces and draw in natural light. To facilitate Pacific Islands marine biology research, a suite of laboratory and support areas was designed over two floors on the southwest side of Building 176. A loading dock on the first floor gives access to the necropsy suites, with the main body of laboratory accommodations serving chemistry, biology, and molecular biology disciplines located immediately above. Figure 6.6 shows the level-one, level-two, and level-three floor plans of NOAA IRC main building. Figure 6.7 presents the cardinal front elevations of the building. Figure 6.8 delineates the cross and long sections of the building.

6.2.3   DAYLIGHT AND THERMAL DESIGN

The building is ideally oriented on the north-east axis, with long facades facing south-east and north-west to maximize daylight and ventilation and minimize the solar heat gain.

The vast footprints of 222.5 m by 82.3 m (730 ft by 270 ft) of the existing hangar structures meant that daylight through perimeter windows could illuminate only a fraction of the floor area. A building occupant, for example, could sit as far as 38.1 m (125 ft) away from a window or clerestory. Physical and virtual computer modeling aided to develop a design solution that drives daylight deep into the building.

The design solution was a grid of 110 skylight apertures 6.09–7.62 m (20–25 ft) apart, each about 0.37 m² (4 ft²), sprinkled across the roof. Under the apertures, reflective-lined plastic-resin solar tubes direct sunlight into the workspaces through finned diffusers that drop from the ceiling

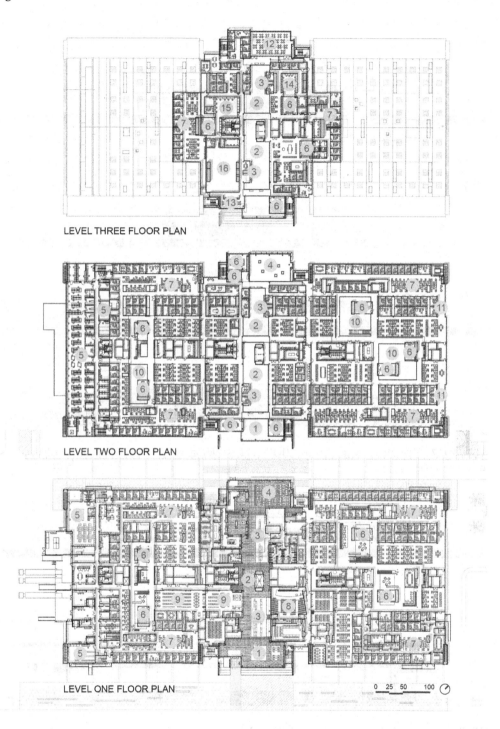

LEVEL THREE FLOOR PLAN

LEVEL TWO FLOOR PLAN

LEVEL ONE FLOOR PLAN

0 25 50   100

FIGURE 6.6 Floor plans, Image courtesy HOK. 1. Entry/reception; 2. central gallery/atrium; 3. NOAA exhibits/displays; 4. dining hall; 5. research laboratory and support; 6. collaborative gathering spaces/internal courtyards; 7. administrative spaces/offices; 8. auditorium; 9. library; 10. open to below; 11. open team space; 12. roof terrace; 13. kitchen/pantry; 14. Pacific Tsunami Warning Center; 15. information technology; 16. data center.

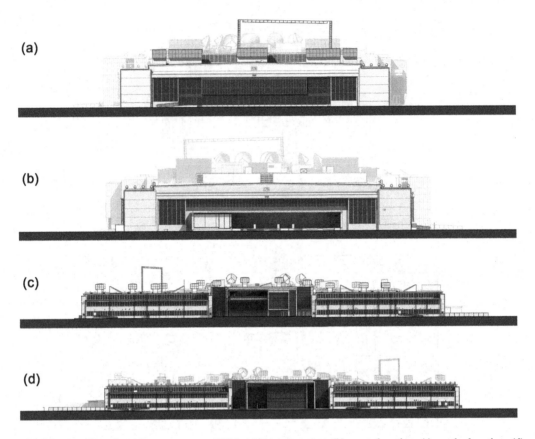

FIGURE 6.7 Elevations, Image courtesy HOK. (a) East elevation; (b) west elevation; (c) north elevation; (d) south elevation.

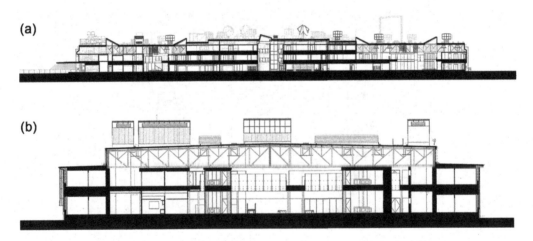

FIGURE 6.8 Sections, Image courtesy HOK. (a) Longitudinal section; (b) transverse section.

like enormous square Slinkys™. Those plastic-resin translucent fins, which are connected by steel cables, serve two purposes: scattering the sunlight to reduce glare and reflecting it to the ceiling, where it is further diffused (Figure 6.9). The entire ceiling becomes a luminaire and largely eliminates the need for electric lights during the day without the typical heat associated with direct overhead lighting. When clouds cover the sun, the interior space takes on a completely different

FIGURE 6.9 Skylight apertures across the roof directing sunlight into the workspace via solar tubes. Image courtesy HOK

feeling: ambient, glowing light from diffuse sources (i. e. the sky and the earth) still powered by the sun. This natural cycle of illumination provides a consistent spread of 323 lux/m^2 (30 lux/ft^2) across the sprawling floor plates. The floor plates feature translucent partitions, bringing light where little came before and reducing electric lighting loads by 50%. Daylight autonomy achieved is 57%, which ensures that regularly occupied area receives at least 300 lux, 50% of the annual occupied hours; Figure 6.10 shows the daylight simulations (computer modeling).

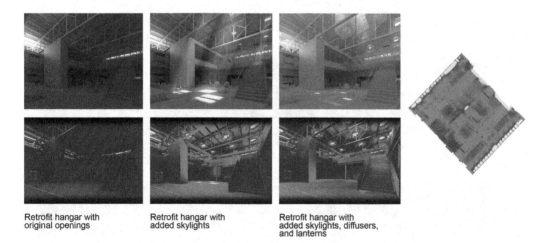

Retrofit hangar with
original openings

Retrofit hangar with
added skylights

Retrofit hangar with
added skylights, diffusers,
and lanterns

FIGURE 6.10 Simulation models showcasing abundant natural lighting via added skylights, diffusers, and lanterns. Image courtesy HOK

6.2.4 ENERGY SYSTEMS

The morphology of native *Albizia Saman* (Monkeypod) tree – principles of natural lighting and cooling – was one of HOK's design bio-inspirations (biomimicry) for the NOAA IRC (Woolford and Todd 2018). The tree canopy grows wider than the tree is tall; it has multiple layers and deflects the heat of the sun, while filtering natural light toward the ground. The dense canopy of foliage provides shade and thermal comfort during the hot, humid days (Figure 6.11). The trees protect themselves from hot temperatures by pulling soil moisture through the vascular structure into the leaves. Moisture released as water vapor creates a cooling effect around the canopy. Then when it's overcast and raining, the leaves curl up into a narrow, conic shape and allow the breezes and rain to reach the ground. The grass is greener under the monkeypod tree because it allows the rain to reach the soil.

Based on indigenous principles, Hawaii's first hydronic passive downdraft cooling system of its size and design uses deep seawater, naturally cold at 14.44–15°C (58–59°F), pumped from the geothermal well into roof cool coils. Prevailing sea breezes, which is warmer than 18.3°C (65°F) in the day time at Ford Island, are captured by 38 wind scoops, each roughly of 3.66 m × 7.32 m (12 ft × 24 ft) located on the building rooftop. The incoming air is cooled by passing through cold water 'reverse radiators' in the vertical shafts 'thermal chimneys' (passive downdraft). The moisture content of the intake air is reduced to 60% humidity by cooling it and condensing out some moisture. The naturally cold seawater is mechanically chilled to an even colder 9.44–10°C (49–50°F) so it can cool the indoor air. The downdraft in each shaft drops up to 84.95 m³ (3000 ft³) per minute of cooler, denser air to the base of the building; from there, cool air is supplied to occupied spaces via a raised access concrete floor utilizing 'displacement' principles. After being supplied at low levels, the air draws heat from occupants, equipment, lights, and solar, and, as its temperature rises, it becomes less dense and more buoyant, rising through the building via interconnected light wells and atria before being exhausted to atmosphere. With natural stack and Venturi effects, the exhaust ventilation system is also 100% passive, though few powered fans are needed for airflow, and the point-chilling does require conventionally generated energy. When outside temperatures dip below 18.3°C (65°F), the system reverses itself, assisted by heating, instead of cooling coils. The buoyancy-driven ventilation system moves only 100% fresh outside air cleaned through ultra-violet and dust and pollen filters. Figure 6.12 illustrates the passive downdraft cooling system installed in NOAA IRC.

The geothermal well is 396.24 m (1300 ft) deep, with a 762 mm (30 inches) diameter and pumps up to 14,195.29 liters (3750 gallons) of seawater per minute to the surface and is primarily used

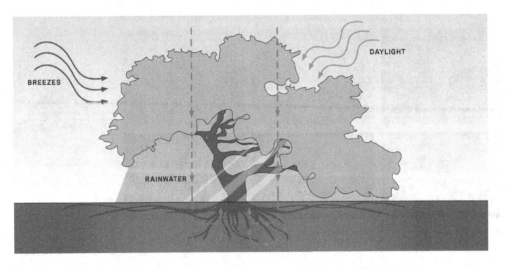

FIGURE 6.11 Natural principle of Monkeypod tree as inspiration for the passive downdraft system. Image courtesy HOK

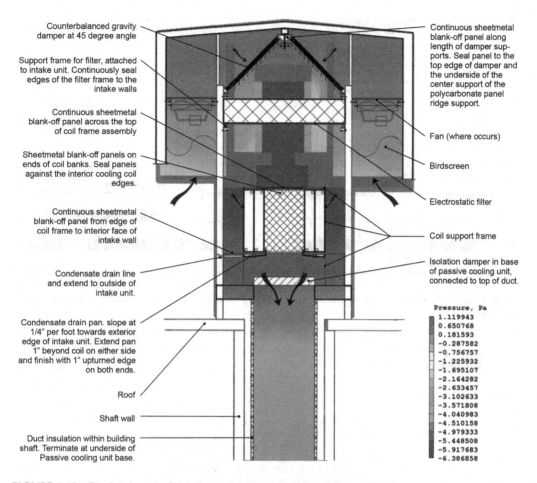

FIGURE 6.12 Passive downdraft cooling unit at the roof. Image courtesy HOK

to supply the outdoor marine animal tanks, seawater ports in the research laboratories, and an aquaria room.

By bringing in 100% outside air through the ventilation system and large amounts of daylight through the skylights, occupants are given fresh air and natural light throughout the working day and provide 50% savings on energy. Ninety-five percent of the occupants can control their light levels. The underfloor air distribution system enables all occupants to control their thermal comfort by adjusting floor diffusers. Figure 6.13 illustrates the principle of passive downdraft cooling and displacement ventilation system. The commissioning process provided many lessons in regulating the temperature throughout the building. Passive systems, daylighting, and a backup generator serving the central part of the building allow for three days of operations in case of emergency. The design captures renewable energy through solar thermal systems. Figure 6.14 illustrates the energy reduction matrix from the base case to the design case.

6.2.5 SUSTAINABLE THINKING

Material selections were informed by maximizing efforts to reuse the existing structure and local materials (such as ohia wood and basalt stone), long-term durability, compliance with the hurricane, and anti-terrorism requirements. The project complied with NAVFAC's affirmative procurement program, which mandates compliance with EPA's comprehensive procurement guidelines to

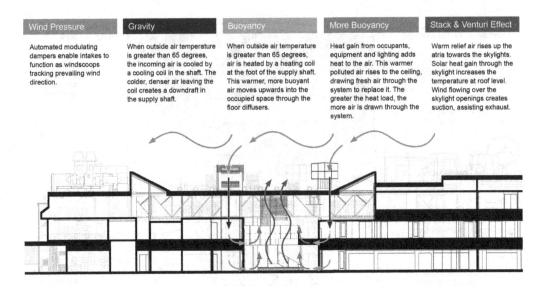

Wind Pressure	Gravity	Buoyancy	More Buoyancy	Stack & Venturi Effect
Automated modulating dampers enable intakes to function as windscoops tracking prevailing wind direction.	When outside air temperature is greater than 65 degrees, the incoming air is cooled by a cooling coil in the shaft. The colder, denser air leaving the coil creates a downdraft in the supply shaft.	When outside air temperature is greater than 65 degrees, air is heated by a heating coil at the foot of the supply shaft. This warmer, more buoyant air moves upwards into the occupied space through the floor diffusers.	Heat gain from occupants, equipment and lighting adds heat to the air. This warmer polluted air rises to the ceiling, drawing fresh air through the system to replace it. The greater the heat load, the more air is drawn through the system.	Warm relief air rises up the atria towards the skylights. Solar heat gain through the skylight increases the temperature at roof level. Wind flowing over the skylight openings creates suction, assisting exhaust.

FIGURE 6.13 Principle of passive downdraft cooling and displacement ventilation system. Image courtesy HOK

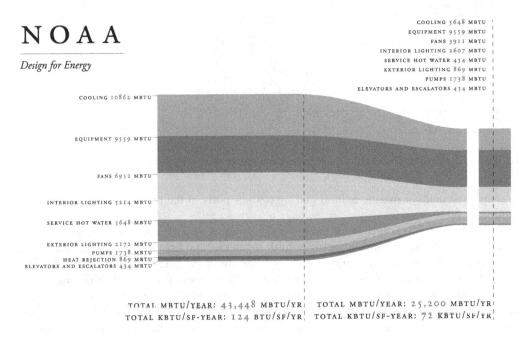

FIGURE 6.14 Energy reduction from base case to design case. Image courtesy HOK

measure, report, and reduce greenhouse gas pollution from agency operations and to reduce waste, increase efficiencies, and cut costs.

To maximize the use of recovered materials, the team specified construction products including structural fiberboard, laminated paperboard, plastic pipe and fittings, geotextiles, cement/concrete containing ground-granulated blast furnace slag, carpet, and floor tiles. Eighteen percent of building materials are from recycled content.

Low-emitting adhesives and sealants, paints and coatings, carpet systems, and composite wood all exceed LEED requirements.

1941

2018
18% Building materials from recycled content
95% Construtcion waste diverted from landfill

FIGURE 6.15   Resource-efficient design from recycled content. Image courtesy HOK

A construction waste management program reduced waste by 95% through reuse, recycling, and supplier take-back of materials. Figure 6.15 shows the material selection matrix.

The design capitalizes on opportunities for water conservation and reuse. The entire diaphragm of the roof is a water capture device, with each drain linked to a channel leading to a dedicated rainwater retention tank. The harvested (non-potable) water supplies the toilet flushing requirements for plumbing systems.

A second strategy relies on the capture and use of the graywater waste from electrical water coolers and condensate from the rooftop mechanical equipment. The building lavatories and showers are fitted with low-flow fixtures. Passive cooling units on the roof capture up to 37,854.12 liters (10,000 gallons), a day of condensate from the chilled coils of water, recycling it into a graywater system used to flush toilets and irrigate the 12.14 hectare (30 acres) site's native landscaping; no graywater leaves the site.

The percentage of water consumed on-site from rainwater capture is 20%. The percentage of water consumed on-site from graywater/blackwater capture and treatment is 53%. The percentage of rainwater that can be managed on-site from maximum anticipated 24-hour, 2-year storm event is 40% (Figure 6.16). Table 6.2 summarizes the design profile of NOAA IRC, Honolulu, Hawaii.

6.3   STANFORD UNIVERSITY CENTRAL ENERGY FACILITY, STANFORD (ZONE 3C WARM MARINE, COTE 2017)

6.3.1   DESIGN INTENTIONS

At the heart of Stanford University's transformational, campus-wide energy system is a new, technologically advanced Central Energy Facility (CEF) designed in response to Stanford's long-range

20% Water consumption from rainwater   53% Water consumption from graywater   40% Condensate manged on site

FIGURE 6.16   Water consumption strategies for rainwater, graywater, and condensate from AHUs. Image courtesy HOK

Energy and Climate Action Plan and overall Stanford Energy System Innovations (SESI) program (Figure 6.17). The facility replaces an aging 100% fossil-fuel (natural gas) powered central cogeneration heat and power plant with grid-sourced electricity-powered 'regeneration' plant – 65% of which comes from renewable energy sources and a first-of-its-kind innovative heat recovery system that captures nearly two-thirds of waste heat generated by the campus cooling system (chilled water return loop) to produce 93% of hot water required for the heating system. The facility is projected to improve campus energy performance by 75% and reduce campus-wide potable water usage by 18%, and expected energy savings over the life cycle of 35 years is estimated at $425 million.

The net-positive-energy facility of 11,669.9 m² (125,614 ft²) comprises five distinct components: an entry court and a high-performance office building, a heat recovery chiller (HRC) plant with two large cold storage tanks, the California State Office of Health Planning and Development (OSHPD) cooling and heating plant, a service yard, and a new campus-wide main electrical substation. The AIA COTE Top Ten 2017 jury comments:

> This project fulfills a carbon-neutral strategy for Stanford and houses a central plant and facilities building. The facility demonstrates a long-range climate and energy plan in action. It transforms what would be a typical unappreciated energy plant into a classroom and a moment of architectural joy. A naturally ventilated, daylit work environment is provided for facilities staff who would normally be in a windowless basement. It sets a high bar for a university to provide national environmental leadership and design excellence.

> **(AIA 2017b)**

Designed by ZGF Architects in partnership with Affiliated Engineers Inc. CEF integrates its state-of-the-art technological advances and the ultra-energy-efficient district heating and cooling system (Figure 6.18), with the architectural expression of lightness, vivid colors, dramatic material contrasts, transparency, and sustainability to express the facility's purpose: resonating the grand-old radicalism of the Centre Pompidou. Stanford has made the facility part of its campus tours, alongside the original mission-style architecture and mascot-inspiring evergreens.

6.3.2   CLIMATE AND SITE

This part of California is characterized by a year-round warm marine climate and moderate rainfall. The average annual rainfall is 378.3 mm in the region. August is commonly the hottest month when the mean daily maximum temperature is 26.6°C, and the mean daily minimum temperature is 15.3°C. January is the coldest month when the mean daily maximum temperature is 13.8°C, and the mean daily minimum temperature is 5.9°C. Throughout the year the morning maximum relative humidity is above 77% and the afternoon minimum relative humidity ranges between 49% and 63%.

Throughout the year cloud cover ranges between 42% and 58%. Sunshine hours of the year are 3035.9 h. Wind speed ranges between 2.4 and 3.6 m/s, and the north is the prevailing wind direction along with north-west for a few months. The global solar radiation goes as high as 8.1 kWh/m². The climatic data of San Jose are given in Table 6.3, and the sun-path diagram is in Figure 6.19.

TABLE 6.2
Design Profile NOAA Daniel K. Inouye Regional Center, Honolulu, Hawaii

Building Profile	Location	Ford Island, Honolulu, Hawaii, US Project Site: Historic Structure or District
	Principal use	Laboratories, offices, visitor center
	Employees/occupants	715
	Expected (design) occupancy	850 (residents, occupants, visitors)
	Site area	117,336.5 m² (1,263,000 ft²)
	Built-up area	32,516 m² (350,000 ft²)
	No. of stories	Three
	Distinctions/awards	AIA/COTE Top Ten Green Projects Award (2017), LEED Gold, IES Special Citation Award for energy and environmental lighting design, San Francisco AIA – Merit Award – 2015, IIDA Northern California Chapter – Merit Award – 2015, Historic Hawaii Foundation – Preservation Award – 2014
	Total cost of construction (excluding furnishing)	$157,000,000
	Substantial completion/occupancy	2014
Building Team	Owner/representative	National Oceanic and Atmospheric Administration (NOAA)
	Architect	HOK, Ferraro Choi and Associates
	General contractor	The Walsh Construction
	Mechanical/electrical/plumbing engineer, energy modeler	WSP Flack + Kurtz
	Structural engineer	SOHA Engineers
	Civil engineer	Kennedy Jenks Consultants
	Environmental consultant	Kleinfelder
	Landscape architect	Ki Concepts
	Lighting design	WSP
	LEED consultant	Brightworks
	Commissioning agent	Glumac Engineers
Solar Design Profile	Latitude	N 21° 19′
	Longitude	W 155° 57′
	Altitude	2 m
	Heating degree days (base ~18.3°C–65°F)	4
	Cooling degree days (base ~18.3°C–65°F)	4594
	Annual hours operation	2080
	Orientation	Northwest
Performance Profile	Annual energy use intensity (EUI) (Site) consumption	12.54 kWh/m² (42.8 kBtu/ft²)
	Electricity (grid purchase)	9.03 kWh/m² (30.8 kBtu/ft²)
	Natural gas	3.52 kWh/m² (12 kBtu/ft²)
	Annual on-site renewable energy exported	0
	Annual energy cost index	$27.115/m² (2.52/ft²)
	Annual load factor	36.5%
	Savings vs. Standard 90.1-2007 Design Building	42.6%
	Energy star rating	96

(Continued)

TABLE 6.2 (CONTINUED)
Design Profile NOAA Daniel K. Inouye Regional Center, Honolulu, Hawaii

Water Efficiency	Annual water use	3,393,966.14 liters (896,591 gallons)
Key Sustainability Features	Water conservation	Rainwater harvesting; condensate capture; graywater used for irrigation and toilet flushing
	Recycled materials	Reuse of structural steel
	Daylighting	Large skylights with diffusers to bring daylight into the deep floor plate
	Individual controls	Underfloor air distribution vents can be controlled by occupants
	Carbon reduction strategies	100% natural ventilation in the office/public spaces
	Transportation mitigation strategies	Organization-funded transit
	Renewable energy	Solar thermal
	Other major sustainable features	Hydronic passive downdraft HVAC system
Building Envelope	Roof	Type: Metal deck, rigid insulation, bituminous Overall R-value 2.94301 m^2 K/W (16.7 °F ft^2 h/Btu) Reflectivity 0.89
	Walls	Type: Curtain wall-cast-in-place concrete Overall R-value 1.93851 m^2 K/W (11°F ft^2 h/Btu) Glazing percentage of 0.65
	Windows	Effective U-factor for Assembly 5.399 W/m^2 K (0.951 Btu/h ft^2°F) Solar heat gain coefficient (SHGC) 0.391 Visual light transmittance 0.57

Source: Adapted from www.hpbmagazine-digital.org/hpbmagazine/winter_2018/MobilePagedReplica.action?pm=2&folio=6#pg8

FIGURE 6.17 Stanford Central Energy Facility with photovoltaic (PV) trellis as entrance feature. © Tim Griffith

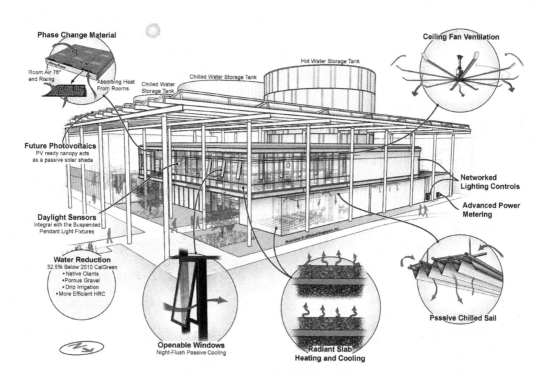

FIGURE 6.18   Sustainable design strategies. Image courtesy ZGF Architects

As an infrastructural facility with a large electrical transformer field, the project was deliberately sited on a 1.21 hectare (3 acres) site on the edge of the Stanford campus edge by converting a portion of the university practice golf course (Figure 6.20). The facility is strategically sited to integrate with future academic quad expansions respecting Olmstead's original axial master campus plan. The project boundary and building footprint were deliberately modified to avoid impacting native California tiger salamander habitat, as well as conserving several existing heritage oak trees on-site.

The facility takes advantage of California's warm marine climate by creating an indoor–outdoor workplace. The main entrance is on the prominent eastern edge, facing the central campus, while the electrical substation is located on the western edge to minimize its visual presence from the central campus, and a landscaped berm screens it from the adjacent golf course.

A welcoming photovoltaic (PV) trellis, supported by steel pipe columns, unifies the front entrance and wraps around the corner. To the northeast is a two-story, 929.03 m² (10,000 ft²) L-shaped climate-responsive office building for staff and visitors. On its ground level, a large flex room, which often serves as the starting point for tours of the facility, doubles as an extra classroom. A paved and landscaped courtyard within the office building offers another prime congregational space for visitors and tour groups, while the courtyard is shielded from the sun by a trellis topped with steel grating. A grand staircase serves as the entry to the second floor offices for the 16-person staff, outdoor room, conference room, and staff kitchen and lounge, and also imaginatively functions as theater bleacher seating for tours and lectures, with the hot water tank painted 'Stanford Red' as the stage backdrop (Figure 6.21).

The heat recovery plant (HRP), the key player in this ultra-efficient energy loop system, sits along the prominent eastern edge of the complex, immediately south of the office building and shares the overhead PV trellis as an organizing and unifying architectural element. The two main volumes that house the heat recovery equipment are simple steel-framed structures with cast-in-place concrete walls.

TABLE 6.3

Climatic Data, San Jose, California

	Latitude N 37° 22'				Longitude W 121° 55'					Altitude 16 m			
	Climate Warm Marine				ASHRAE 3C					Köppen Csb			
Months	**Jan**	**Feb**	**Mar**	**Apr**	**May**	**Jun**	**Jul**	**Aug**	**Sep**	**Oct**	**Nov**	**Dec**	**Year**
Sunshine h[a]	185.9	207.7	269.1	309.3	325.1	311.4	313.3	287.4	271.4	247.1	173.4	160.6	3061.7
Cloud (%)	54.8	56.9	63.1	57.6	38.5	28.4	34.8	31.7	30.6	43.0	44.0	50.6	44.5
Solar irradiation daily average (Wh/m²)													
Global	2253	2954	3452	6066	7311	8102	7721	6933	5674	4069	2701	2169	4950
Diffuse	1072	1331	1750	2126	2249	2220	2143	1937	1733	1449	1054	900	1664
Relative humidity (%)													
Morning	91	91	79	83	79	78	87	84	85	79	77	85	83.2
Evening	58	58	56	49	45	45	51	45	45	42	48	54	49.7
Dry-bulb temperature (°C)													
Max	13.8	14.0	18.1	18.5	22.1	24.0	24.7	26.6	24.6	23.5	16.8	14.7	20.1
Min	5.9	6.4	10.4	9.3	12.0	13.8	13.8	15.3	13.9	13.0	9.0	7.1	10.8
Mean	9.9	10.2	14.3	13.9	17.1	18.9	19.3	21.0	19.3	18.3	12.9	10.9	15.5
Neutrality	20.9	21.0	22.2	22.1	23.1	23.7	23.8	24.3	23.8	23.5	21.8	21.2	22.6
Upper limit	23.4	23.5	24.7	24.6	25.6	26.2	26.3	26.8	26.3	26.0	24.3	23.7	25.1
Lower limit	18.4	18.5	19.7	19.6	20.6	21.2	21.3	21.8	21.3	21.0	19.3	18.7	20.1
Rain (mm)[b]	75.9	84.3	51.8	26.9	9.9	2.3	0	0	5.8	19.8	47.8	53.8	378.3
Prec (mm)[c]	299	332	204	106	39	9	0	0	23	78	188	212	1490
Wind (m/s)	2.4	2.8	2.7	3.5	3.6	4.1	3.6	3.0	2.8	2.5	2.5	2.8	3.0
HDD	270	227	119	136	53	15	11	2	16	40	173	234	1296
CDD	0	0	0	0	7	25	26	59	22	19	0	0	158

(Continued)

TABLE 6.3 (CONTINUED)
Climatic Data, San Jose, California

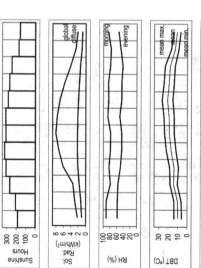

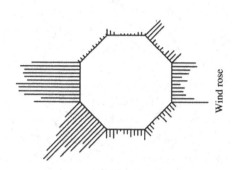

Wind rose

| | Average diurnal range (°K) | 9.3 |
| | Annual mean range (°K) | 20.7 |

Recommended 'design' conditions[d]

Summer		
	DBT (°C)	31.4
	MCWB (°C)	19.0
	WBT (°C)	20.1
	MCDB (°C)	29.1
Winter		
	DBT (°C)	3.2

Sources of data: https://energyplus.net/weather, [a]WMO (2010), [b]http://worldweather.wmo.int/en/city.html?cityId=852, [c]NOAA (2017), [d]ASHRAE(2009), Sunshine hours for San Francisco, California (48.4 miles)

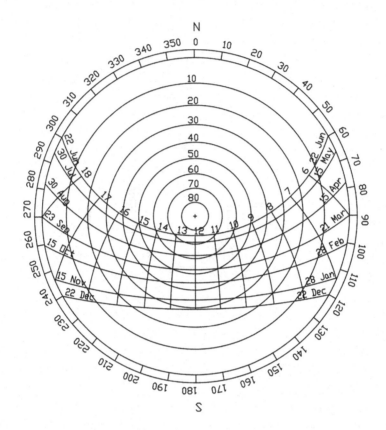

FIGURE 6.19 A stereographic sun-path diagram for San Jose, latitude N 37° 22′.

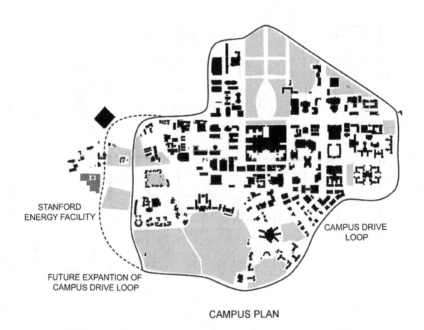

CAMPUS PLAN

FIGURE 6.20 Campus layout and location of the Stanford Central Energy Facility. Image courtesy ZGF Architects

FIGURE 6.21 Central courtyard with a grand staircase serving as entrance and theater seating. © Tim Griffith

The cooling and heating plant (CHP) is physically separated from the rest of the energy complex to meet the requirements of OSHPD (California's Office of Statewide Health Planning and Development) since it serves the energy loads of School of Medicine and the Lucille Packard Children's Hospital. The plant houses three boilers with room for future expansion of two boilers, six chillers, and a main electrical room. The glazed east wall of the boiler area also serves as a visual connection to the entry court and the office building, allowing the equipment to be showcased as part of the overall storytelling of innovative energy technology used throughout the complex.

The centerpiece of the facility is a 9.46 million liters (2.5-million-gallons) hot water tank, nearly 21.34 m (70 ft) in diameter and protected by a 19.51 m (64 ft) tall screen whose stainless steel and aluminum-perforated panels shimmer in the sun. The two large, cold water thermal storage tanks are part of the massing of the plant and are lowered 7.62 m (25 ft) below grade and visually shielded by a perforated box-ribbed aluminum screen wall that extends beyond the height of the plant itself to reduce their visual impact.

The substation brings electricity from the Stanford Solar Generating Station (SSGS), developed under a partnership between Stanford and Sun Power, as well as other grid generating sources to power the CEF and other campus buildings.

The site landscaping utilizes native plants to lower water demand, porous gravel to assist in recharging the groundwater table and bioswales with bio-retention basins to collect stormwater runoff. Drip irrigation serves native and low-water landscaping.

Figure 6.22 shows the plans of the entire facility, while Figure 6.23 shows floor plans of the office building. Figure 6.24 presents the sections of the facility.

6.3.3 DAYLIGHT AND THERMAL DESIGN

The warm marine climate allows floor-to-ceiling glass curtain wall with operable windows on both sides of narrow 8.23 m (27 ft) floor plate of the office building. It maximizes daylight, natural ventilation, and a clear view of the surrounding landscape (Figure 6.25). Sun angles and shading mask

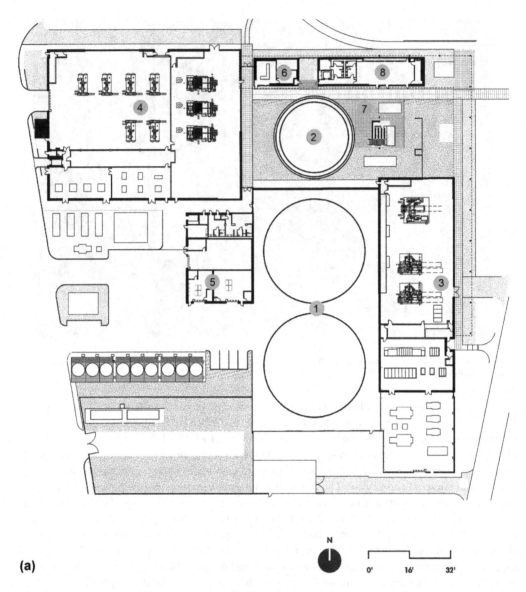

(a)

N

0'   16'   32'

FIGURE 6.22 Floor plans, Image courtesy ZGF Architects. (a) Ground floor plan; (b) first floor plan; Legend. 1. Chilled water storage tanks; 2. hot water storage tank; 3. heat recovery chillers; 4. OSHPD plant; 5. workshops; 6. control room; 7. entry courtyard; 8. conference room; 9. balcony; 10. office

analysis determined the optimal size and spacing of the opaque shading devices to prevent excess solar gain and prevent glare. Even the more industrial areas housing chiller and boilers have generous daylighting with a series of translucent skylights and ground-level curtain walls designed to provide ambient light during daytime hours for worker circulation (Figure 6.26). Hundred percent floor area achieves adequate daylight levels without the use of electric lighting. User-controlled operable windows allow for natural ventilation when outside conditions warrant in this year-round warm marine climate.

The building envelope is heavily insulated (minimum R-34 in the walls and R-30 in the roof) and detailed to provide at least 76.2 mm (3 inches) of continuous insulation on the exterior of the metal studs, metal roof deck, and elevated floor deck (Figure 6.27). This layer of continuous insulation prevents thermal bridging, a common issue with conventional building envelopes.

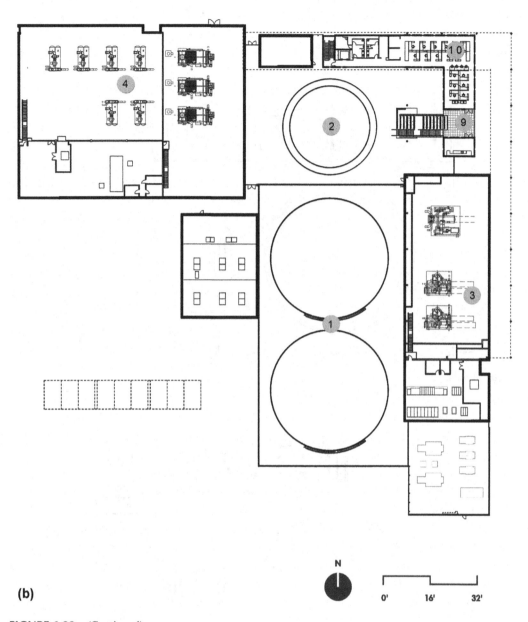

(b)

N

0'   16'   32'

FIGURE 6.22 (Continued)

6.3.4 ENERGY SYSTEMS

The office building utilizes an energy-efficient radiant floor system throughout for radiant heating and cooling, with additional climate control provided by ancillary systems since the response time of the radiant floor system is too long to provide comfort conditions quickly enough. The radiant floor system, in this case, is a 'single-pipe' system embedded in a concrete slab, which requires a 'dead band' in the operation between the heating and cooling modes. This generally occurs, for example, when foggy mornings give way to warm and sunny afternoons, an occasional event in this microclimate.

The first ancillary 'system,' used in the conference room on the first floor, is ceiling-mounted *chilled sails*, to respond to the sudden increase in cooling load caused by a large increase in the number of occupants. Installed directly below the ceiling, the folded metal chilled sails effectively

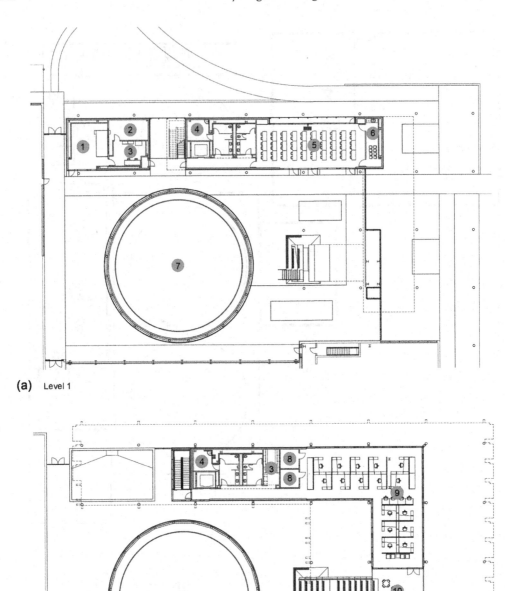

(a) Level 1

(b) Level 2

FIGURE 6.23 Level plans of entry courtyard and office block, Image courtesy ZGF. (a) Level 1; (b) Level 2. Legend. 1. Control room; 2. server; 3. workroom; 4. janitor closet; 5. flex space; 6. storage 7. hot water tank; 8. conference room; 9. open office; 10. terrace; 11. staff lounge

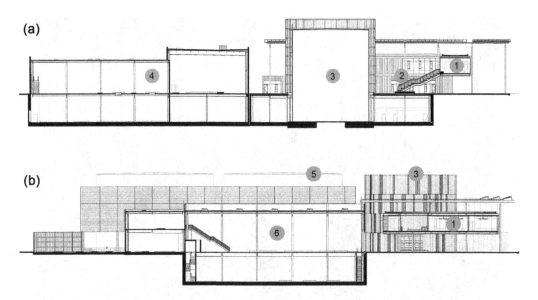

FIGURE 6.24 Sections, Image courtesy ZGF. (a) Longitudinal section; (b) transverse section. Legend 1. Office; 2. entry courtyard; 3. hot water storage tank; 4. OSHPD plant; 5. chilled water storage tanks; 6. heat recovery chillers.

FIGURE 6.25 Curtain wall maximizing natural light for interior spaces. © Tim Griffith

provide an instant response by utilizing very effective convective heat transfer with the room air. Like chilled beams, chilled sails use chilled water and passive air movement to provide the desired thermal comfort conditions in the space. On peak cooling days, the chilled sail system is backed up by a fan-coil unit, which is used only under those more extreme conditions.

The second ancillary 'system,' used throughout the office building, is a blanket-type bio phase change material (PCM) as a passive cooling technique, acting as thermal mass to absorb heat at

FIGURE 6.26 Glass facade serving as the view to outdoor spaces and light to indoor spaces. © Tim Griffith

a constant temperature rather than causing an increase in room air temperature. These PCMs are located above and on the suspended ceiling, where they can interact with the room environment. These materials absorb heat from the space above a certain temperature (called the 'Q-value') and gradually become liquified. To perform the same function on the following day, the material must be exposed to a lower temperature and returned to a solid state during the night. This is accomplished by using a small exhaust fan at night in the plenum space only, where the PCM is located, to chill down that space so that the PCM is fully re-charged for the next day.

When daylighting is insufficient, LED lighting is used throughout and controlled through occupancy sensors and daylight dimming controls. The facility design exceeds state energy guidelines.

Initially, the large cross-laminated wood canopy structure placed above the office building was meant for shading purposes; later the solar PV system totaling 175 kW was installed on the canopy and started to produce power that could be utilized by the building. The system produced almost 290 MWh of energy over the course of the year from 5/2017 to 4/2018, which is almost double the actual annual energy use of 151.6 MWh measured for the building during the period from 5/2016 to 4/2017. The building EUI of 15.86 kWh/m^2 (54.1 kBtu/ft^2 per year) includes not only the office building but also the intensive process loads of the central plant control room and server room that are associated with larger campus operations. The building easily achieved ZNE (Zero-Net-Energy), and the over-production of energy is put into the general campus power grid.

A patented technology developed by Stanford continuously monitors the plant's equipment and predicts campus energy loads, grid prices, and weather, steering the system to optimal efficiency. The automated software also reviews its performance.

6.3.5 SUSTAINABLE THINKING

An industrial facility, the CEF uses local, durable, and sustainable materials to ensure a long service life under demanding conditions while taking cues from Stanford's rich collection of historical and contemporary buildings. The materials selection maximizes recycled contents and minimizes additional finish materials layers or trim.

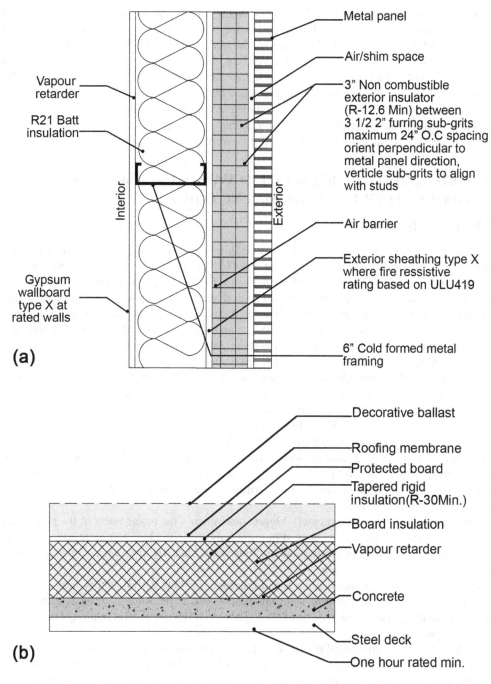

FIGURE 6.27 Building envelope, Image courtesy ZGF. (a) Wall section (metal panel on metal studs); (b) roof section (membrane over the insulation on concrete deck).

Warm, board-formed concrete contains local aggregate and accentuates mass walls within the public view and relates to the sandstone on campus. Corten steel panels have 85% recycled content, and natural patina references the red tile terra cotta roofs. Structural steel elements exceed 80% recycled content and, along with aggregates, were locally sourced. Exposed concrete floors reduced the need for a carpet and support radiant thermal cooling and heating. Cross-laminated structural panels provide the decking under the PV array, creating a visible reference to the wood ceilings of arcades on campus. Other wood used in accents throughout the facility is reclaimed from a hangar at the nearby Moffett Federal Airfield, which had housed dirigibles. FSC certified reclaimed wood accents bench seating and a café wall. Low-VOC and low-odor paints advance a healthy indoor environment quality, and carpeting minimizes indoor air contaminants.

The Central Energy Facility building features low-flow fixtures to minimize potable water use. The project is designed to connect to a future municipal system that will provide non-potable water for process cooling and irrigation of native landscaping – the largest water demands.

Table 6.4 collate the design profile of Stanford University Central Energy Facility.

6.4 EDITH GREEN–WENDELL WYATT (EGWW) FEDERAL BUILDING, PORTLAND (ZONE 4C MIXED MARINE, COTE 2016)

6.4.1 DESIGN INTENTIONS

The General Services Administration's (GSA) Edith Green–Wendell Wyatt (EGWW) Federal Building Project modernized an existing 18-story, 47,610.4 m^2 (512,474 ft^2) office tower in downtown Portland, Oregon, which was originally built in 1974. The project intentions included upgrading building's outdated and worn out MEP (mechanical, electrical and plumbing) and other systems, updating work environments and improving accessibility, while also meeting the stringent energy and water conservation requirements of the EISA, complying with federal standards for blast resistance, and providing new code compliant egress stairs, entries, and restrooms. The building received funding under the American Recovery and Reinvestment Act (ARRA), which entailed goals of energy savings of 30% over ASHRAE/IESNA Standard 90.1-2007 and a 55% reduction in fossil-fuel energy over 2003 Commercial Buildings Energy Consumption Survey Baseline (CBECS).

The building is named after Congresswoman Edith Green, who represented Oregon's Third Congressional District from 1955 to 1974, and Congressman Wendell Wyatt, who represented Oregon's First Congressional District from 1964 to 1975.

The design by SERA and Cutler Anderson Architects completely transformed the existing office building from an aging, uncomfortable energy hog into highly efficient Leadership in Energy and Environment (LEED) Platinum-certified modern office environment for 16 federal agencies and was completed within 39 months.

Implementation of one of the project's biggest innovation – the replacement of the precast concrete cladding with a high-performance skin – not only made the building more energy-efficient but also allowed for a radiant heating and cooling system that freed up approximately 2880 m^2 (31,000 ft^2) extra net rentable space.

Besides, this building modernization sets a precedent for how a mid-century high rise can be adapted and reused to new functions and requirements and at the same time meet stringent sustainability standards. It has become one of the highest performance federal buildings in the GSA's portfolio (Figure 6.28).

The AIA COTE recommended the EGWW for the Top Ten Plus award in 2016 for exceptional post-occupancy performance data, as the building achieved 55% energy use reduction and 65% water use reduction with increased occupant satisfaction; the jury comments on the project:

This project transforms a generic concrete office building into a high-performance, environmentally responsive, comfortable place to work. There are a lot of existing, low-performance buildings out there that don't contribute much to the urban fabric. In terms of impact, these are the buildings we need to address. This sets a great precedent for re-use and upgrade, and demonstrates the potential for creative, green reuse projects.

(AIA 2016a)

TABLE 6.4
Design Profile Stanford University Central Energy Facility, Stanford

	Location	Stanford, California
Building profile	Principal use	Infrastructure
	Number of residents, occupancy	42
	Site area	50,076.5 m² (534,019 ft²)
	Built-up area	11,668.6 m² (125,600 ft²)
		Office 929.03 m² (10,000 ft²)
		Gross-conditioned floor area: 885.37 m² (9530 ft²)
		Gross-unconditioned floor area: 10,784.56 m² (116,084 ft²)
	Distinctions/awards	AIA San Francisco Design Awards (2016)
		Institute Honor Awards for Architecture (2017)
		AIA/COTE Top Ten Green Projects Award (2017)
	Cost of construction excluding furnishing	$175,000,000
	Cost per square foot	$1393
	Total project cost	$485,000,000
	Substantial completion/occupancy	March 2015
Building team	Owner/representative	Stanford University
	Architect	ZGF Architects LLP, Portland
	General contractor	The Whiting Turner Contracting Company
	Mechanical/electrical/plumbing Engineer	Affiliated Engineers Incorporation (AEI), San Francisco
	Structural engineer	Rutherford + Chekene, San Francisco
	Civil engineer	BKF Engineers
	Landscape architect	Tom Leader Studio, San Francisco
	Lighting design	Pivotal lighting design
	Acoustical design	Colin Gordon Associates, Inc.
	Commissioning agent	Affiliated Engineers Incorporation
Solar design profile	Latitude	N 37° 22′
	Longitude	W 121° 55'
	Altitude	16 m
	Heating degree days (base ~ 18.3 °C-65°F)	2410
	Cooling degree days (base ~ 18.3 °C-65°F)	171
	Annual hours operation	8760
	Orientation	Along east-west axis
Performance profile	Annual energy use intensity (EUI) (site) consumption	12.02 kWh/m² (38.1 kBtu/ft²)
	Electricity (on-site solar or wind installation)	32.5 kWh/m² (102.9 kbtu/ft²)
	Savings vs. Standard 90.1-2007 Design Building	100%
	Energy star rating	100
	Percentage of power represented by renewable energy certificates	100%
Water efficiency	Annual water use	6114.95 kl (1615,400 gallons ~ lower the 18% of annual consumption)
Key sustainability features	Water conservation	Dual setting aerator, native plants, drip irrigation, porous gravel, and more efficient HRC
	Recycled materials	Reclaimed wood soffits

(Continued)

TABLE 6.4 (CONTINUED)
Design Profile Stanford University Central Energy Facility, Stanford

	Daylighting	Natural daylighting, daylight sensors, networked lighting controls
	Individual controls	Operable windows, ceiling fan ventilation, occupant accessible lighting control
	Carbon reduction strategies	Use of renewable energy (65%)
	Transportation mitigation strategies	Shuttle, regional rail, bike with shower facility
	Other major sustainable features	Phase change materials in the ceiling, radiant heating and cooling, night-flush passive cooling, passive chilled sail
Building envelope	Roof	Metal roof deck, Overall R-value: 5.287 m² K/W (30°F ft² h/Btu)
	Walls	Metal panels, metal/glass curtain wall, precast concrete overall R-value: 5.992 m² K/W (34°F ft² h/Btu)
	Windows	Curtain wall overall R-value: 5.992 m² K/W (34°F ft² h/Btu)

www.districtenergy.org/HigherLogic/System/DownloadDocumentFile.ashx?DocumentFileKey=61a8df0d-ff3a-03c1-17 55-028616438442&forceDialog=1

FIGURE 6.28   Massive concrete structure modified into a lightweight and energy-efficient EGWW building. ©Nic Lehoux

6.4.2   Climate and Site

The mixed marine climate (zone 4C) of Portland is characterized by a narrow annual range of temperature, dampness in the atmosphere nearly throughout the year, and high rainfall. The average annual rainfall is 914.7 mm in Portland. August is the hottest month when the mean daily maximum temperature is 27.2°C, and the mean daily minimum temperature is 13.4°C at Portland. January is the coldest month when the mean daily maximum temperature is 7.0°C, and the mean daily minimum temperature is 4.1°C at Portland. Throughout the year the morning maximum relative humidity is above 77%, and the afternoon minimum relative humidity ranges between 34% and 68%.

Throughout the year cloud cover ranges between 23% and 81%. During the period July to September, clear or lightly clouded skies prevail generally. Sunshine hours of the year are 2340.9 h. Wind speed ranges between 1.9 and 7.2 m/s, and the north is the prevailing wind direction along with the west for a few months. The global solar radiation goes as high as 6.2 kWh/m². The climatic data of Portland are given in Table 6.5, and the sun-path diagram is in Figure 6.29.

The existing site is within the city's central business district and adjacent to the city hall and the federal courthouse. It's also less than two blocks from the downtown bus mall and within walking distance of four light rail lines (Figure 6.30).

Significant efforts like high-performance green building workshops followed by virtual modeling in BIM (Building information modeling), intensive energy modeling technique, and physically testing in the lighting lab were involved to analyze which energy and Indoor Environment Quality (IEQ) measures delivered the best value. After completing the analyses, the data and constraints were translated into a synthesized aesthetic expression, whose focus was to communicate sustainability on an emotional and physical level, both inside and out; Figure 6.31 is the entry-level plan, and Figure 6.32 is the tenant-level plan.

Two big changes to the area surrounding EGWW occurred during construction that altered its microclimate and required adaptive responses by the building team: the removal of a building to the south, and the addition of a building to the east. With the removal of a building to the south, the zone immediately adjacent to the window became warmer.

6.4.3 Daylight and Thermal Design

The passive and energy-efficient design of the building entailed transforming the existing, uninsulated facade into a high-performance curtain wall with elevation-specific shading devices (Figure 6.33). To evaluate the optimum combination of shading and daylighting, a parametric analysis evaluated peak cooling loads for each orientation to find shading requirements. Three glazing percentages (40%, 50%, and 57%) with and without shading were modeled for a typical space. This parametric analysis led to the following high-performance design requirements: 40–42% vision glazing on the tower, maximizing glazing where shading minimizes solar gain. After determining which facades needed shading (southwest, southeast, and northwest) and which did not (northeast), the next step was to determine the percentage of time each facade would need to be shaded. The depth and spacing of the horizontal and vertical shading devices were varied by the designers to arrive at both the desired performance metrics and the architectural expression. Thus, every face of the building presents a different set of aluminum reflecting and shading elements that were modeled to respond to year-round variations of the sun. Horizontal panels reflect light and provide shade on the northwest and southeast, while the southwest-facing full height vertical 'reeds' provide relief from the low-angle sun – all without obstructing views (Figure 6.34).

One design strategy that has proven to benefit building occupants is the use of light shelves and reflectors below the window sill to maximize daylight penetration (Figure 6.35). In addition to energy savings, the effects of the daylighting helped with circadian entrainment.

Additionally, the building program was tuned to optimize performance with an ideal plan, optimized at 70% open offices/ 30% closed offices, provided for each agency. Agencies with greater requirements for enclosed space are located lower in the building where surrounding buildings provide additional shading benefits. Daylighting at levels that allow lights to be off during daylight hours is 51%.

Double-height public space at the base of the building brings daylight from the upper, plaza level to the lower, ground level (Figure 6.36). A large canopy on the top of the building provides additional shading for the taller 18th floor, as well as supporting optimally angled photovoltaics and providing a water collection area.

TABLE 6.5
Climatic Data, Portland, Oregon

| | Latitude N 45° 32' | | | Longitude W 122° 24' | | | Altitude 11 m | | |
| Climate | Mixed marine | | | ASHRAE 4C | | | Köppen Csb | | |
Months	Jan	Feb	Mar	Apr	May	Jun	Jul	Aug	Sep	Oct	Nov	Dec	Year
Sunshine h[a]	85.6	116.4	191.1	221.1	276.1	290.2	331.9	298.1	235.7	151.7	79.3	63.7	2340.9
Cloud (%)	71.9	58.6	78.0	67.1	71.7	55.8	34.8	33.5	23.9	63.5	81.0	73.3	59.4
Solar irradiation daily average (Wh/m^2)													
Global	1237	2084	2736	4117	4958	5960	6213	5651	4626	2350	1269	969	3514
Diffuse	737	1047	1714	2106	2714	2769	2439	2199	1698	1226	835	659	1679
Relative humidity (%)													
Morning	80	77	80	83	85	84	87	85	80	83	93	81	83.2
Evening	68	58	58	51	50	49	42	37	34	57	77	73	54.5
Dry-bulb temperature (°C)													
Max	7.0	10.6	10.4	15.7	17.6	23.2	26.6	27.2	25.4	16.9	10.0	7.2	16.5
Min	4.1	5.2	4.5	7.2	8.7	12.8	14.2	13.4	10.8	10.2	6.0	5.2	8.5
Mean	5.6	7.9	7.5	11.5	13.2	18.0	20.4	20.3	18.1	13.6	8.0	6.2	12.5
Neutrality	20.9	20.9	20.9	20.9	21.9	23.4	24.1	24.1	23.4	22.0	20.9	20.9	21.7
Upper limit	23.4	23.4	23.4	23.4	24.4	25.9	26.6	26.6	25.9	24.5	23.4	23.4	24.2
Lower limit	18.4	18.4	18.4	18.4	19.4	20.9	21.6	21.6	20.9	19.5	18.4	18.4	19.2
Rain (mm)[b]	124.0	93.0	93.5	69.3	62.7	43.2	16.5	17.0	37.3	76.2	143.0	139.4	915.1
Prec (mm)[c]	488	366	368	273	247	170	65	67	147	300	563	549	3603
Wind (m/s)	7.2	4.4	3.8	2.5	2.0	2.2	2.0	1.9	2.6	3.3	2.4	6.2	3.4
HDD	396	298	336	206	152	41	5	8	49	161	317	373	2342
CDD	0	0	0	0	1	36	71	67	30	0	0	0	205

(Continued)

TABLE 6.5 (CONTINUED)
Climatic Data, Portland, Oregon

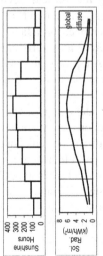
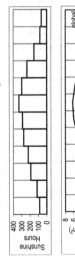
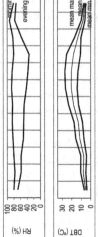
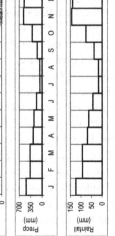

| | | Average diurnal range (°K) | 8.0 |
| | | Annual mean range (°K) | 23.1 |

Wind rose

Summer	DBT (°C)	30.6
	MCWB (°C)	19.2
	WBT (°C)	19.9
	MCDB (°C)	29.2
Winter	DBT (°C)	−1.9

Sources of data: https://energyplus.net/weather, [a]WMO (2010), [b]http://worldweather.wmo.int/en/city.html?cityId=810, [c]NOAA (2017), [d]ASHRAE(2009)

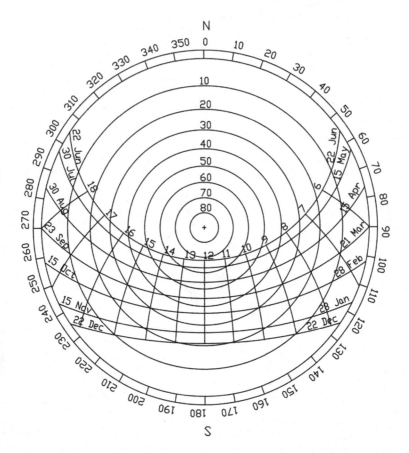

FIGURE 6.29   A stereographic sun-path diagram for Portland, latitude N 37° 22′

In addition to providing the necessary glazing shading, the staggered vertical 'reeds' on the northwest facade also support plant growth and provide a native ecosystem around the building. Selected from the bio-region for their qualities of beauty, drought tolerance, soil adaptability, and compatibility with security guidelines, the planting scheme is conceived as a vertical landscape that provides a unique setting for the re-birth of this urban building. A tapestry of climbing vines ascends the reeds, unique to each solar exposure, connects the facade to the ground plane and surrounding landscape, and communicates GSA's green commitment.

The plants used on the building are a mixture of evergreen and deciduous vines, which creates different habitats for a variety of species. The majority of vines are deciduous which allow winter sun and provide shading in autumn with a pop of color. Because all plantings are located above a below-ground parking garage, the depth of plantings to be considered was limited. A lightweight soil matrix supports vigorous growth and allows drainage to the structure below. The soil is 508 mm (20 inches) deep and consists of a blend of organics, pumice, and local sandy topsoil. Plants were selected, sourced, container-grown, and developed for an additional season to target a 3.66–4.57 m (12–15 ft) height at planting. Vines are spaced 914.4 mm (3 ft) apart on both sides of the screen, and the species are varied with each exposure and building character and provide fragrance at the entrance and exit.

6.4.4   Energy Systems

The EGWW is predicted to achieve a 60% reduction in energy use compared to the existing building and a 45% reduction in energy use intensity (EUI), exceeding the EISA performance goals which are in alignment with the AIA's 2030 Commitment.

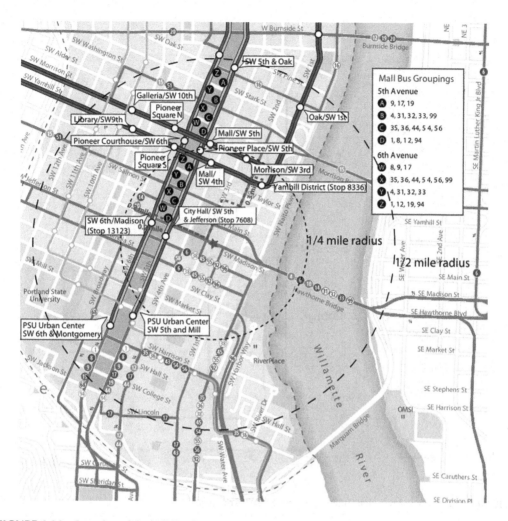

FIGURE 6.30   Location of the building in the city's central business district.

These savings result directly from an integrated design process (IDP) that prioritized thermal comfort for the occupants and energy performance.

The hydronic radiant heating ceiling panels with DOAS was projected to a $2 million reduction in lifecycle costs compared to GSA's traditional variable air volume (VAV) systems. This cost reduction was based on an estimated annual energy savings of $228,000. In the two years since its completion, EGWW has boasted an average savings of $306,450 annually.

Although the system provides a high level of thermal comfort and indoor air quality (IAQ), it was predicted that occupants accustomed to noisily blown cool or warm air (as in traditional systems) would perceive a problem linking minimal air movement to a lack of cooling. One way this was corrected during the post-occupancy evaluation was to adjust the range of temperature allowed to improve the balance between energy savings and occupant comfort. A thermal comfort study provided a solution to warm perimeter zone by adding an active radiant panel to the perimeter soffit. This not only addressed the added thermal load at the perimeter but also provided a higher degree of thermal comfort and more uniform temperatures throughout the depth of the space.

Solar thermal panels are estimated to provide for 30% of the building's domestic hot water. A 1207.7 m^2 (13,000 ft^2) solar roof that would produce 3% of the building's electrical energy requirements annually. Modernized energy efficient elevators that generate power as they descend. Incorporation of a server farm in the basement of the building, which allowed for the collection of

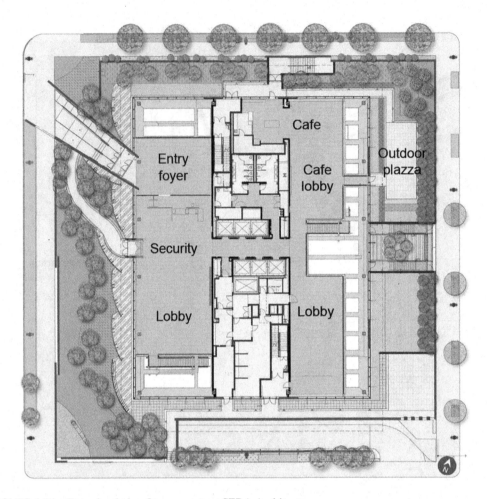

FIGURE 6.31   Entry-level plan, Image courtesy SERA Architects

waste heat for warming other areas of the building. This strategy results in additional energy savings compared to the sprinkling of servers on multiple floors.

The building also provides enhanced indoor air quality (IAQ) through the use of a 100% dedicated outdoor air system (DOAS), resulting in above-code ventilation with excellent filtration. A large portion of those savings will come from having eliminated forced-air fans.

In six months of occupancy, the team has incorporated a series of 'aftercare' measures to monitor energy use and help building operators tune the building to achieve its goals – 39% energy cost savings and 46% energy use savings compared to ASHRAE 90.1-2007

The radiant panel ceiling system included 203.2 mm (8 inches) inactive sections in addition to the standard 609.6 × 1219.2 mm (2 ft × 4 ft) active sections located in alignment with the building's exterior mullions. The system accommodates the insertion of walls on a 1524 mm (5 ft) grid and is critical for long-term flexibility and adaptability (Figure 6.37). The radiant panels were also designed with disconnects that allow for the building mechanical zones to be reconfigured. The active sections with piping attached are designed for radiant heating and cooling; while the inactive sections without piping are configured to support light fixtures, sprinklers, and other building infrastructure. The team validated the design methodology, building out a prototypical tenant space after the original ceiling installation was completed.

Energy-efficient electric lighting systems with advanced controls that intend to reduce light energy usage by 40% compared to Oregon code: high-efficiency task lighting, appliances, and A/V equipment. Lighting power density (LPD) is 6.46 W/m^2 (0.60 W/ft^2).

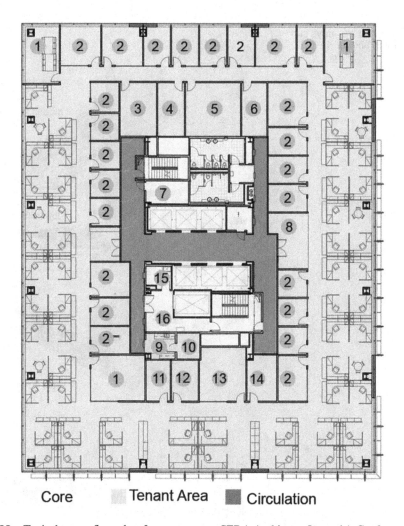

FIGURE 6.32   Typical tenant floor plan, Image courtesy SERA Architects. Legend 1. Conference; 2. office; 3. storage; 4. copy/work; 5. library; 6. store; 7. toilet; 8. reception; 9. break; 10. tele; 11. IT; 12. kitchen; 13. mail; 14. cust; 15. elevated vestibule.

6.4.5   SUSTAINABLE THINKING

The building's 3337 tons of precast concrete cladding was replaced with a relatively light glass and aluminum curtain wall system. This modification lightened the structure and simplified the seismic retrofit of the building to meet today's more stringent earthquake code without requiring a major structural upgrade – typically one of the costliest expenses in a renovation of this scale. The curtain wall also features special 'blast-resistant glass,' which GSA mandates for 21st-century buildings. To limit materials being landfilled, the project team initially focused on resource conservation and material reuse. Careful demolition eliminated over $1,000,000 in contingency, which was used to buy additional sustainable design features. Additionally, 3337 tons of precast concrete was crushed and reused as roadbed, and 3500 tons of materials and products were given new lives, including:

- Two drinking fountains and five doors were given to an inner-city church.
- Grab bars were donated to special needs individuals.
- 30 solid core doors were sent to a village in Africa.
- Mahogany strips were made into bicycle fenders by a local craftsman.

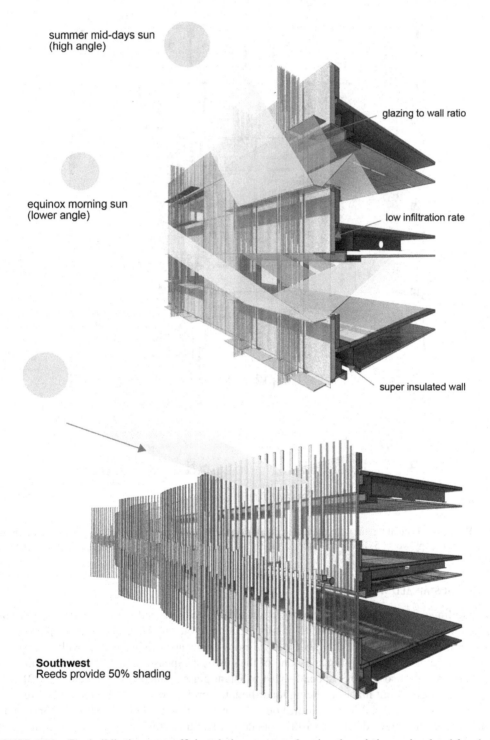

summer mid-days sun
(high angle)

equinox morning sun
(lower angle)

glazing to wall ratio

low infiltration rate

super insulated wall

Southwest
Reeds provide 50% shading

FIGURE 6.33   The building's energy-efficient design was transforming the existing, uninsulated facade to a high-performance curtain wall with elevation-specific shading devices. Image courtesy SERA Architects

SOUTH-WEST **NORTH-WEST** **NORTH-EAST** **SOUTH-EAST**

FIGURE 6.34 Building elevations. Image courtesy SERA Architects

In addition to 'Buy-American' requirements and durability, the project team focused on selecting regional materials (11.9%) with high recycled content (29.8%). Indoor air quality (IAQ) was also considered through the selection of low-emitting adhesives, floor systems, and composite wood and agri-fiber products.

To further reduce the building's environmental impacts, the property management team has incorporated the GSA's green leasing and operations policy and has created a program to educate tenants (present and future) about the building's green features and the impacts they have.

The project's detailed water usage modeling predicted greater than 60% water savings to be achieved through a combined strategy of incorporating water-conserving plumbing fixtures and a rainwater catchment and reuse system.

The EGWW water conservation strategy started with an analysis of the existing building's historical water usage (Figure 6.38). This analysis showed that 87% of the building's water usage is for domestic uses and 13% is used for irrigation of surrounding vegetation. Because of this large interior use, the primary strategy focused on rainwater reuse for non-potable flush fixtures. A water model predicted that the building would harvest approximately 2,044,122.36 liters (540,000 gallons) of water annually – a 60% reduction in water use compared to a typical building. However, in the first two years of use, the actual rainwater collected annually was 2,371,727.04 liters (626,544 gallons), or a 65% reduction in water use. Landscape irrigation water usage is reduced by over 50% through the use of drought resistance landscaping and the incorporation of subsurface irrigation. A 750,104.85 liters (165,000 gallons) storage tank, created by repurposing an old firearm target range in the basement, allows rainwater to be stored and reused for toilet flushing, irrigation, and mechanical cooling tower makeup water. Over the course of a year, an impressive 231,727 liters (626,544 gallons) of rainwater has been collected and reused, the equivalent of seven competition-size high school swimming pools. During winter months, rainwater accounts for up to 85% of the total building water use. The tank also supports another project goal: mitigating the negative effects of urban runoff.

The gas and electric utility bills demonstrate a 45% energy savings in the first two years of use, compared to a building built to code. Besides, the water collecting 'canopy' supports a 180-kW solar PV array that provides 4% of the building's total energy.

A tenant-led waste diversion program was carried out for employees. For the fiscal year 2015, EGWW had an impressive 51.33% waste diversion rate. In addition to standard recycling of paper, plastic, glass, etc., the building also participates in the offsite composting of landscape organics. Construction waste divergence rate of 87%.

One year after moving into EGWW, tenants indicated increased satisfaction compared to their temporary quarters in a survey for the Center for the Built Environment.

Table 6.6 presents the design profile of Edith Green–Wendell Wyatt (EGWW) Federal Building Portland.

FIGURE 6.35 Integrated light reflector for diffused sunlight. Image courtesy SERA Architects. (a) Light shelves at the still level; (b) northwest facade.

6.5 NATIONAL RENEWABLE ENERGY LABORATORY, GOLDEN, COLORADO (ZONE 5B COOL DRY, COTE 2011)

6.5.1 DESIGN INTENTIONS

The Department of Energy's (DOE) Research Support Facility (RSF) at the National Renewable Energy Laboratory (NREL) is a physical manifestation of the goal to transform innovative research in energy efficiency and renewable energy into market-viable technologies and practices. The building is meant to serve as a long-term replicable blueprint to achieve a substantial reduction in

FIGURE 6.36   Double-height public space to bring daylight from the upper plaza level. ©Nic Lehoux

building energy use and to adopt a net-zero energy approach for large-scale commercial buildings without increasing the cost.

The foundation of this blueprint is a national design competition and performance-based design-build project delivery model developed by NREL. The design-build team started the four-month competition phase with a multi-day design charrette involving all the integrated disciplines that focused on design solutions addressing the LEED Platinum requirement and the owner's desire for a net-zero-energy building – while meeting the functional, programmatic and cost requirements of the project. The project pursued three of NREL's four definitions of net-zero energy: site energy, source energy, and energy emissions (net-zero energy cost was not pursued) (Hootman et al. 2012). Designed by RNL Design, Denver, Colorado, the project was the AIA COTE top ten winner in 2011.

Integrated into an existing NREL campus on the foothills of the Rocky Mountains, the RSF provides a large-scale 20,624.5 m² (222,000 ft²) office space to more than 800 staff, who had been previously scattered in different locations. The building also houses a data center that serves the entire NREL campus. Figure 6.39 shows the view of the entrance to RSF at day time and brightly lit at night time.

6.5.2   CLIMATE AND SITE

The cool dry climate (zone 5B) of Denver is characterized by seasonal and diurnal temperature swings, generally dry throughout the year with an average humidity of about 20–45% and a moderate annual rainfall of 363.2 mm.

February is the coldest month when the mean minimum temperature is about −5.2°C and the mean maximum temperature is about 6.1°C. June is the hottest month with an average maximum temperature of about 30.5°C, and the mean minimum temperature of about 14.9°C. Climate of Denver is dry with low relative humidity round the year. September month is slightly low humid when relative humidity is around 20%, and April month is slightly highly humid when relative humidity is around 45%.

RADIANT SYSTEM ZONES

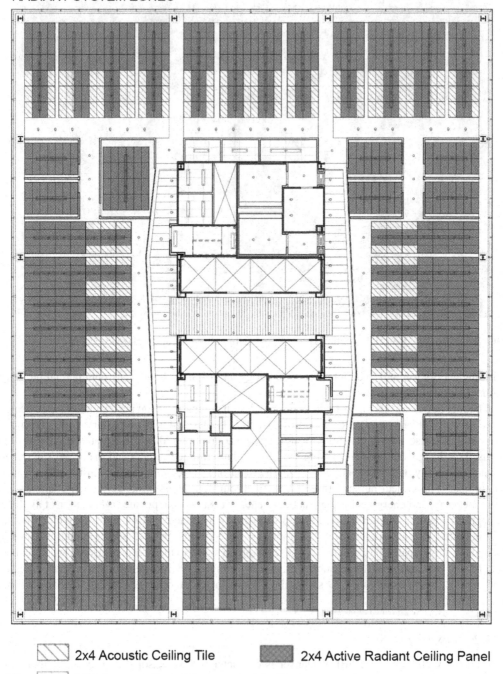

2x4 Acoustic Ceiling Tile   2x4 Active Radiant Ceiling Panel

(a)   5/8" Gypsum Board Suspended Ceiling

FIGURE 6.37   Radiant heating system. Image courtesy SERA Architects. (a) Radiant panel layout; (b) panel mechanism.

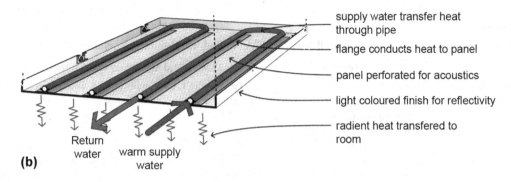

supply water transfer heat through pipe

flange conducts heat to panel

panel perforated for acoustics

light coloured finish for reflectivity

radient heat transfered to room

Return water

warm supply water

(b)

FIGURE 6.37   (Continued)

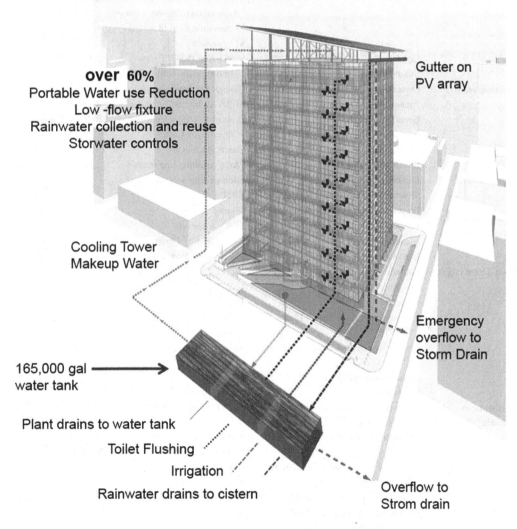

RAINWATER REUSE SYSTEM

over 60%
Portable Water use Reduction
Low -flow fixture
Rainwater collection and reuse
Storwater controls

Gutter on PV array

Cooling Tower Makeup Water

165,000 gal water tank

Plant drains to water tank

Toilet Flushing

Irrigation

Rainwater drains to cistern

Emergency overflow to Storm Drain

Overflow to Strom drain

FIGURE 6.38   Rainwater reuse mechanism. Image courtesy SERA Architects

TABLE 6.6

Design Profile Edith Green–Wendell Wyatt (EGWW) Federal Building, Portland, Oregon

	Location	1220 SW 3rd Ave, Portland, OR 97204, USA
Building profile	Principal use	Federal offices: GSA offices includes a central data center, federal courts, and office space
	Employees/occupants	1520
	Expected (design) occupancy	1530
	Percent occupied	99%
	Built-up area	47,610.39 m^2 (512,474 ft^2) conditioned space 40,779.97 m^2 (438,952 ft^2)
	Distinctions/awards	2013, LEED 2009-NC, Platinum; 2014 GSA Design Honor Award; 2014 Tall Building in America Award I Council on Tall Buildings and Urban Habitat; 2014 AIA COTE Top Ten Award; 2013 AIA Technology in Architectural Practice (TAP)/Building Information Modelling (BIM) Winner I Construction Owners Association of America Owners' Choice Award: Renovation/Retrofit Award
	Total cost	$136 million
	Cost	$2657.7/m^2 ($247/ft^2)
	Substantial completion/occupancy	2013 (originally built in 1974)
Building team	Owner/representative	General Services Administration
	Architect	SERA Architects/Cutler Andersen Architects
	General contractor	Howard S. Wright Construction, a Balfour Beatty Construction Company
	Mechanical Engineer	Stantec/McKinstry
	Electrical Engineer	PAE Consulting Engineers
	Plumbing Engineer	Interface Engineering
	Structural Engineer	KPFF
	Energy Modeler, Commissioning Agent	Glumac
	Environmental, LEED Consultant	SERA Architects
	Lighting Design	Luma Lighting Design
	Landscape Architect	Place Studio, Portland
Solar design profile	Latitude	45.5145° N
	Longitude	122.6770° W
	Altitude	11 m
	Heating degree days (base ~ 18.3 °C-65°F)	4214
	Cooling degree days (base ~ 18.3 °C-65°F)	433
	Annual hours occupied	2800
	Orientation	City block (NSEW)
Performance profile	Annual energy use intensity (EUI) (site) consumption	11.29 kWh/m^2 (35.8 kBtu/ft^2)
	Electricity (grid purchase)	7.51 kWh/m^2 (23.82 kBtu/ft^2), Natural gas 3.15 kWh/m^2 (10.74 kBtu/ft^2)
	Electricity (on-site solar installation)	0.38 kWh/m^2 (1.2 kBtu/ft^2)
	Annual energy cost index	$0.72/ft^2
	Annual source (primary) energy	27.13 kWh/ m^2 (86 kBtu/ft^2)
	Savings vs. Standard 90.1-2007 Design Building	46%
	Energy star rating	97

(Continued)

TABLE 6.6 (CONTINUED)

Design Profile Edith Green–Wendell Wyatt (EGWW) Federal Building, Portland, Oregon

Water efficiency	Annual water use	4.85 million liters (1.28 million gallons) (61% reduction in potable water use)
		Annual water use (measured): 10,014,522.64 liters (2,645,557 gallons)
Key sustainability features	Water conservation	A 624,592.9 liters (165,000 gallons) rainwater cistern (a former rifle range) holds water funneled from the roof. Predicted water savings of 61% compared to code. Stormwater mitigation strategies.
	Recycled materials	Reduction of 92% in paper used for architectural contract documents. Construction waste divergence rate of 87%. Nearly 100% of the structural elements saved.
	Daylighting	Light shelves, reflectors, daylight sensors (40% reduction)
	Transportation mitigation strategies	Two blocks from major public transit line
	Other major sustainable features	Rooftop photovoltaic panels (184), a 100% dedicated outdoor air system, heat recovery ventilator, hydronic radiant heating/cooling ceilings, demand–control ventilation
Building envelope	Roof	Built-up 152.4 mm (6 inch) concrete with continuous insulation, overall R 7.05 m^2 K/W (40°F ft^2 h/Btu), high reflectivity
	Walls	Insulated spandrel with continuous insulation overall R 4.4 m^2 K/W (25°F ft^2 h/Btu), glazing percentage 41%
	Basement/foundation	Slab edge insulation R 3.52 m^2 K/W (20°F ft^2 h/Btu) Basement wall insulation R 2.5 m^2 K/W (14.2°F ft^2 h/Btu)
	Windows	Effective U-factor for assembly 2.21 W/m^2 K (0.39 Btu/h ft^2°F), solar heat gain coefficient (SHGC) 0.27, visual light transmittance VLT (51%)

Source: Foreman and Lowen (2015)

Throughout the year cloud cover ranges between 38% and 56%. Sunshine hours of the year are 3106.6 h. Wind speed ranges between 3 and 5.3 m/s, and the south is the prevailing wind direction along with the north and south-west for a few months. The global solar radiation goes as high as 7.2 kWh/m^2. The climatic data of the nearest city Denver are given in Table 7.16, and the sun-path diagram is in Figure 7.15.

The campus is well connected to Denver city by shuttles and city buses, while on-site car parking is minimized and bike parking is maximized to encourage the employees to use alternate modes of transportation.

The campus planning protects the local herd of deer and existing wildlife as well as respond to the surrounding community life. As desired by the local community the building height is kept lower than zoning codes to maintain view corridor and the campus is planned to maintain trail access. Part of the project included the dedication of land area on the campus's South Table Mountain mesa as permanent open space to extend an existing conservation easement.

The sustainable site design is achieved by natural stormwater management techniques, open space preservation, permeable paving systems, native landscape integration, use of high-albedo (reflectance) pavements, and the innovative use of on-site excavated rock for gabion walls. The RSF has been selected as one among 150 projects to take part in an international pilot project program to evaluate the new Sustainable Sites Initiative (SITES) rating system for sustainable landscapes.

FIGURE 6.39   RSF, NREL, Golden, Colorado, a net-zero-energy building embedded with the concepts of bioclimatic design. (a) Daytime view © C. Kabre; (b) nighttime view, NREL, public domain.

The stormwater management strategies include a series of rain gardens for roof drainage collection, porous paving within the courtyards, and bioswales to connect water collection points to the existing natural arroyo system of the campus consistent with its pre-developed hydrology.

The building is a simple architectural response to the climate, site, and ecosystem that envelops it and the desire for a flexible, high-performance workplace within. The building evolved in the form of lazy 'H' shape with its long and narrow office wings oriented along the east/west axis that is 15° from being parallel to each other and connected by a central connector space consisting of the lobby and conference facilities. This shape also provided two exterior courtyards, which have become popular amenities. Figure 6.40 is an aerial view of the RSF, and Figure 6.41 shows the site plan.

The open office plan resulted in a higher-density workplace, reducing the building footprint per person. Figure 6.42 presents the ground floor and first floor plans, and Figure 6.43 shows the second floor and third floor plans.

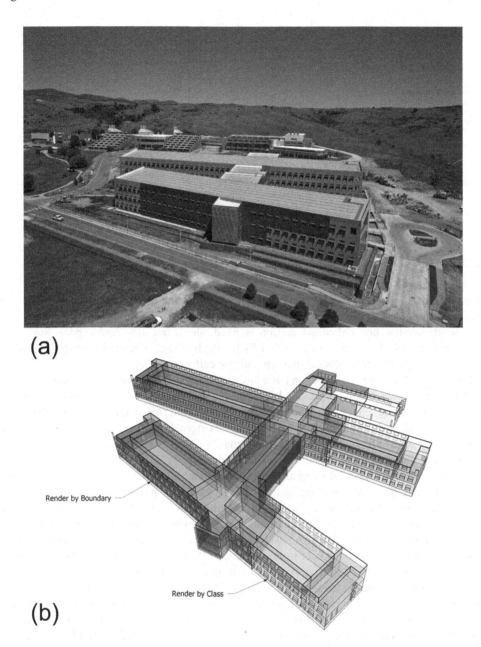

(a)

(b)

FIGURE 6.40 a/b Climate optimized 'H' form building with long and narrow office wings connected by a central spine, NREL, public domain.

6.5.3 Daylight and Thermal Design

The primary building section addresses design strategies to harness passive energy such as daylighting, natural cross-ventilation, and solar heat gain and glare control (Figure 6.44). The resulting section is 18.29 m (60 ft) wide for the two office wings. Building orientation and geometry minimizes east and west windows, while north and south windows are optimally sized with average window-to-wall ratios (WWR) of 27%, still harvesting daylight in 92% of all regularly occupied spaces.

The array of primary windows across the south facade is the pinnacle of the climate-responsive architecture (Figure 6.45). These south windows have two distinct divisions, the lower one is view

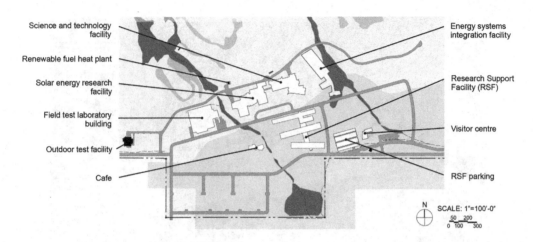

Science and technology facility

Renewable fuel heat plant

Solar energy research facility

Field test laboratory building

Outdoor test facility

Cafe

Energy systems integration facility

Research Support Facility (RSF)

Visitor centre

RSF parking

N   SCALE: 1"=100'-0"

FIGURE 6.41   Site plan, NREL, public domain.

section and the upper one is daylight section (Figure 6.46). Figure 6.47 explains the concept of daylighting. Figure 6.48 is a view of the daylit interior.

The south window upper daylight section is fixed, shade-free has double pane low-E glazing with high visible light transmittance (VLT) for daylighting. A reflective daylighting device (LightLouver™ system), redirects sunlight up onto the ceiling deep into space.

The south window lower view section has a low U-value, solar heat gain coefficient (SHGC), and VLT. The window view section is provided with the external three-sided shade to control solar heat gain in the summer and reduce glare most of the year, so no internal shading is needed. This lower view section's triple and Low-E glazing with individual external sunshades contributes to improved thermal performance. Thermo-chromic glass or smart window is introduced in the western wall, which becomes tinted on the application of a small voltage.

Thermal comfort is addressed using an integrated system of cross-ventilation and thermal mass with night flushing for passive cooling and thermal labyrinth for passive space heating. The view portion of the window has an operable part, with two-thirds of the operable part being manual and one-third automated. A small icon appears on occupants' computers informing them when conditions are favorable for opening the windows, particularly in spring and fall. The geometry of the building was studied in a wind tunnel to fine-tune the effectiveness of the operable windows and cross-ventilation, as well as pedestrian comfort in the exterior courtyards.

During the summer diurnal temperature range is high (15.6 K), which is the most favorable condition for passive cooling by the integration of thermal mass and night-flush cooling. The building's interior has a significant amount of exposed thermal mass, which can absorb much of the internal summer heat gains. The automated operable low-level view windows on the south facade and high-level automated windows on the north facade are opened to allow the thermal mass to flush and replace with the cool night air.

The RSF building includes a large crawl space under the two main office wings. Dark-colored perforated sheet metal attached to south facade functions as a solar collector; the heated air behind the dark corrugated metal cladding is drawn by fans located in the building's crawl space, a thermal mass labyrinth. The crawl space functions as a thermal battery, storing thermal energy and allowing the ventilation air for the building to be passively preheated during the heating season. The labyrinth also serves as a heat sink for reject heat from the data center, dramatically lowering the cooling load of the data center year-round.

RSF adopted modular construction with the two primary exterior wall assemblies including a precast concrete assembly and a steel stud assembly. The precast concrete walls include continuous rigid insulation and use a low-conductivity connector between the interior and exterior concrete layers. The steel stud wall uses stamped openings within the web, which reduces the thermal bridging.

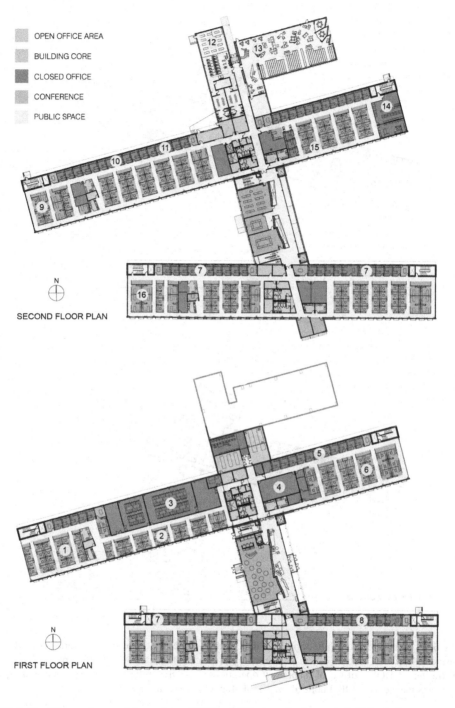

FIGURE 6.42 First and second floor plans, NREL, public domain. Legend. 1. ISO - Information systems office; 2. ISO - Information Systems and networks; 3. NREL/GO server room; 4. Emergency operations center; 5. MRI - finance; 6. Environment, safety and health & quality; 7. DOE - office of commercialization and project management; 8. DOE - office of acquisition and financial assistance; 9. D-systems integration; 10. H-ISO; 11. Education programs; 12. Fitness center; 13. Information commons; 14. ISO-IR-library and pubs/ pix/prints; 15. Human resources; 16. DOE-OMA.

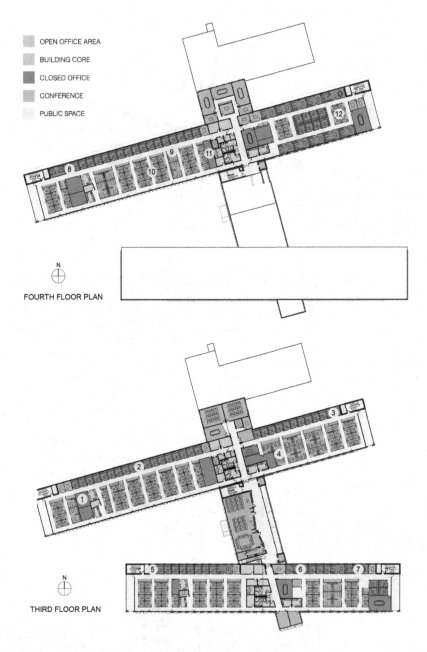

FIGURE 6.43 Third and fourth floor plans, NREL, public domain. Legend. 1. Site operations; 2. Contracts and business services; 3. Legal office; 4. Technology transfer office; 5. DOE - office of management and administration; 6. DOE - office of laboratory operations; 7. DOE - office of the manager; 8. Finance; 9. Requirements management; 10. Public affairs; 11. Lab dev; 12. Executives.

In addition to high R-value wall and roof assemblies, careful attention was paid to the intersections of assemblies to reduce thermal bridging (Hootman 2012).

6.5.4 Energy Systems

Lighting is an integrated system of daylighting, daylight control systems, occupancy controls, and high-efficiency lighting fixtures. The ambient daylight levels for office areas with 270 lux (25 footcandles)

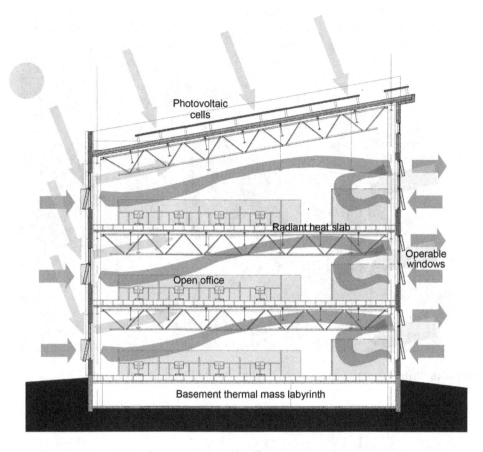

FIGURE 6.44   Building section of passive design strategies, NREL, public domain.

FIGURE 6.45   Arrays of the window on the south facade for view and daylight. © C. Kabre

FIGURE 6.46   Triple glazed, operable windows with individual sunshades. © C. Kabre

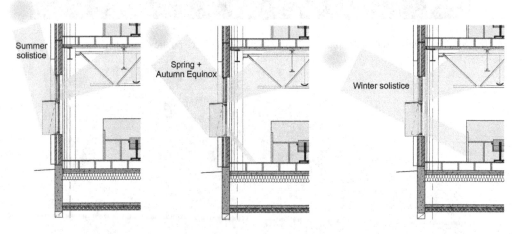

FIGURE 6.47   Optimized window design to cut summer sunlight and harness winter sunlight, NREL, public domain.

are supplemented by electric lighting of the primary fluorescent fixture utilizing two 25 W T-8 lamps throughout the offices. Each workstation is fitted with a 6 W LED task lamp. The electric lighting control approach is manual-on, manual-off, with automatic-off backup. The lights are always off unless manually turned on by an occupant. When manually switched on, lights are automatically dimmed, stepped down, and turned off based in response to daylight based on photocells (sensors) and a time clock. In private offices and conference rooms, vacancy sensors also shut off lights when the space is unoccupied. The RSF is realizing up to 85% savings in lighting energy use during sunny midday hours. During a typical sunny day, the light power density (LPD) is often at or below 1.614 W/m² (0.15

FIGURE 6.48 Daylit interior workplaces, NREL, public domain.

W/ft²), significant savings compared to typical code minimum office building with an LPD of 10.76 W/m² (1 W/ft²), and the installed LPD of 6.78 W/m² (0.63 W/ft²).

Space conditioning is an integration of climate-responsive design optimizing heating and cooling load, and a low-energy HVAC system can condition the space when needed. Strategies include decoupled ventilation air and space conditioning, low-pressure drop design, equipment efficiency, and leveraging free sources of heating (and cooling) like air-to-air heat recovery and IT equipment.

The ventilation system is decoupled from the space conditioning system using hydronic radiant slab ceilings and a dedicated outdoor air system (DOAS) for ventilation and dehumidification. The hydronic radiant tubing is cast into the structural floor and roof decks to condition the space below. Ventilation air is distributed via 300 mm (12 inches) raised access floors. This decoupled approach eliminates reheat energy and much of the fan power associated with a more conventional all-air VAV reheat system.

CO_2 sensors allow the active ventilation system to ramp down when occupancy is low or natural ventilation is in use.

Non-office spaces such as conference rooms are conditioned with a more traditional VAV reheat system to provide quick response to large changes in occupancy and to accommodate high occupant densities. CO_2-based demand–control ventilation is provided to minimize fan, cooling, and reheat energy. Hydronic radiant panels provide heat during unoccupied hours without engaging air handlers and reheat energy during periods of low use. The data center's dedicated cooling system is built around free cooling that exports useful heat to the rest of the building.

The photovoltaic systems totaling approximately 1.6 MW for the project consists of three systems. One photovoltaic system of 449 kW on roof and 524 kW system installed over visitor parking and 706 kW array installed over staff parking. The PV on the roof of the RSF was procured using a Power Purchase Agreement (PPA). The remaining PV for the project was purchased using ARRA funds.

The building's extremely detailed energy model predicts an energy use intensity (EUI) of 10.41 kWh/m² (33 kBtu/ft²/year). The on-site photovoltaic system is sized to meet net-zero site energy at 11.04 kWh/m² (EUI of 35 kBtu/ft²/year).

6.5.5 Sustainable Thinking

The material choices for the building were driven by the desire to have flexible and durable materials with low health impacts and reduced resource consumption. Construction waste diversion (75%), low-emitting materials, recycled-content materials (34%), regional materials (13%), and certified wood materials (59%) were all incorporated based on LEED compliance. Several highly visible, innovative materials were also incorporated into the project, including the use of reclaimed natural gas pipe as structural columns and beetle kill pine used as decorative wall elements throughout and as a multi-story feature wall in the lobby.

Although rainwater cannot be captured, as per Colorado water laws, the project still used the roof area of the project to determine a water use budget. In an average year, a little more than 3,016,973.2 liters (797,000 gallons) of waterfalls on the roof of the building. The building and site water uses are modeled at just over 2,994,260.72 liters (791,000 gallons) per year. A combination of highly efficient plumbing fixtures, native and xeriscape vegetation and drip irrigation with a satellite-based smart irrigation controller was used to reduce building water use by 55% and irrigation water use by 84% compared to LEED baselines.

All building occupants and visitors can see in real-time daily and annual energy use summary, broken down by end uses, and total energy on two large video displays in the lobby. This is part of educating the building users about net-zero energy future.

Table 6.7 presents the design profile of the RSF, NREL, Golden, Colorado.

6.6 UNIVERSITY OF WYOMING – VISUAL ARTS FACILITY, LARAMIE, WYOMING (ZONE 6B – COLD DRY, COTE 2016)

6.6.1 Design Intentions

The Visual Arts Facility (VAF) at the University of Wyoming consolidates the University's fine arts program from its scattered locations throughout the campus, addressing the limitations of the former facilities while providing students and faculty and enriching teaching-learning environment that is both state-of-the-art in occupational safety and its concern for environmental aspects and impacts. Sited at the east end of campus, Figure 6.49 shows how the VAF visually integrates with the adjacent iconic Centennial complex designed by Antoine Predock as well as the industrial architecture of the University power plant. The building with a floor area of 7432 m^2 (80,000 ft^2) on 4 hectares (10 acres) site, is designed to be state-of-the-art, competitive, accessible, and sufficiently flexible in accepting program and technological evolution.

The building design responds to the campus sustainability committee mission 'to advance sustainability at the University of Wyoming and the broader community by promoting sustainability awareness, using the campus as a living laboratory, and providing strategies for sustainable operations.'

The integrated design process combined with energy efficiency and outdoor water management garnered a LEED platinum award as well as a Portland 2030 Challenge Design award based on a reduction in carbon dioxide emissions (among other design criteria). Thus, the building fulfills the University's policy to construct LEED-certified buildings on the campus as part of the American College and University Presidents' Climate Commitment. Designed by Hacker architects, the project was winner of the AIA COTE Top Ten award in 2016, the jury comments:

> A contextually appropriate design that fits the landscape very well. The jury was impressed by the attention paid to the health and well-being of the building occupants, the way the design addressed air quality in the studios, and the way this was expressed by the ventilation stacks on the exterior. Art materials contain many toxic chemicals; this project is a model for how to do this type of facility.

(AIA 2016b)

TABLE 6.7
Design Profile Research Support Facility, National Renewable Energy Laboratory, Golden, Colorado

	Building Name	US Department of Energy's (DOE) Research Support Facility at the National Renewable Energy Laboratory (NREL)
Building profile	Location	1617 Cole Boulevard, Golden Colorado 80401, United States
	Principal use	Laboratory for research and development of renewable energy and energy efficiency technologies
	Employees/occupants	822
	Expected (design) occupancy	1040*
	Percent occupied	79%
	Site area	12.14 hectares (30 acres)
	Built-up area	20,624.5 m² (222,000 ft²)
	Distinctions/awards	LEED-NC Platinum 2011
		Sustainable Sites Pilot, AIA COTE Top 10 Green project (2011), Green Gov Presidential Award, Green Innovation (RSF Data Centre) 2011, McGraw Hill Construction Outstanding Green Building, 2010, American Institute of steel construction IDEAS2 Award 2011, Design-Build Institute of America, Merit Award, 2011
	Total cost	$64,000,000
	Cost per square foot	$259
	Substantial completion/occupancy	2010 (June)
Building team	Owner/representative	Department of Energy's National Renewable Energy Laboratory, Federal Government
	Architect	RNL Design
	General contractor	Haselden Construction, LLC
	Mechanical engineer	Stantec Consulting
	Electrical engineer	Stantec Consulting
	Structural engineer	KL&A Engineering, Golden Office
	Civil engineer	Martin/Martin, Inc.
	Landscape architect	RNL
	Lighting design	RNL
	Commissioning agent	RNL
Solar design profile	Latitude	39° 44′28″ N
	Longitude	105° 13′14″ W
	Altitude	1650 m
	Heating degree days (base ~ 18.3 °C-65° F)	6220
	Cooling degree days (base ~ 18.3 °C-65°F)	1154
	Annual hours occupied	50 hours per person per week; and 60 visitors per week, 2 hours per visitor per week
	Orientation	Along east-west axis
Performance profile	Annual energy use intensity (EUI) (site) consumption	11.17 kWh/m² (35.4 kbtu/ft²)
	Electricity (grid purchase)	0.31 kWh/ m² (1.01 kBtu/ft²)
	Electricity (on-site solar or wind installation)	7.6 kWh/ m² (24.2 kBtu/ft² /y)

(Continued)

TABLE 6.7 (CONTINUED)

Design Profile Research Support Facility, National Renewable Energy Laboratory, Golden, Colorado

	Annual net energy use intensity	3.53 kWh/m² (11.2 kBtu/ft²)
	Annual source (primary) energy	5.36 kWh/m² (17 kBtu/ft²) (including on-site PV)
	Savings vs. Standard 90.1-2007 Design Building	31.6% *
	Energy star rating	100
Water efficiency	Annual water use	2,995,014.02 liters (791,199 gallons)
Key sustainability features	Water conservation	Dual flush water closets, waterless urinals, low-flow lavatories and showers, drip irrigation, irrigation zones based on water frequency, satellite-based smart irrigation, roof drainage irrigates rain garden
	Recycled materials	Wood from pine trees killed by beetles;recycled runway materials from Denver's closed Stapleton Airport; and reclaimed steel gas piping (About 75% of construction waste materials have been diverted from landfills)
	Daylighting	Narrow floor plate 18.6 m (60 ft), daylight controls, Louvers designed for south-facing daylight windows, 92% daylit spaces
	Individual controls	Operable windows, underfloor ventilation air, individual task lights, occupant accessible lighting control
	Carbon reduction strategies	Creation of outdoor pedestrian spaces, use of renewable energy
	Transportation Mitigation Strategies	Minimizing on-site parking and to promote the use of public means (shuttles and city buses)
	Other major sustainable features	Thermal mass interiors, natural ventilation and night purging, thermal labyrinth crawl space, transpired solar collectors, radiant heating and cooling, underfloor ventilation, office recycling/composting program
Building envelope	Roof	Photovoltaics on standing seam roof, overall R-value: 5.82 m²K/W (33°F ft² h/Btu)
	Walls	Precast wall assembly, overall R-value: 2.64 m² K/W (15°F ft² h/Btu)
		Glazing Percentage: 27%
	Basement/foundation	Grade beam/foundation R-value: R 1.76 m²K/W (10°F ft² h/Btu)
		0.85 m²K/W (0.15°F ft² h/Btu)
		Floor over crawl space/labyrinth; R-value: 3.17 m²K/W (18°F ft² h/Btu)
	Windows	Typical view window, U-value: 0.965 W/m²K (0.17 Btu/h ft² °F), SHGC 0.23, VLT 0.43
		Typical daylight window U-value: 1.64 W/m²K (0.29 Btu/h ft² °F) SHGC: 0.38, VLT: 0.70

Source: Hootman et al. (2012).

FIGURE 6.49 The Visual Arts Facility (VAF) consolidates the fine arts program from its scattered locations throughout the campus. © Lara Swimmer

6.6.2 CLIMATE AND SITE

The cold dry climate of Laramie is characterized by long, cold winters with freezing temperatures, driving snow, and relentless winds. The average annual rainfall is 404.7 mm in Laramie. December is the coldest month when the mean minimum temperature is about −5.9°C and mean maximum temperature is about −0.4°C. July is the hottest month with an average maximum temperature of about 26.7°C, and the mean minimum temperature of about 11.5°C. The atmosphere is generally dry, July is slightly low humid when relative humidity is around 30%, and the month of December is slightly high humid when relative humidity is around 56%.

Throughout the year cloud cover ranges between 37% and 68%. Sunshine hours of the year are 248.5 h. Wind speed ranges between 4 and 6.3 m/s, and the west is the prevailing wind direction along with the south and the south-west for a few months. The global solar radiation goes as high as 7.3 kWh/m². The climatic data of Laramie are given in Table 6.8, and the sun-path diagram is in Figure 6.50.

The site lies at the edge, where the university's development joins the alluvial wash of the Laramie Mountains and the majestic Veda woo rock within. The art faculty and students find great inspiration in Wyoming's extraordinary environment of big sky, dynamic winds, rocky formations, the native grasses, and plants placed around the site provide several unique 'outdoor studio' opportunities; Figure 6.51 shows the site plan.

Massing density and materiality of built form relate explicitly to the harsh winter winds. The VAF building houses offices, faculty studios, classrooms, and labs for foundations of art, ceramics, sculpture, painting, drawing, small metalworking, printmaking, graphic design, and digital arts. Public gathering spaces, critique and display areas, and an exhibition gallery that converts to a black box address the additional needs of the art department. Figure 6.52 presents the ground floor and first floor plans, and Figure 6.53 shows four cardinal elevations of the building. Figures 6.54, 6.55, and 6.56 are the building sections.

TABLE 6.8
Climatic Data, Laramie, Wyoming

		Latitude N 41° 19' Climate Cold Dry				Longitude W 105°40' ASHRAE 6B				Altitude 2215 m Köppen BSk			
Months	**Jan**	**Feb**	**Mar**	**Apr**	**May**	**Jun**	**Jul**	**Aug**	**Sep**	**Oct**	**Nov**	**Dec**	**Year**
Sunshine h[a]	190.7	202.6	253.1	271.9	291.9	303.2	317.5	297.4	262.3	237.0	178.8	175.4	248.5
Cloud (%)	68.4	66.9	37.0	63.6	49.3	52.9	41.4	45.0	59.0	48.8	46.0	40.8	51.6
Solar irradiation daily average (Wh/m²)													
Global	2112	3257	5171	5308	6444	7314	6670	5830	4075	3400	2376	1877	4486
Diffuse	723	940	1167	1848	2115	2123	2197	1913	1688	1097	817	588	1435
Relative humidity (%)													
Morning	68	79	68	70	77	74	75	71	79	68	62	69	71.7
Evening	43	43	33	33	41	27	30	25	40	34	40	56	37.1
Dry-bulb temperature (°C)													
Max	1.1	2.1	9.7	11.3	15.2	21.2	26.7	25.7	19.3	12.1	5.2	-0.4	12.4
Min	-5.9	-6.8	-2.6	0.3	3.6	6.6	11.5	10.8	7.7	0.3	-2.1	-6.1	1.4
Mean	-2.4	-2.4	3.6	5.8	9.4	13.9	19.1	18.3	13.5	6.2	1.6	-3.3	6.9
Neutrality	20.9	20.9	20.9	20.9	20.9	22.1	23.7	23.5	22.0	20.9	20.9	20.9	20.9
Upper limit	23.4	23.4	23.4	23.4	23.4	24.6	26.2	26.0	24.5	23.4	23.4	23.4	23.4
Lower limit	18.4	18.4	18.4	18.4	18.4	19.6	21.2	21.0	19.5	18.4	18.4	18.4	18.4
Rain (mm)[b]	8.4	11.9	26.7	45.2	59.4	59.4	55.6	49.5	37.6	23.6	15.0	12.4	404.7
Prec (mm)[c]	27	34	58	107	169	154	143	123	111	80	54	32	1092
Wind (m/s)	6.3	5.5	6.2	6.3	5.7	5.7	4.7	4	5.1	5.9	6.3	5.4	5.6
HDD	656	591	466	367	262	104	5	15	148	379	527	690	4210
CDD	0	0	0	0	0	4	60	29	0	0	0	0	93

(Continued)

TABLE 6.8 (CONTINUED)
Climatic Data, Laramie, Wyoming

Average diurnal range (°K)	11.0
Annual mean range (°K)	33.5

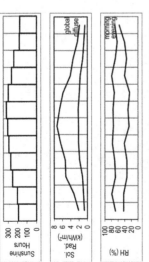
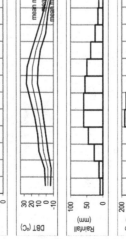
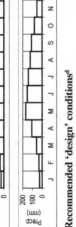
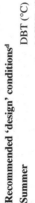

Wind rose

Recommended 'design' conditions[d]

Summer		
	DBT (°C)	27.8
	MCWB (°C)	12.5
	WBT (°C)	14.3
	MCDB (°C)	23.4
Winter		
	DBT (°C)	−19.0

Sources of data: https://energyplus.net/weather, [a]WMO (2010), [b]http://worldweather.wmo.int/en/city.html?cityId=736, [c]NOAA (2017), [d]ASHRAE(2009) Sunshine hours and rainfall for Cheyenne, Wyoming (50.6 miles)

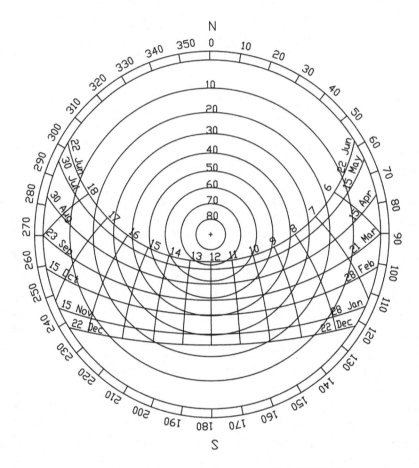

FIGURE 6.50 A stereographic sun-path diagram for Laramie, latitude N 37° 22′.

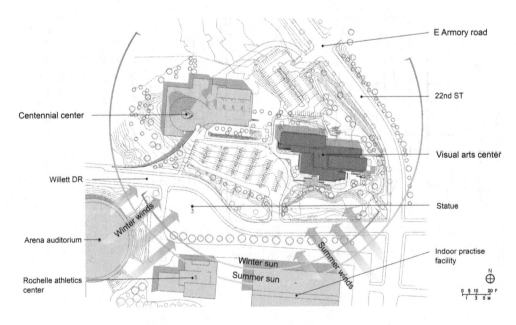

FIGURE 6.51 Site plan with climate analysis. Image courtesy Hacker architects.

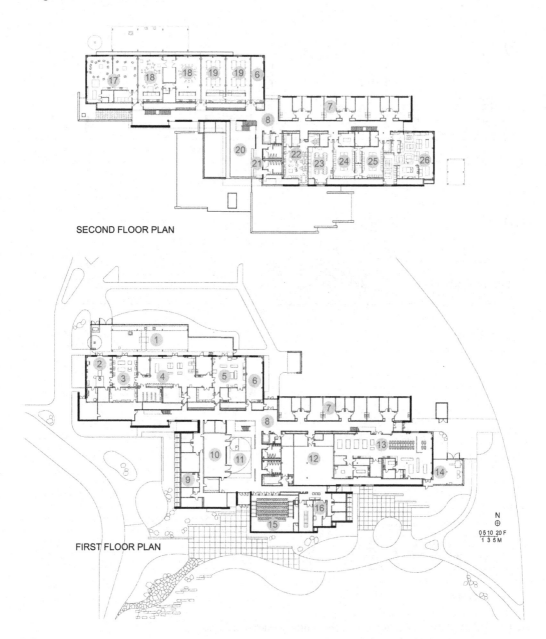

FIGURE 6.52   First and second floor plans; Image courtesy Hacker architects. Legend. 1. Sculpture yard; 2. foundry; 3. steel fab; 4. structure lab; 5. wood lab; 6. senior studios 7. faculty studios; 8. corridor; 9. administration 10. galleries; 11. lobby; 12. mechanical; 13. ceramics lab; 14. ceramics work yard; 15. classroom; 16. art history; 17. painting lab; 18. drawing lab; 19. foundations lab; 20. open below; 21. cyber lounge; 22. small metals lab; 23. graphic design work room; 24. graphic design computer lab; 25. digital media lab; 26. print making lab.

Low impact development (LID) design approach allows the natural flow of water on the site through infiltration, filtering, ponding, and evaporating. The VAF site allows water to permeate the ground by decreasing concrete and increasing native landscaping, treating 100% of the pollutants associated with the stormwater runoff. Site water flows from the north around the building to the east before entering the retention swale between the street approach and the front entrance demonstrating the process in full view.

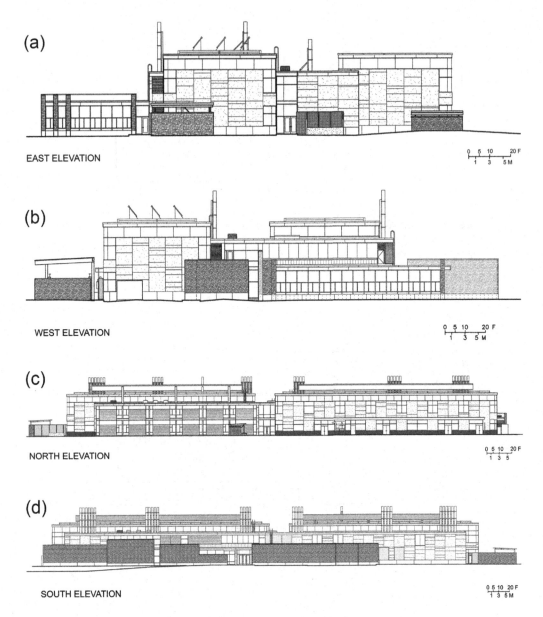

FIGURE 6.53　Building elevations. Images courtesy Hacker architects.

6.6.3　Daylight and Thermal Design

Many art-making processes are enhanced by the controlled use of natural lighting. The building orientation and form is based on the study of the sun's interaction with interior spaces, simultaneously distributing reflected daylight while eliminating solar heat gain. The design team used computer modeling to test specific formal and glazing decisions for maximum daylight benefit. Figure 6.57 shows the top lighting strategies for the art studio. Figure 6.58 shows daylit foyer. The VAF utilizes high-performance windows and a well-insulated, airtight-building envelope to minimize cooling or heating loads and to ensure interior thermal comfort. The building form places large swaths of solid walls clad in the traditional campus stone to break the cold winter winds.

Laramie's seasonal winds are harnessed for natural ventilation to cool common spaces and corridors through a system of intake exterior louvers and exhaust fans positioned through CFD analysis.

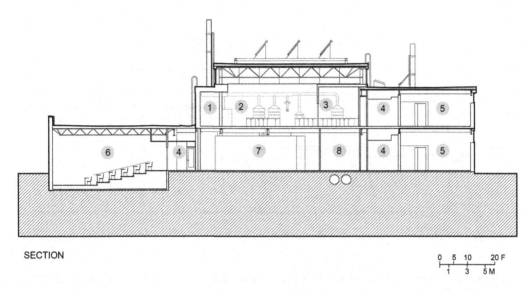

SECTION

FIGURE 6.54 Transverse section (east side). Image courtesy Hacker architects. Legend. 1. Mechanical; 2. small metals lab; 3. investment; 4. corridor 5. faculty studio; 6. classroom; 7. mechanical room; 8. electrical.

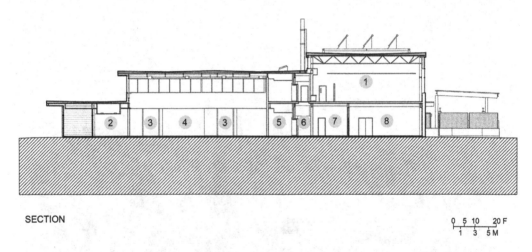

SECTION

FIGURE 6.55 Transverse section (east side). Image courtesy Hacker architects. Legend. 1. Foundations; 2. vestibule; 3. open to gallery; 4. common 5. corridor; 6. mechanical 7. raw materials; 8. wood lab.

Office spaces utilize operable windows and ceiling fans for occupant comfort to eliminate mechanical ventilation and cooling.

6.6.4 Energy Systems

The VAF's electric lighting utilizes automatic daylight sensors for common areas and exterior lighting and manual timer switches preset to coincide with class scheduling for offices, classrooms, and studios.

The VAF building achieves energy conservation in several ways, which include centralized evaporative cooling in arid summer, natural ventilation, displacement air distribution, exhaust air process control, exhaust heat recovery, hydronic in-slab radiant heating, and other innovative approaches. It is projected to save energy and produce 54% lower carbon dioxide compared to the US national average.

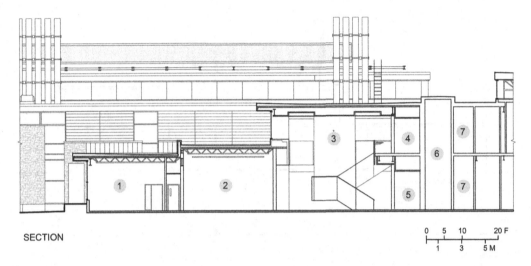

SECTION

FIGURE 6.56 Longitudinal section. Image courtesy Hacker architects. Legend: 1. Adjunct; 2. galleries; 3. lobby ; 4. cyber lounge 5. hall; 6. elevator; 7. electrical and mechanical.

FIGURE 6.57   Natural light intake and its suitability for studio works. © Lara Swimmer

Laboratory studios are usually among the least energy-efficient buildings because of their high ventilation demands and. Each laboratory studio and classroom is served with an independent, variable air system with low-velocity displacement diffusers system controlled by demand–control sensors. This system provides only the necessary air at the necessary time, resulting in significant energy savings. Process ventilation is critical to the safety of the artist students. Fifteen unique process ventilation types all run through heat recovery units before exiting to the atmosphere. Where possible, the VAF uses user-controlled, timed process exhaust, a simple innovative energy-saving approach. Heat recovery coil units on much of the building's exhaust air systems capture

FIGURE 6.58   Daylit foyer. © Lara Swimmer

heat leaving the building and use it to preheat the outdoor air at the central air-handling units. Heat recovery is a major contributor to energy performance due to the lab-style 100% exhaust ventilation throughout the studios. Slot vents on countertops extract fumes before people can breathe them, and a user-friendly, easily accessible filter system signals an alarm when vents and filters need service.

Fully 70% of the building's electrical power is provided from renewable resources by contract. The building also harnesses renewable solar energy from an on-site 357.73 kWh/year (1134 MBtu/year) through a glycol-based evacuated solar tube collector array (the dark, angled structures visible on top of the building) that produces hot water for the in-slab hydronic radiant heating as well as domestic hot water and pre-treatment of outside air for ventilation. The German-made Oventrop system's solar heat collectors are 127 mm (5 inches) diameter, 1524 mm (5 ft) long evacuated tubes that cover two sections of the roof is one of the largest solar evacuated tube installations in the United States. For initial start-up, the collectors were installed at night so as not to overwhelm the system. This system provided 37% of the building's heating in the first reporting year, and tuning is continued to adjust for full efficiency.

Through its first year of reporting, post-commissioning, the VAF achieved a measured EUI of 23.03 kWh/m^2 (73 kBtu/ft^2/yr), 72% better than the 2030 baseline. ME Engineers also designed a system to provide heated air at a head level even though ceilings reach up to 4.88 m (16 ft) on the first floor.

6.6.5   Sustainable Thinking

The refined palette of materials including quarried stone, weathering steel, precast concrete, zinc, and oak are chosen to blend harmoniously with both the existing historical campus and the Wyoming landscape. Figure 6.59 shows weathering steel as shading elements. Materials used in the construction of the VAF went through multi-faceted analysis including the study of embodied energy, recycled content and recyclability, durability, and VOC content. All of the paints, adhesives, and carpeting installed in the building are low/no VOC. The construction of the VAF involves over 20% of the materials extracted and manufactured within a 500-mile radius of the site. The stone is

FIGURE 6.59   Use of salvaged iron as shading element. © Lara Swimmer

FIGURE 6.60   Regular structural and glazing bays for future flexibility. © Lara Swimmer

TABLE 6.9

Design Profile University of Wyoming Visual Arts Facility, Laramie, Wyoming

Building Name	University of Wyoming Visual Arts Facility
Building profile Location	2129 Willet Drive, Laramie, Wyoming 82072, United States
Principal use	Education – College/University (campus-level)
Site area	7432.3 m² (80,000 ft²)
Built-up area	7198.5 m² (77,484 ft²)
Distinctions/awards	LEED Platinum (2012), 2030 Compliant, AIA COTE Top Ten Green Project (2016), Honorable Mention, AIA Portland COTE 2030 Challenge Award (2011)
Total cost	$27,000,000.00
Cost per square foot	$348.5*
Substantial completion/occupancy	2012 (January)
Building team Owner/representative	University of Wyoming
Architect	Hacker Architects & Malone Belton Able PC
General contractor	GE Johnson Construction Company
Mechanical engineer	M-E Engineers
Electrical engineer	M-E Engineers
Structural engineer	Martin/Martin Inc.
Civil engineer	Trihydro Corporation
Landscape architect	EDAW
Solar design profile Latitude	41.3°N
Longitude	105.6°W
Altitude	+7333 ft elevated
Heating degree days (base ~18.3°C–65 °F)	4210
Cooling degree days (base ~18.3°C–65°F)	93
Annual hours occupied	4560 (6 am–10 pm)*
Orientation	East-west axis with large north-facing facade
Performance profile Annual energy use intensity (EUI) (site) consumption	36.93 kWh/m² (126 kBtu/ft²)
Electricity (grid purchase)	4.42 kWh/m² (14 kBtu/ft²*)
Electricity (on-site solar or wind installation)	35.33 kWh/m² (112 kBtu/ft²)
Savings vs. Standard 90.1-2007 Design Building	72%
Key sustainability features Water conservation	Efficient plumbing fixtures, low impact development approach, swales, and native landscaping
Recycled materials	Use of 50% recycled and salvaged construction waste
Daylighting	Automatic daylight sensors, manual timer switches, clerestory windows
Individual controls	Process specific adjustable exhaust snorkels
Carbon reduction strategies	A solar thermal system, cycling, and public transit
Other major sustainable features	Solar evacuated tubes, hydronic radiant slab, domestic hot water, evaporative cooling, custom exhaust systems

(Continued)

TABLE 6.9 (CONTINUED)

Design Profile University of Wyoming Visual Arts Facility, Laramie, Wyoming

Building envelope	Roof	Type: insulated roof overall *R*-value: 42
	Walls	Stone veneer wall: R 13.2 insulation, U 0.070 for assembly
		Precast wall: R 19.8 insulation, U 0.048 for assembly
		Metal siding wall: R 22.3 insulation, U 0.042 for assembly
		Vertical glazing: 34%
	Windows	High-performance windows

*Computed hours: (16 hours/day) × 285 working days.

the same stone that has made up every campus building since the founding, having been quarried from the campus's quarry at the edge of the city; making the campus literally of the site.

A cast-in-place, steel-framed building; the building envelope combines architectural precast concrete, high-performance aluminum curtain wall, zinc panels, glazing and quarried stone (Figure 6.60).

A full-scale mock-up was constructed to illustrate part of the building envelope, allowing construction managers to review construction sequencing and evaluate the integrity of the building envelope in the project's quest for a hardy, airtight design. 'Building information modeling' (BIM) streamlined the design and construction process for the building envelope as well as for the mechanical, electrical, steel frame, and other elements. The construction process maximized the recycled or salvaged rate of construction waste to the extent of 50% or greater.

Water use within the building is reduced through water-efficient low-flow plumbing fixtures and encouraged conservation behavior, and the education of occupants contributed to an overall 41% reduction in regulated potable water. Maximum LEED points were achieved in water use reduction by these measures. Sediment traps on all studio sink capture and contains any toxic chemicals, minerals, and solids from exiting the building and entering the waste stream.

An interactive touch screen dashboard in the lobby of VAF educates visitors and occupants about the building's energy efficiency and other sustainable features; exemplify the concept of 'buildings that teach.'

Measurement and verification of the project's energy conservation strategies are supported by metering to provide concrete real-time data. There are 63 different meters installed in the VAF to record electrical loads, water usage, gas usage, and HVAC systems to verify and improve performance. Table 6.9 delineates design profile of University of Wyoming -Visual Arts Facility, Laramie, Wyoming.

REFERENCES

AIA (2011) The National Renewable Energy Laboratory Research Support Facility. The American Institute of Architects, the Committee on the Environment (COTE), http://www.aiatopten.org/node/103, accessed on 18 January 2020.

AIA (2016a) The Edith Green – Wendell Wyatt Federal Building. The American Institute of Architects, the Committee on the Environment (COTE), https://www.aiatopten.org/node/494, accessed on 19 January 2020.

AIA (2016b) The University of Wyoming-Visual Arts Facility. The American Institute of Architects, the Committee on the Environment (COTE), https://www.aiatopten.org/node/425, accessed on 19 January 2020.

AIA (2017a) The NOAA Daniel K. Inouye Regional Center. The American Institute of Architects, the Committee on the Environment (COTE), https://www.aia.org/showcases/76911-noaa-daniel-k-inouye-regional-center, accessed on 19 January 2020.

AIA (2017b) The Stanford Central Energy Facility. The American Institute of Architects, the Committee on the Environment (COTE), https://www.aia.org/showcases/76996-stanford-university-central-energy-facility, accessed on 19 January 2020.

Amelar S (2015) From Sea to Shining Sea: NOAA Inouye Regional Center. *Architectural Record*, 1 August, https://www.architecturalrecord.com/articles/7358-noaa-inouye-regional-center, accessed on 4 January 2020.

Bennett H (2015) Something Old, New, Borrowed and Green. *Hawaii Business*, 1 July.

Foreman C, Lowen C (2015) Optimizing a Landmark: Edith Green-Wendell Wyatt Federal Building. High Performing Buildings, ASHRAE, Summer, pp. 49–58. http://www.hpbmagazine.org/Case-Studies/Edith-Green-Wendell-Wyatt-Federal-Building-Portland-OR/, accessed on 2 November 2019.

Hootman T, Okada D, Pless S, Sheppy M, Torcellini P (2012) Net Zero Blueprint: National Renewable Energy Laboratory Research Support Facility. High Performing Buildings, pp. 20–33. http://www.hpbmagazine.org/attachments/article/12170/12F-Department-of-Energys-National-Renewable-Energy-Laboratory-Research-Support-Facility-Golden-CO.pdf, accessed on 18 January 2020.

https://www.uwyo.edu/sustainability/leed_buildings/

https://www.hok.com/projects/view/national-oceanic-and-atmospheric-administration-noaa-daniel-k-inouye-regional-center/, accessed January 4, 2020.

https://www.architectmagazine.com/project-gallery/stanford-universitys-central-energy-facility_o

https://worldarchitecture.org/articles/cgnzh/aia_honor_awards_2017_recognize_excellence_in_architecture_interior_and_regional_urban_design.html

https://www.architonic.com/en/project/zgf-architects-llp-stanford-university-central-energy-facility/5103246

https://www.architecturalrecord.com/articles/11538-stanford-university-central-energy-facility

https://architectureprize.com/winners/winner.php?id=2533

https://issuu.com/constructionbusinessmedia/docs/1705_nzb

https://www.zgf.com/project/stanford-university-central-energy-facility/

https://idesignawards.com/winners/zoom.php?eid=9-10659-16

Knittel T (2014) How Reverse Engineering Can Spur Design Innovation. *Fast Co. Exist*, 9 January, accessed on 4 January 2020.

Manahane S (2013) NOAA IRC Dedication at Joint Base Pearl Harbor-Hickham. *US Navy News, Hawaii*, 19 December, accessed on 4 January 2020.

The Rafu Shimpo, Los Angeles Japanese Daily News (2013) NOAA Dedicates Inouye Regional Center in Honolulu. 17 December. Available at http://www.rafu.com/2013/12/noaa-dedicates-inouye-regional-center-in-honolulu/, accessed on 4 January 2020.

Woolford P, See T (2018) Adaptive Reuse for Hawaii Hangars, NOAA Daniel K. Inouye Regional Center, High Performing Buildings, pp. 6–16. http://www.hpbmagazine-digital.org/hpbmagazine/winter_2018/MobilePagedArticle.action?articleId=1257380, accessed on 4 January 2020.

7 Climate Data and Sun-Path Diagrams

7.1 INTRODUCTION

The synergistic design process will play a major role in the creation of next-generation sustainable buildings that integrate natural processes and programmatic requirements and synthesize into architectural design. In any project development today, when the potentials of passive cooling, passive heating, natural ventilation, and daylighting are fully exhausted – depending upon the climate and other site conditions – then only the relatively low energy (hybrid) design strategies and active design strategies available should be explored to meet occupants' comfort demand.

Every geographical location has its climate, its own set of natural resources, and conditions manifested by the interplay of climatic elements. The climatic matrix does not differ so much from point to point within a given region, but it differs considerably from region to region. In those differences of climatic matrix lie the origins of environmentally sustainable building design, which has become a larger concern for architecture's avant-garde and mainstream alike.

Today's designers can access an enormous amount of climate data that is light-years ahead in its accuracy and precision. However, the real challenge lies in extracting and assessing the pertinent climate data to respond with intelligent design decisions.

This chapter aims to present a narrative summation of macro climate data, wind roses, and sun path diagrams for the major US cities, which can be used to conduct climate analysis; then the summary moves into specific data to inform sustainable building design and renewable energy system application. The objective is to provide climatic data for the capital city of each of the 50 states in the United States (Figure 7.1). Table 7.1 presents the list of states with their respective capital cities and alternative nearest cities, for which climatic data are available, are mentioned for Kentucky, Maryland, and national capital Washington, DC. The selection of locations was based on two criteria:

1. The intention to obtain a fairly even distribution of points over the country
2. The availability of data

The basic climatic data are taken from the following sources:

(i) The Department of Energy (DOE), US Government, where the nation's energy and climate research efforts are centered for building applications, the EnergyPlus website publishes climatic data for 1020 locations in the United States, including Guam, Puerto Rico, and US Virgin Islands, derived from the 1991 to 2005 period of record. All 1020 locations in the Typical Meteorological Year 3 (TMY3) use comprehensive weather data over several decades to determine a 'typical' year hourly weather data (Wilcox and Marion 2008); each station has a summary data STAT file.
(ii) The National Oceanic and Atmospheric Administration – NOAA's National Climatic Center in Ashville, North California, collects and summarizes data from major weather stations around the nation, each one recording detailed (hourly or three-hourly) data on numerous weather elements.
(iii) The World Meteorological Organization (WMO).
(iv) The American Society of Heating, Refrigerating and Air-Conditioning Engineers

The climatic data for the 50 US cities are presented in Tables 7.2, 7.2, 7.3, 7.4, 7.5, 7.6, 7.7, 7.8, 7.9, 7.10, 7.11, 7.12, 7.13, 7.14, 7.15, 7.16, 7.17, 7.18, 7.19, 7.20, 7.21, 7.22, 7.23, 7.24, 7.25, 7.26, 7.27, 7.28, 7.29,

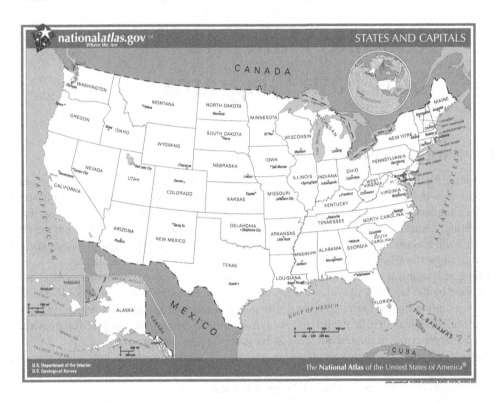

FIGURE 7.1 Map of the United States showing 50 states and national capital Washington, DC. Source: Wikimedia Commons (2017), public domain

7.30, 7.31, 7.32, 7.33, 7.34, 7.35, 7.36, 7.37, 7.38, 7.39, 7.40, 7.41, 7.42, 7.43, 7.44, 7.45, 7.46, 7.47, 7.48, 7.49, 7.50 and 7.51. Each table gives summary of the basic climatic data for 12 months: mean sunshine hours, cloud cover (%), daily average solar irradiation global and diffuse on horizontal surface (W/m²), maximum and minimum relative humidity (%), maximum and minimum dry bulb temperature (°C), rainfall (mm), precipitation (mm), wind speed (m/s), heating degree days (HDDs), and cooling degree days (CDDs). The annual average is computed for cloud cover, irradiation, relative humidity, temperature, wind speed from the data of 12 months and shown in the last column 'year'. While the annual sum is computed for sunshine hours, rainfall, precipitation, HDD and CDD from the data of 12 months and shown in the last column 'year.' The average diurnal range and annual mean range, thermal neutrality, and its upper and lower limits are computed from the temperature data. The heating degree days (HDDs) and cooling degree days (CDDs) are for the base temperature of 18°C. The wind direction is graphically represented in the wind rose diagram. The sunshine hours, solar irradiations, relative humidities, dry bulb temperatures, and rainfall are also shown graphically for better understanding. Also the design conditions as per ASHRAE 2009 are given at the end of the table. The second set of Tables 7.52 to 7.101 gives the hourly dry-bulb temperature for all 12 months, with overheated, comfortable, and underheated period delineated in different shades for the design of solar control. The second set of tables also gives the hourly direct and diffuse solar radiation for the hottest and coldest months. Chapter 2 explains the origin and derivation of the climatic data presented in this chapter. The sources of data in each case are indicated below the first table. The tables for cities are listed alphabetically. Stereographic sun-path diagrams for 50 locations shown in Figures 7.2 to 7.51 are drawn using the software Winshade (Kabre 1999). Shadow protractor is shown in Figure 7.52. The second set of Tables 7.52 to 7.101 and Figures 7.2 to 7.52 are available on the publishers' web site.

In addition to macroclimate data presented in this chapter, the designers may understand microclimate variations and measure them with compact recording units developed for use on the actual design sites, which will allow finely tuned design responses.

TABLE 7.1
List of US States and Capital City

S. No.	State	Capital City
1.	Alabama	Montgomery
2.	Alaska	Juneau
3.	Arizona	Phoenix
4.	Arkansas	Little Rock
5.	California	Sacramento
6.	Colorado	Denver
7.	Connecticut	Hartford
8.	Delaware	Dover
9.	Florida	Tallahassee
10.	Georgia	Atlanta
11.	Hawaii	Honolulu
12.	Idaho	Boise
13.	Illinois	Springfield
14.	Indiana	Indianapolis
15.	Iowa	Des Moines
16.	Kansas	Topeka
17.	Kentucky	Frankfort/Louisville (alternative city)
18.	Louisiana	Baton Rouge
19.	Maine	Augusta
20.	Maryland	Annapolis, Washington DC/Baltimore (alternative city)
21.	Massachusetts	Boston
22.	Michigan	Lansing
23.	Minnesota	Minneapolis
24.	Mississippi	Jackson
25.	Missouri	Jefferson City
26.	Montana	Helena
27.	Nebraska	Lincoln
28.	Nevada	Carson city
29.	New Hampshire	Concord
30.	New Jersey	Trenton
31.	New Mexico	Santa Fe
32.	New York	Albany
33.	North Carolina	Raleigh
34.	North Dakota	Fargo
35.	Ohio	Columbus
36.	Oklahoma	Oklahoma city
37.	Oregon	Salem
38.	Pennsylvania	Harrisburg
39.	Rhode Island	Providence
40.	South Carolina	Columbia
41.	South Dakota	Pierre
42.	Tennessee	Nashville
43.	Texas	Austin
44.	Utah	Salt Lake City
45.	Vermont	Montpelier
46.	Virginia	Richmond
47.	Washington	Olympia
48.	West Virginia	Charleston
49.	Wisconsin	Madison
50.	Wyoming	Cheyenne

TABLE 7.2

Climatic Data, Albany, New York

	Jan	Feb	Mar	Apr	May	Jun	Jul	Aug	Sep	Oct	Nov	Dec	Year
Latitude N 42° 45'							**Longitude** W 73° 48'				**Altitude** 84 m		
Climate Cool Humid							**ASHRAE** 5A				**Köppen** Bwh		
Months	Jan	Feb	Mar	Apr	May	Jun	Jul	Aug	Sep	Oct	Nov	Dec	Year
Sunshine h[a]	141.1	158.5	200.3	218.9	248.9	262.2	289.2	253.2	210.5	168.8	100.7	108.3	2360.6
Cloud (%)	70.2	67.1	64.8	66.2	69.5	63.3	58.2	62.1	63.9	59.0	60.2	70.5	64.6
Solar irradiation daily average (Wh/m²)													
Global	1646	2732	3798	4561	5741	6114	6115	5341	4169	2836	1733	1487	3856
Diffuse	997	1516	2056	1997	2671	2753	2969	2470	1957	1399	955	937	1890
Relative humidity (%)													
Morning	79	77	71	72	81	89	86	95	89	93	81	76	82.4
Evening	67	58	49	45	47	54	53	53	55	60	59	60	55
Dry-bulb temperature (°C)													
Max	−4.2	−1.0	6.7	13.8	20.5	23.2	27.1	25.6	21.3	14.2	7.9	0.9	13
Min	−8.7	−6.2	−0.5	5.0	10.7	13.5	17.0	15.4	11.8	6.4	2.3	−3.7	5.3
Mean	−6.5	−3.6	3.1	9.4	15.6	18.4	22.1	20.5	16.6	10.3	5.1	−1.4	9.1
Neutrality	20.9	20.9	20.9	20.9	22.6	23.5	24.6	24.2	22.9	21.0	20.9	20.9	20.6
Upper limit	23.4	23.4	23.4	23.4	25.1	26.0	27.1	26.7	25.4	23.5	23.4	23.4	23.1
Lower limit	18.4	18.4	18.4	18.4	20.1	21.0	22.1	21.7	20.4	18.5	18.4	18.4	18.1
Rain (mm)[b]	65.8	55.9	81.5	80.5	91.7	96.3	104.6	87.9	83.8	93.5	83.6	74.4	999.5
Prec (mm)[c]	259	220	321	317	361	379	412	346	330	368	329	293	3935
Wind (m/s)	4.8	4.5	5.6	5.0	4.1	4.4	3.8	3.4	3.4	3.4	3.7	5.4	4.3
HDD	767	610	479	267	92	32	1	7	81	254	402	611	3603
CDD	0	0	0	2	21	46	131	85	29	1	0	0	315

(Continued)

TABLE 7.2 (CONTINUED)
Climatic Data, Albany, New York

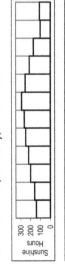

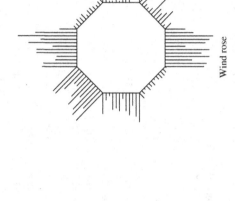

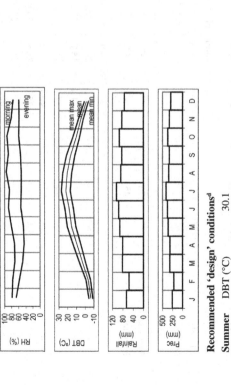

| | Average diurnal range (°K) | 7.8 |
| | Annual mean range (°K) | 35.8 |

Wind rose

Recommended 'design' conditions[d]

Summer	DBT (°C)	30.1
	MCWB (°C)	21.9
	WBT (°C)	23.4
	MCDB (°C)	27.9
Winter	DBT (°C)	−16.2

Sources of data: https://energyplus.net/weather, [a]WMO (2010), [b]http://worldweather.wmo.int/en/city.html?cityId=712, [c]NOAA(2017), [d]ASHRAE(2009).

TABLE 7.3

Climatic Data, Atlanta, Georgia

Latitude	N 33° 37'					**Longitude**	W 84° 25'			**Altitude**	308 m		
Climate	Warm Humid					ASHRAE	3A			Köppen	Cfa		
Months	**Jan**	**Feb**	**Mar**	**Apr**	**May**	**Jun**	**Jul**	**Aug**	**Sep**	**Oct**	**Nov**	**Dec**	**Year**
Sunshine h[a]	164	171.7	220.5	261.2	288.6	284.8	273.8	258.6	227.5	238.5	185.1	164.0	2783.3
Cloud (%)	59.4	55.9	54.9	43.2	57.0	46.5	51.8	53.0	44.9	46.5	50.0	53.4	51.4
Solar irradiation daily average (Wh/m²)													
Global	2777	3328	4649	5741	6320	6731	6306	5815	4361	3836	2953	2525	4612
Diffuse	1181	1249	1860	2300	2827	2936	2889	2773	2018	1277	1267	1089	1972
Relative humidity (%)													
Morning	77	70	72	81	87	84	91	90	86	93	81	78	82.5
Evening	52	45	41	42	48	44	52	53	57	53	43	48	48.2
Dry-bulb temperature (°C)													
Max	7.8	12.6	18.7	23.1	26.2	30.3	31.1	31.1	26.6	21.4	17.8	12.6	21.6
Min	0.8	3.7	9.2	11.4	15.7	19.1	21.3	22.5	18.9	11.3	7.1	3.5	12.0
Mean	4.3	8.2	14.0	17.3	21.0	24.7	26.2	26.8	22.8	16.4	12.5	8.1	16.8
Neutrality	20.9	20.9	22.1	23.1	24.3	25.5	25.9	26.1	24.9	22.9	21.7	20.9	23.0
Upper limit	23.4	23.4	24.6	25.7	26.8	28.0	28.4	28.6	27.4	25.4	24.2	23.4	25.5
Lower limit	18.4	18.4	19.6	20.7	21.8	23.0	23.4	23.6	22.4	20.4	19.2	18.4	20.5
Rain (mm)[a]	106.7	118.6	122.2	85.3	93.2	100.3	133.9	99.1	113.5	86.6	104.1	99.1	1262.6
Prec (mm)[b]	480	501	498	365	405	476	502	431	456	345	422	416	5297
Wind (m/s)	4.6	5.2	4.9	4.0	3.5	3.9	3.2	3.4	3.1	4.0	4.2	4.7	4.1
HDD	435	435	435	435	435	435	435	435	435	435	435	435	5220
CDD	0	0	0	0	0	0	0	0	0	0	0	0	0

(Continued)

TABLE 7.3 (CONTINUED)
Climatic Data, Atlanta, Georgia

Average diurnal range (°K)	9.6
Annual mean range (°K)	30.3

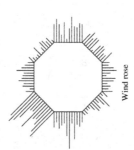

Wind rose

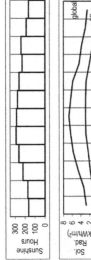

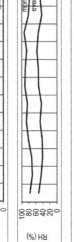

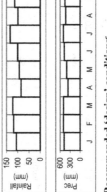

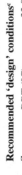

Recommended 'design' conditions[c]

Summer		
DBT (°C)	33.0
MCWB (°C)	23.3
WBT (°C)	24.5
MCDB (°C)	30.3
Winter	DBT (°C)	−3.5

Sources of data: https://energyplus.net/weather, [a]WMO (2010), [b]http://worldweather.wmo.int/en/city.html?cityId=268, [c]NOAA(2017), [d]ASHRAE(2009).

TABLE 7.4
Climatic Data, Augusta, Maine

Latitude	N 44° 19'			Longitude	W 69° 48'				Altitude	107 m		
Climate	Cold Humid			ASHRAE	6A				Köppen	Dfb		
Months	Jan	Feb	Mar	Apr	May	Jun	Jul	Aug	Sep	Oct	Nov	Dec	Year
Sunshine h[a]	164.8	172.8	205.2	213.5	243.2	259.1	282.2	267.6	229.1	195.7	138.7	140.9	2512.8
Cloud (%)	60.4	54.8	58.5	72.8	69.5	84.8	65.8	59.3	62.2	67.1	73.2	67.9	66.4
Solar irradiation daily average (Wh/m²)													
Global	1471	2075	4076	4063	5070	4679	4757	4560	3806	2128	1560	1114	3280
Diffuse	686	1027	1428	2006	2496	2857	2398	2129	1604	1118	870	622	1603
Relative humidity (%)													
Morning	69	66	74	83	84	93	89	89	94	89	81	76	82.3
Evening	54	43	47	53	53	65	61	61	59	69	63	55	56.9
Dry-bulb temperature (°C)													
Max	−8.8	−4.3	7.9	10.4	16.5	19.8	23.4	24.6	23.2	12.8	6.9	0.2	11.1
Min	−14.7	−11.9	−0.2	3.7	8.4	13.3	15.9	16.9	15.1	7.4	2.3	−5.3	4.2
Mean	−11.8	−8.1	3.9	7.1	12.5	16.6	19.7	20.8	19.2	10.1	4.6	−2.6	7.6
Neutrality	20.9	20.9	20.9	20.9	21.7	22.9	23.9	24.2	23.7	20.9	20.9	20.9	20.2
Upper limit	23.4	23.4	23.4	23.4	24.2	25.4	26.4	26.7	26.2	23.4	23.4	23.4	22.7
Lower limit	18.4	18.4	18.4	18.4	19.2	20.4	21.4	21.7	21.2	18.4	18.4	18.4	17.7
Rain (mm)[b]	66.3	61.7	85.6	96.3	93.7	90.4	86.9	84.1	95.0	110.7	110.7	82.3	1063.7
Prec (mm)[c]	261	243	337	379	369	356	342	331	374	436	436	324	4188
Wind (m/s)	4.6	4.4	4.5	4.8	4.3	3.7	3.9	2.5	4.2	3.4	4.3	4	4.1
HDD	930	739	449	335	177	64	11	13	39	258	411	655	4081
CDD	0	0	0	0	2	16	56	85	58	2	0	0	219

(Continued)

TABLE 7.4 (CONTINUED)
Climatic Data, Augusta, Maine

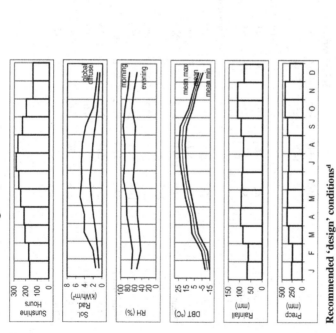

| | Average diurnal range (°K) | 6.8 |
| | Annual mean range (°K) | 39.3 |

Wind rose

Recommended 'design' conditions[d]

Summer	DBT (°C)	28.7
	MCWB (°C)	20.6
	WBT (°C)	22.0
	MCDB (°C)	26.6
Winter	DBT (°C)	−17.3

Sources of data: https://energyplus.net/weather, [a]WMO (2010), [b]http://worldweather.wmo.int/en/city.html?cityId=1959, [c]NOAAA (2017), [d]ASHRAE (2009), Sunshine hours for Portland, Maine (59 miles).

TABLE 7.5

Climatic Data, Austin, Texas

Latitude	N 30° 17'				Longitude			W 97° 44'		Altitude		213 m	
Climate	Hot humid				ASHRAE			2A		Köppen		Cfa	
Months	Jan	Feb	Mar	Apr	May	Jun	Jul	Aug	Sep	Oct	Nov	Dec	Year
Sunshine h[a]	163.8	169.3	205.9	205.8	227.1	285.5	317.2	297.9	233.8	215.6	168.3	153.5	220.3
Cloud (%)	62.8	63.0	61.0	68.7	63.9	48.3	45.5	45.5	43.9	64.7	67.0	49.4	56.97
Solar irradiation daily average (Wh/m²)													
Global	2505	3340	4485	4888	5582	6412	6987	6074	5351	3786	2703	2506	4552
Diffuse	1131	1420	1654	2184	2587	2385	2188	2113	1723	1449	1341	1083	1772
Relative humidity (%)													
Morning	82	76	81	89	93	95	90	96	84	94	83	79	86.8
Evening	54	46	51	56	52	51	46	47	46	65	55	50	51.6
Dry-bulb temperature (°C)													
Max	15.6	18.2	21.9	26.2	29.1	32.2	34.0	33.6	30.1	25.9	21.6	17.5	25.5
Min	7.5	9.7	13.0	17.9	18.9	22.1	23.8	22.0	20.2	18.7	14.8	8.8	16.5
Mean	11.6	14.0	17.5	22.1	24.0	27.2	28.9	27.8	25.2	22.3	18.2	13.2	21.0
Neutrality	21.4	22.1	23.2	24.6	25.2	26.2	26.8	26.4	25.6	24.7	23.4	21.9	24.3
Upper limit	23.9	24.6	25.7	27.1	27.7	28.7	29.3	28.9	28.1	27.2	25.9	24.4	26.8
Lower limit	18.9	19.6	20.7	22.1	22.7	23.7	24.3	23.9	23.1	22.2	20.9	19.4	21.8
Rain (mm)[b]	57.2	57.4	77.0	56.4	97.3	115.6	91.2	53.8	76.5	99.8	74.9	62.7	919.8
Prec (mm)[c]	200	204	209	212	199	198	211	224	211	212	207	204	2491
Wind (m/s)	4.1	3.7	4.0	3.8	3.3	3.4	3.1	2.9	2.9	3.2	4.0	3.2	3.5
HDD	224	132	74	3	4	0	0	0	0	9	64	184	694
CDD	5	5	38	109	175	267	339	288	199	125	47	9	1606

(Continued)

TABLE 7.5 (CONTINUED)
Climatic Data, Austin, Texas

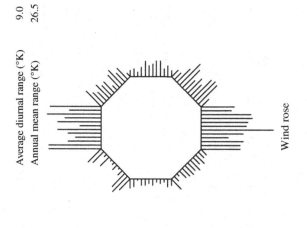

| Average diurnal range (°K) | 9.0 |
| Annual mean range (°K) | 26.5 |

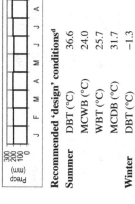

Wind rose

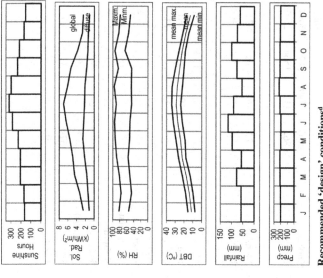

Recommended 'design' conditions[d]

Summer	DBT (°C)	36.6
	MCWB (°C)	24.0
	WBT (°C)	25.7
	MCDB (°C)	31.7
Winter	DBT (°C)	−1.3

Sources of data: https://energyplus.net/weather, [a]WMO (2010), [b]http://worldweather.wmo.int/en/city.html?cityId=719 [c]NOAAA (2017), [d]ASHRAE(2009).

TABLE 7.6
Climatic Data, Baltimore, Maryland

Latitude			N 39° 10'			Longitude			W 76°40'			Altitude			45 m	
Climate			Mixed Humid						4A			Köppen			Dfa	
Months	Jan	Feb	Mar	Apr	May	Jun	Jul	Aug	Sep	Oct	Nov	Dec	Year
Sunshine h[a]	155.4	164	215	230.7	254.5	277.3	290.1	264.4	221.8	205.5	158.5	144.5	2581.7
Cloud (%)	57.0	58.9	57.3	60.1	55.9	52.8	52.5	68.5	57.8	54.2	50.0	59.5	57.05
Solar irradiation daily average (Wh/m²)													
Global	2017	2749	3875	5094	5629	6460	5983	5257	4301	3449	2220	1825	4072
Diffuse	862	1173	1564	2102	2397	2579	2503	2509	1937	1367	974	880	1737
Relative humidity (%)													
Morning	72	69	61	74	84	86	80	88	89	78	81	68	77.5
Evening	58	50	38	46	53	46	53	50	63	47	52	48	50.3
Dry-bulb temperature (°C)													
Max	2.1	5.7	12.1	17.7	23.2	28	29.3	28.9	24.6	17.3	13.3	5.3	17.3
Min	−2.4	−0.4	3.8	9.1	13.3	17.5	21.4	19.8	16.8	9	5.5	−0.8	9.4
Mean	−0.2	2.7	8.0	13.4	18.3	22.8	25.4	24.4	20.7	13.2	9.4	2.3	13.3
Neutrality	20.9	20.9	20.9	22.0	23.5	24.9	25.7	25.3	24.2	21.9	20.9	20.9	21.9
Upper limit	23.4	23.4	23.4	24.5	26.0	27.4	28.2	27.8	26.7	24.4	23.4	23.4	24.4
Lower limit	18.4	18.4	18.4	19.5	21.0	22.4	23.2	22.9	21.7	19.4	18.4	18.4	19.4
Rain (mm)[b]	74.2	77.5	101.6	83.8	107.2	94.2	86.6	93.7	95.5	83.8	90.2	89.4	1077.7
Prec (mm)[c]	305	290	390	319	399	346	407	329	403	333	330	337	4188
Wind (m/s)	4.3	5	4.8	4.2	3.7	3.7	3.4	2.8	3.7	3.6	3.8	4.1	3.9
HDD	574	442	314	144	53	3	0	0	11	172	273	494	2480
CDD	0	0	4	0	63	151	228	195	84	6	0	0	731

(Continued)

TABLE 7.6 (CONTINUED)
Climatic Data, Baltimore, Maryland

| Average diurnal range (°K) | 7.9 |
| Annual mean range (°K) | 31.7 |

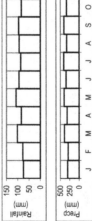

Wind rose

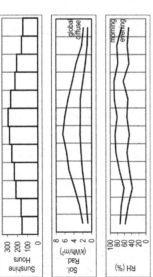

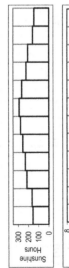

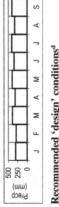

Recommended 'design' conditions[d]

Summer	DBT (°C)	32.9
	MCWB (°C)	23.5
	WBT (°C)	24.9
	MCDB (°C)	30.3
Winter	DBT (°C)	−8.2

Sources of data: https://energyplus.net/weather, [a]WMO (2010), [b]http://worldweather.wmo.int/en/city.html?cityId=720[c]NOAAA (2017), [d]ASHRAE(2009).

TABLE 7.7

Climatic Data, Baton Rouge, Louisiana

Latitude	N 30° 31′				Longitude			W 91° 9′		Altitude		20 m	
Climate	Hot Humid				ASHRAE			2A		Köppen		Cfa	
Months	Jan	Feb	Mar	Apr	May	Jun	Jul	Aug	Sep	Oct	Nov	Dec	Year
Sunshine h[a]	153.0	161.5	219.4	251.9	278.9	274.3	257.1	251.9	228.7	242.6	171.8	157.8	2648.9
Cloud (%)	61.3	63.6	65.2	44.0	57.7	50.0	53.1	41.0	40.4	40.2	51.4	47.3	51.3
Solar irradiation daily average (Wh/m²)													
Global	2444	3323	4217	5394	5619	5962	5811	5653	4535	4330	2825	2564	4390
Diffuse	1031	1601	2029	2269	2670	2624	2575	2230	2035	1550	1169	999	1899
Relative humidity (%)													
Morning	84	88	93	95	88	89	94	91	91	95	91	90	90.8
Evening	62	57	55	50	59	49	61	59	55	51	61	53	56.0
Dry-bulb temperature (°C)													
Max	14.2	15.0	22.0	25.0	28.4	31.3	31.4	31.0	29.7	25.9	20.0	15.0	24.1
Min	7.5	7.5	12.5	14.5	20.2	21.2	23.3	22.9	20.4	14.6	11.8	6.0	15.2
Mean	10.9	11.3	17.3	19.8	24.3	26.3	27.4	27.0	25.1	20.3	15.9	10.5	19.6
Neutrality	21.2	21.3	23.1	23.9	25.3	25.9	26.3	26.2	25.6	24.1	22.7	21.1	23.9
Upper limit	23.7	23.8	25.6	26.4	27.8	28.4	28.8	28.7	28.1	26.6	25.2	23.6	26.4
Lower limit	18.7	18.8	20.7	21.4	22.8	23.4	23.8	23.7	23.1	21.6	20.2	18.6	21.4
Rain (mm)[a]	145.3	128.0	112.0	113.3	124.2	162.8	126.0	147.8	115.3	119.4	104.1	142.2	1540.4
Prec (mm)[b]	572	504	441	446	489	641	496	582	454	470	410	560	6065
Wind (m/s)	3.6	3.8	3.8	3	2.7	3.1	2.4	2.4	2.6	3.3	3.2	2.7	3.1
HDD	252	200	69	26	0	0	0	0	4	33	108	257	949
CDD	14	3	39	78	194	247	281	273	206	78	25	1	1439

(Continued)

TABLE 7.7 (CONTINUED)
Climatic Data, Baton Rouge, Louisiana

Average diurnal range (°K)	8.9
Annual mean range (°K)	25.4

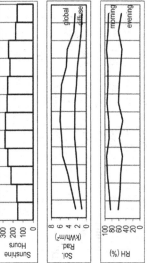

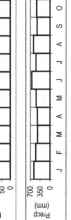

Wind rose

Recommended 'design' conditions[c]

Summer	
DBT (°C)	33.8
MCWB (°C)	25.1
WBT (°C)	26.5
MCDB (°C)	31.1
Winter	
DBT (°C)	−0.5

Sources of data: https://energyplus.net/weather, [a]WMO (2010), [b]http://worldweather.wmo.int/en/city.html?cityId=721 [c]NOAA (2017), [d]ASHRAE(2009), Sunshine hours for New Orleans, Louisiana (81.7 miles).

TABLE 7.8

Climatic Data, Boise, Idaho

		Latitude N 43° 37'					Longitude W 116° 12'				Altitude 701 m			
Climate		Cool Dry					ASHRAE 5B				Köppen BSk			
Months	Jan	Feb	Mar	Apr	May	Jun	Jul	Aug	Sep	Oct	Nov	Dec	Year
Sunshine h[a]	109.3	151.9	238.6	281.4	335.5	351.6	399.8	358.8	303.6	238.1	119.6	105.2	2993.4
Cloud (%)	73.3	65.4	64.6	67.4	44.0	28.5	17.2	34.1	21.8	33.9	65.4	71.4	48.92
Solar irradiation daily average (Wh/m²)													
Global	1685	2490	3652	5454	6307	7555	7613	6633	4995	3402	1837	1408	4419
Diffuse	1037	1215	1437	2173	2215	2105	1782	1443	1285	1160	851	688	1449
Relative humidity (%)													
Morning	88	84	73	71	73	60	52	55	61	61	75	76	69.1
Evening	72	53	37	35	36	25	18	20	27	30	52	57	38.5
Dry-bulb temperature (°C)													
Max	1.9	7.6	11	17.3	19.8	26.3	32.4	29.1	23.9	17.9	8.5	5.7	16.8
Min	-3.9	-0.3	1.6	5.8	9.1	12.3	16.3	13.4	11.6	6.3	1.6	-0.2	6.1
Mean	-1.0	3.7	6.3	11.6	14.5	19.3	24.4	21.3	17.8	12.1	5.1	2.8	11.5
Neutrality	20.9	20.9	20.9	21.4	22.3	23.8	25.3	24.4	23.3	21.6	20.9	20.9	21.4
Upper limit	23.4	23.4	23.4	23.9	24.8	26.3	27.8	26.9	25.8	24.1	23.4	23.4	23.9
Lower limit	18.4	18.4	18.4	18.9	19.8	21.3	22.9	21.9	20.8	19.1	18.4	18.4	18.9
Rain (mm)[b]	31.5	25.1	35.3	31.2	35.3	17.5	8.4	6.1	14.7	19.1	34.3	39.4	297.9
Prec (mm)[c]	124	99	139	123	139	69	33	24	58	75	135	155	1173
Wind (m/s)	3	3.3	3.7	4	3.1	3.7	3.2	3.6	3.3	3.3	2.9	3.4	3.4
HDD	610	428	370	194	119	35	1	19	54	199	419	500	2948
CDD	0	0	0	1	14	88	204	119	36	3	0	0	465

(Continued)

TABLE 7.8 (CONTINUED)
Climatic Data, Boise, Idaho

Average diurnal range (°K)	10.7
Annual mean range (°K)	36.3

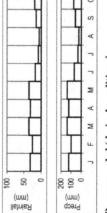

Wind rose

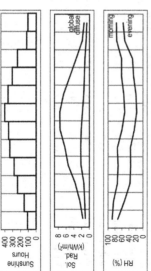

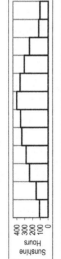

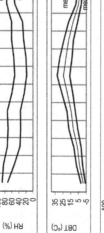

Recommended 'design' conditions[d]

Summer		
	DBT (°C)	35.0
	MCWB (°C)	17.3
	WBT (°C)	18.3
	MCDB (°C)	32.2
Winter	DBT (°C)	−11.9

Sources of data: https://energyplus.net/weather, [a]WMO (2010), [b]http://worldweather.wmo.int/en/city.html?cityId=725 [c]NOAA (2017), [d]ASHRAE(2009).

TABLE 7.9

Climatic Data, Boston, Massachusetts

Latitude		N 42° 22'				Longitude		W 71° 1'			Altitude		6 m	
Climate		Cool Humid				ASHRAE		5A			Köppen		Dfa	
Months	Jan	Feb	Mar	Apr	May	Jun	Jul	Aug	Sep	Oct	Nov	Dec	Year
Sunshine hᵃ	163.4	168.4	213.7	227.2	267.3	286.5	300.9	277.3	237.1	206.3	143.2	142.3	2633.6
Cloud (%)	59.2	53.8	61.8	63.2	59.5	56.7	56.8	65.5	51.0	56.8	59.5	54.5	58.2
Solar irradiation daily average (Wh/m²)													
Global	1791	2776	3608	4512	5572	5764	6068	5347	4258	2985	2007	1528	3851
Diffuse	881	1373	1711	2160	2463	2562	2782	2492	1849	1393	997	743	1784
Relative humidity (%)													
Morning	65	61	67	76	77	77	80	86	81	77	77	65	74.1
Evening	55	54	51	60	61	58	53	60	54	53	64	53	56.3
Dry-bulb temperature (°C)													
Max	-0.8	1.9	6.7	11.2	17.8	22.3	26.8	25	21.9	15.8	8.3	4.7	13.5
Min	-4.6	-2.5	0.9	6.0	11.9	16.1	19.9	18.6	14.6	9.5	4.4	0.4	7.9
Mean	-2.7	-0.3	3.8	8.6	14.9	19.2	23.4	21.8	18.3	12.7	6.4	2.6	10.7
Neutrality	20.9	20.9	20.9	20.9	22.4	23.8	25.0	24.6	23.5	21.7	20.9	20.9	21.1
Upper limit	23.4	23.4	23.4	23.4	24.9	26.3	27.5	27.1	26.0	24.2	23.4	23.4	23.6
Lower limit	18.4	18.4	18.4	18.4	19.9	21.3	22.5	22.1	21.0	19.2	18.4	18.4	18.6
Rain (mm)ᵃ	85.3	82.6	109.7	95.0	88.6	93.5	87.1	85.1	87.4	100.1	101.3	96.0	1111.7
Prec (mm)ᵇ	336	325	432	374	349	368	343	335	344	394	399	378	4377
Wind (m/s)	5.8	5.9	6	6.5	5.6	4.9	5	5.3	4.6	5.3	4.9	6.1	5.5
HDD	650	519	441	289	116	45	1	2	31	184	352	491	3121
CDD	0	0	0	6	20	71	168	116	34	5	0	0	420

(Continued)

TABLE 7.9 (CONTINUED)
Climatic Data, Boston, Massachusetts

| | | | Average diurnal range (°K) | 5.5 |
| | | | Annual mean range (°K) | 31.4 |

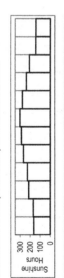

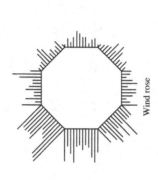

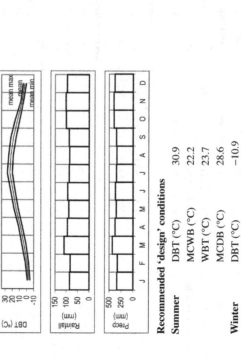

Wind rose

Recommended 'design' conditions

Summer	DBT (°C)	30.9
	MCWB (°C)	22.2
	WBT (°C)	23.7
	MCDB (°C)	28.6
Winter	DBT (°C)	−10.9

Sources of data: https://energyplus.net/weather; [a]WMO (2010), [b]http://worldweather.wmo.int/en/city.html?cityId=273[c]NOAA (2017), [d]ASHRAE (2009).

TABLE 7.10

Climatic Data, Charleston, West Virginia

Latitude	N 38° 22'				Longitude		W 81° 34'			Altitude		310 m		
Climate	Mixed Humid				ASHRAE		4A			Köppen		Dfa		
Months	Jan	Feb	Mar	Apr	May	Jun	Jul	Aug	Sep	Oct	Nov	Dec	Year
Sunshine h[a]	179.3	186.7	243.9	275.1	294.8	279.5	287.8	256.7	219.7	224.5	189.5	171.3	2808.8
Cloud (%)	79.8	73.8	65.5	59.1	58.8	55.4	52.0	62.6	56.0	54.5	63.8	71.5	62.7
				Solar irradiation daily average (Wh/m²)										
Global	1921	2632	3712	4997	5904	6251	6003	5361	4271	3224	2261	1783	4027
Diffuse	1243	1420	1779	2128	2654	2723	2773	2514	1782	1451	1236	1017	1893
					Relative humidity (%)									
Morning	74	80	69	83	85	92	96	92	95	91	86	79	85.2
Evening	59	60	44	43	49	56	59	66	56	50	52	55	54.1
					Dry-bulb temperature (°C)									
Max	1.7	4.5	14.1	17.7	23.0	27.1	29.4	26.6	24.2	18.2	12.6	6.2	17.1
Min	−2.7	−0.9	5.2	7.2	12.5	16.9	19.6	19	14.9	8.1	4.3	−0.1	8.7
Mean	−0.5	1.8	9.7	12.5	17.8	22.0	24.5	22.8	19.6	13.2	8.5	3.1	12.9
Neutrality	20.9	20.9	20.9	21.7	23.3	24.6	25.4	24.9	23.9	21.9	20.9	20.9	21.8
Upper limit	23.4	23.4	23.4	24.2	25.8	27.1	27.9	27.4	26.4	24.4	23.4	23.4	24.3
Lower limit	18.4	18.4	18.4	19.2	20.8	22.1	22.9	22.4	21.4	19.4	18.4	18.4	19.3
Rain (mm)[b]	76.2	81.0	99.3	82.3	121.9	109.0	125.5	95.0	82.6	67.8	94.7	83.1	1118.4
Prec (mm)[c]	300	319	391	324	480	429	494	374	325	267	373	327	4403
Wind (m/s)	4.1	3.2	3.4	2.6	2.7	1.9	2.1	1.8	2.1	1.4	3.3	2.9	2.6
HDD	588	464	262	183	48	6	0	2	20	175	291	483	2522
CDD	0	0	5	23	38	130	196	146	53	1	0	0	592

(*Continued*)

TABLE 7.10 (CONTINUED)
Climatic Data, Charleston, West Virginia

| | Average diurnal range (°K) | 8.4 |
| | Annual mean range (°K) | 32.1 |

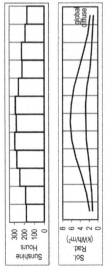
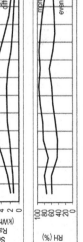
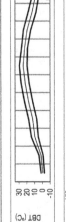
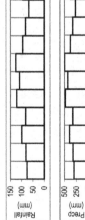

Wind rose

Recommended 'design' conditions[d]

Summer	DBT (°C)	31.6
	MCWB (°C)	22.8
	WBT (°C)	24.1
	MCDB (°C)	29.0
Winter	DBT (°C)	−9.6

Sources of data: https://energyplus.net/weather, [a]WMO (2010), [b]http://worldweather.wmo.int/en/city.html?cityId=773[c]NOAA (2017), [d]ASHRAE(2009).

TABLE 7.11

Climatic Data, Cheyenne, Wyoming

Latitude	N 41° 09'					Longitude				W 104° 48'			Altitude	1867 m
Climate	Cold Dry					ASHRAE				6B			Köppen	BSk
Months	Jan	Feb	Mar	Apr	May	Jun	Jul	Aug	Sep	Oct	Nov	Dec	Year
Sunshine h[a]	190.7	202.6	253.1	271.9	291.9	303.2	317.5	297.4	262.3	237	178.8	175.4	2981.8
Cloud (%)	56.7	58.8	52.1	65.0	68.8	46.1	50.8	49.2	41.8	44.1	53.3	61.1	53.97
Solar irradiation daily average (Wh/m²)													
Global	2247	2896	4417	5060	5719	6959	6697	5912	4759	3684	2412	1919	4390
Diffuse	773	1078	1422	2178	2642	2215	2158	1873	1530	973	934	716	1541
Relative humidity (%)													
Morning	67	65	72	68	71	66	62	70	67	58	55	63	65.3
Evening	44	43	38	37	46	34	29	34	34	30	37	43	37.4
Dry-bulb temperature (°C)													
Max	1.8	1.7	7.2	11.3	13.4	22.0	26.8	25.1	20.6	13.6	5.9	4.2	12.8
Min	−5.8	−6.1	−3.5	1.4	5.6	10.0	14.0	12.7	9.3	3.1	−2.1	−2.7	3.0
Mean	−2.0	−2.2	1.9	6.4	9.5	16.0	20.4	18.9	15.0	8.4	1.9	0.8	7.9
Neutrality	20.9	20.9	20.9	20.9	20.9	22.8	24.1	23.7	22.4	20.9	20.9	20.9	20.2
Upper limit	23.4	23.4	23.4	23.4	23.4	25.3	26.6	26.2	24.9	23.4	23.4	23.4	22.7
Lower limit	18.4	18.4	18.4	18.4	18.4	20.3	21.6	21.2	19.9	18.4	18.4	18.4	17.7
Rain (mm)[b]	8.4	11.9	26.7	45.2	59.4	59.4	55.6	49.5	37.6	23.6	15.0	12.4	404.7
Prec (mm)[c]	33	47	105	178	234	234	219	195	148	93	59	49	1594
Wind (m/s)	7.1	6.3	6.7	5.8	5.1	4.4	4.8	4.3	4.7	6.1	5.6	6.2	5.6
HDD	658	582	516	357	261	77	5	19	116	321	518	566	3996
CDD	0	0	0	0	0	30	85	43	17	0	0	0	175

(Continued)

TABLE 7.11 (CONTINUED)
Climatic Data, Cheyenne, Wyoming

| | Average diurnal range (°K) | 9.8 |
| | Annual mean range (°K) | 32.9 |

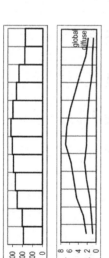

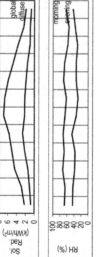

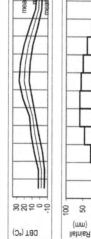

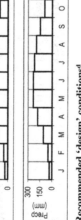

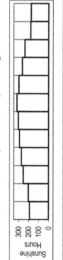

Wind rose

Recommended 'design' conditions[d]

Summer	DBT (°C)	30.2
	MCWB (°C)	14.4
	WBT (°C)	16.5
	MCDB (°C)	24.9
Winter	DBT (°C)	−16.8

Sources of data: https://energyplus.net/weather, [a]WMO (2010), [b]http://worldweather.wmo.int/en/city.html?cityId=736 [c]NOAA (2017), [d]ASHRAE (2009).

TABLE 7.12

Climatic Data, Columbia, South Carolina

	Jan	Feb	Mar	Apr	May	Jun	Jul	Aug	Sep	Oct	Nov	Dec	Year
Latitude	N 33° 57'				Longitude		W 81° 7'			Altitude		65 m	
Climate	Warm Humid				ASHRAE		3A			Köppen		Cfa	
Months	Jan	Feb	Mar	Apr	May	Jun	Jul	Aug	Sep	Oct	Nov	Dec	Year
Sunshine h[a]	172.7	180.7	237.3	269.6	292.9	280	286	263.3	239.8	235	193.8	175	2826.1
Cloud (%)	56.1	59.5	63.5	48.0	52.6	61.4	58.1	55.2	57.1	41.2	49.1	51.5	54.44
Solar irradiation daily average (Wh/m²)													
Global	2544	3228	4420	5893	6029	6421	6108	5435	4748	3685	2703	2377	4466
Diffuse	1131	1235	1888	2015	2539	2842	2700	2547	2000	1484	1085	985	1871
Relative humidity (%)													
Morning	74	83	89	85	91	90	89	96	96	93	91	80	88.1
Evening	42	47	46	35	50	45	46	59	54	56	53	50	48.6
Dry-bulb temperature (°C)													
Max	10.6	15.2	19.9	24.6	27.3	31.7	32.4	30.1	28.9	22.5	18.9	12.0	22.8
Min	2.3	5.2	8.3	10.7	16.3	20.0	21.9	20.9	18.7	11.9	9.3	3.5	12.4
Mean	6.5	10.2	14.1	17.7	21.8	25.9	27.2	25.5	23.8	17.2	14.1	7.8	17.6
Neutrality	20.9	21.0	22.2	23.3	24.6	25.8	26.2	25.7	25.2	23.1	22.2	20.9	23.3
Upper limit	23.4	23.5	24.7	25.8	27.1	28.3	28.7	28.2	27.7	25.6	24.7	23.4	25.8
Lower limit	18.4	18.5	19.7	20.8	22.1	23.3	23.7	23.2	22.7	20.6	19.7	18.4	20.8
Rain (mm)[b]	97.8	95.8	106.9	67.6	76.2	134.4	138.2	123.4	91.4	79.5	78.5	86.4	1176.1
Prec (mm)[c]	363	365	388	269	297	491	518	494	404	332	299	320	4540
Wind (m/s)	3.5	3.2	3.1	3.8	3.4	3.1	3.2	2.2	2.3	2.4	2.1	2.8	2.9
HDD	377	234	153	55	4	0	0	0	0	63	153	349	1039
CDD	0	4	15	44	118	231	285	218	159	35	19	0	1128

(Continued)

TABLE 7.12 (CONTINUED)
Climatic Data, Columbia, South Carolina

| | Average diurnal range (°K) | 10.4 |
| | Annual mean range (°K) | 30.1 |

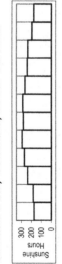
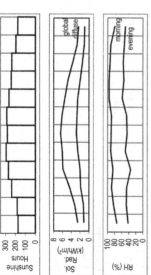
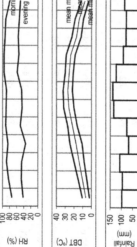
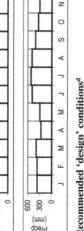

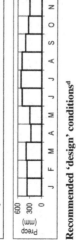

Wind rose

Recommended 'design' conditions[d]

Summer		
	DBT (°C)	34.7
	MCWB (°C)	23.9
	WBT (°C)	25.3
	MCDB (°C)	31.5
Winter		
	DBT (°C)	−3.4

Sources of data: https://energyplus.net/weather, [a]WMO (2010), [b]http://worldweather.wmo.int/en/city.html?cityId=740[c]NOAA (2017), [d]ASHRAE(2009).

TABLE 7.13
Climatic Data, Columbus, Ohio

Latitude	N 39° 58'				Longitude	W 82° 52'				Altitude	247 m		
Climate	Cool Humid				ASHRAE	5A				Köppen	Dfa		
Months	Jan	Feb	Mar	Apr	May	Jun	Jul	Aug	Sep	Oct	Nov	Dec	Year
Sunshine h[a]	110.6	126.3	162	201.8	243.4	258.1	260.9	235.9	212	183.1	104.2	84.3	2182.6
Cloud (%)	62.6	71.4	65.1	58.8	58.1	59.4	57.3	46.1	41.1	49.4	66.8	73.5	59.13
Solar irradiation daily average (Wh/m²)													
Global	1770	2402	3645	4631	5399	6035	5835	5192	4227	3142	1888	1509	3806
Diffuse	939	1458	1896	2316	2670	2869	2775	2394	1813	1436	1150	1021	1895
Relative humidity (%)													
Morning	82	79	83	81	75	87	91	90	92	80	82	77	83.3
Evening	61	62	55	47	50	56	58	57	54	53	57	69	56.6
Dry-bulb temperature (°C)													
Max	−0.1	1.0	9.4	16.2	21.8	26	27.1	26.4	22.4	15.4	10.3	2.3	14.9
Min	−4.5	−4.4	1.4	7.2	12.2	16.5	18.9	17.6	14.2	6.8	4.3	−1.3	7.4
Mean	−2.3	−1.7	5.4	11.7	17.0	21.3	23.0	22.0	18.3	11.1	7.3	0.5	11.1
Neutrality	20.9	20.9	20.9	21.4	23.1	24.4	24.9	24.6	23.5	21.2	20.9	20.9	21.3
Upper limit	23.4	23.4	23.4	23.9	25.6	26.9	27.4	27.1	26.0	23.7	23.4	23.4	23.8
Lower limit	18.4	18.4	18.4	18.9	20.6	21.9	22.4	22.1	21.0	18.7	18.4	18.4	18.8
Rain (mm)[b]	69.3	57.2	76.7	86.4	105.9	101.9	121.7	84.3	72.1	66.3	81.3	75.4	998.5
Prec (mm)[c]	253	197	299	342	441	449	437	339	292	250	318	281	3898
Wind (m/s)	4.2	4.3	3.5	4.0	3.7	3.3	2.1	3.0	2.9	3.2	3.6	4.1	3.5
HDD	634	555	395	188	58	7	0	1	37	226	338	546	2985
CDD	0	0	3	0	38	96	157	122	37	0	0	0	453

(*Continued*)

TABLE 7.13 (CONTINUED)
Climatic Data, Columbus, Ohio

Average diurnal range (°K)	7.4
Annual mean range (°K)	31.6

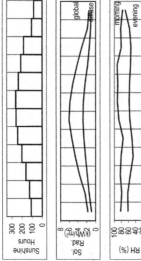
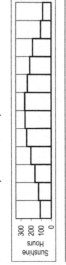
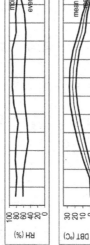
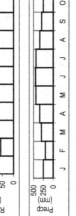
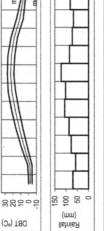

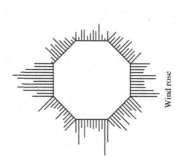

Wind rose

Recommended 'design' conditions[d]

Summer		
DBT (°C)	31.5
MCWB (°C)	22.7
WBT (°C)	24.0
MCDB (°C)	29.2
Winter		
DBT (°C)	−12.7

Sources of data: https://energyplus.net/weather; [a]WMO (2010), [b]http://worldweather.wmo.int/en/city.html?cityId=742 [c]NOAA (2017), [d]ASHRAE (2009).

TABLE 7.14

Climatic Data, Concord, New Hampshire

	Latitude N 43° 12'			Longitude W 71° 30'					Altitude 106m			
	Climate Cold Humid			ASHRAE 6A					Köppen Dfb			
Months	Jan	Feb	Mar	Apr	May	Jun	Jul	Aug	Sep	Oct	Nov	Dec	Year
Sunshine h[a]	162.8	171.8	210.5	223.2	258.4	274.3	295.8	261.9	214.7	183.4	127.8	134.8	2519.4
Cloud (%)	59.1	58.2	61.6	57.2	57.7	61.5	60.4	55.8	58.3	55.3	62.4	73.5	59.0
Solar irradiation daily average (Wh/m²)													
Global	1594	2732	3613	4823	5288	5931	6014	5257	4101	2791	1798	1523	3789
Diffuse	724	1470	1741	2171	2385	2616	2614	2309	1821	1319	936	866	1748
Relative humidity (%)													
Morning	77	84	71	79	87	91	95	93	95	88	82	83	85.4
Evening	47	54	44	43	48	53	49	50	56	49	56	59	50.7
Dry-bulb temperature (°C)													
Max	-1.8	-1	5.8	12.9	19.9	23.4	27.7	25.3	21.9	14.6	6.1	-1.1	12.8
Min	-9.2	-9.7	-2.5	2.5	8.6	13.4	15.8	13.7	10.1	4.0	-0.2	-8.1	3.2
Mean	-5.5	-5.4	1.7	7.7	14.3	18.4	21.8	19.5	16.0	9.3	3.0	-4.6	8.0
Neutrality	20.9	20.9	20.9	20.9	22.2	23.5	24.5	23.8	22.8	20.9	20.9	20.9	20.9
Upper limit	23.4	23.4	23.4	23.4	24.7	26.0	27.0	26.3	25.3	23.4	23.4	23.4	23.4
Lower limit	18.4	18.4	18.4	18.4	19.7	21.0	22.0	21.3	20.3	18.4	18.4	18.4	18.4
Rain (mm)[a]	68.6	66.5	83.1	86.6	93.0	93.7	95.0	80.8	85.9	102.6	94.5	81.3	1031.6
Prec (mm)[b]	270	262	327	341	366	369	374	318	338	404	372	320	4061
Wind (m/s)	3.2	2.7	3.4	3.2	2.9	2.6	2.8	2.3	2.2	2.5	3.6	2.7	2.8
HDD	738	651	508	310	125	33	7	14	87	289	464	705	3931
CDD	0	0	0	0	10	45	125	64	18	0	0	0	262

(Continued)

TABLE 7.14 (CONTINUED)
Climatic Data, Concord, New Hampshire

| | | Average diurnal range (°K) | 9.6 |
| | | Annual mean range (°K) | 37.4 |

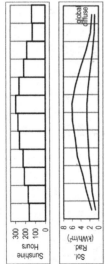

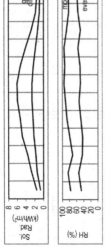

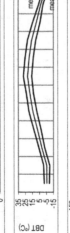

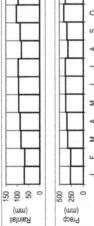

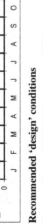

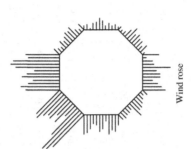

Wind rose

Recommended 'design' conditions

Summer	DBT (°C)	30.5
	MCWB (°C)	21.0
	WBT (°C)	22.7
	MCDB (°C)	27.8
Winter	DBT (°C)	−17.6

Sources of data: https://energyplus.net/weather, [a]WMO (2010), [b]http://worldweather.wmo.int/en/city.html?cityId=743, [c]NOAA (2017), [d]ASHRAE(2009).

TABLE 7.15

Climatic Data, Denver, Colorado

		N 39° 49'			Longitude			W 104°39'			Altitude		1650 m	
Latitude		Cool Dry			ASHRAE			5B			Köppen		BSk	
Climate														
Months	Jan	Feb	Mar	Apr	May	Jun	Jul	Aug	Sep	Oct	Nov	Dec	Year
Sunshine h[a]	215.3	211.1	255.6	276.2	290	315.3	325	306.4	272.3	249.2	194.3	195.9	3106.6
Cloud (%)	41.2	50.6	47.9	74.9	56.9	38.4	42.1	51.6	46.0	48.0	49.0	41.1	49.0
Solar irradiation daily average (Wh/m²)													
Global	2503	3168	4658	4788	6327	7268	6716	5989	5175	3685	2465	2078	4568
Diffuse	738	1108	1373	2018	2358	2099	2199	1974	1617	1123	972	695	1523
Relative humidity (%)													
Morning	58	67	55	74	74	66	72	61	60	76	61	60	65.3
Evening	29	34	23	45	33	24	30	28	20	42	36	31	31.3
Dry-bulb temperature (°C)													
Max	7.5	6.1	12.0	11.1	21.3	30.5	28.8	28.8	26.8	16.5	8.6	8.5	17.2
Min	−3.7	−5.2	1.0	1.1	9.3	14.9	15.4	16.9	12.3	4.9	−0.8	−3.2	5.2
Mean	1.9	0.5	6.5	6.1	15.3	22.7	22.1	22.9	19.6	10.7	3.9	2.7	11.2
Neutrality	20.9	20.9	20.9	20.9	22.5	24.8	24.7	24.9	23.9	21.1	20.9	20.9	21.3
Upper limit	23.4	23.4	23.4	23.4	25.0	27.3	27.2	27.4	26.4	23.6	23.4	23.4	23.8
Lower limit	18.4	18.4	18.4	18.4	20.0	22.3	22.2	22.4	21.4	18.6	18.4	18.4	18.8
Rain (mm)[a]	10.4	9.4	23.4	43.4	53.8	50.3	54.9	42.9	24.4	25.9	15.5	8.9	363.2
Prec (mm)[b]	41	37	92	171	212	198	216	169	96	102	61	35	1430
Wind (m/s)	3.8	3.5	4.4	5.3	3.8	3.4	3	4.5	4.1	3.8	4.2	3.2	3.9
HDD	534	506	368	365	97	1	6	1	36	250	453	514	3131
CDD	0	0	0	0	19	153	138	144	71	3	0	0	528

(Continued)

TABLE 7.15 (CONTINUED)
Climatic Data, Denver, Colorado

Average diurnal range (°K)	12.0
Annual mean range (°K)	35.7

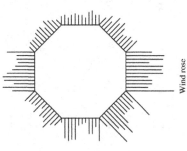

Wind rose

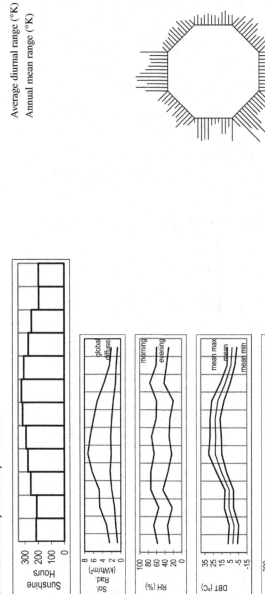

Recommended 'design' conditions

Summer	DBT (°C)	33.2
	MCWB (°C)	15.6
	WBT (°C)	17.6
	MCDB (°C)	27.0
Winter	DBT (°C)	−14.0

Sources of data: https://energyplus.net/weather, aWMO (2010), bhttp://worldweather.wmo.int/en/city.html?cityId=271, cNOAA (2017), dASHRAE(2009).

TABLE 7.16

Climatic Data, Des Moines, Iowa

Latitude	N 41° 31′				Longitude		W 93° 40′			Altitude		292 m	
Climate	Cool Humid				ASHRAE		5A			Köppen		Dfa	
Months	Jan	Feb	Mar	Apr	May	Jun	Jul	Aug	Sep	Oct	Nov	Dec	Year
Sunshine h[a]	157.7	163.3	206	222.2	276	312.1	337.8	297.9	239.8	210	138.5	129.2	2690.5
Cloud (%)	58.0	51.7	61.2	53.4	53.1	48.3	38.6	45.1	40.4	39.6	60.0	61.2	50.86
				Solar irradiation daily average (Wh/m²)									
Global	1965	2854	3816	4999	5890	6710	6425	5707	4740	3306	2127	1704	4187
Diffuse	956	1231	1657	2025	2468	2479	2286	2183	1647	1243	957	845	1665
				Relative humidity (%)									
Morning	69	74	70	72	69	78	85	91	78	71	80	76	76.1
Evening	48	63	52	40	42	49	56	60	48	44	58	62	51.8
				Dry-bulb temperature (°C)									
Max	-2.0	-1.4	7.8	16.1	22.7	27.5	29.3	27.9	24.6	14.8	8.4	-0.6	14.6
Min	-9.3	-8.2	1.0	5.6	12.1	17.8	20.5	18.9	14.9	5.6	1.0	-5.8	6.2
Mean	-5.7	-4.8	4.4	10.9	17.4	22.7	24.9	23.4	19.8	10.2	4.7	-3.2	10.4
Neutrality	20.9	20.9	20.9	21.2	23.2	24.8	25.5	25.1	23.9	21.0	20.9	20.9	21.0
Upper limit	23.4	23.4	23.4	23.7	25.7	27.3	28.0	27.6	26.4	23.5	23.4	23.4	23.5
Lower limit	18.4	18.4	18.4	18.7	20.7	22.3	23.0	22.6	21.4	18.5	18.4	18.4	18.5
Rain (mm)[b]	25.4	32.5	58.4	98	120.4	125.5	113.5	104.9	77.5	67.1	55.6	36.1	914.9
Prec (mm)[c]	92	123	223	371	466	470	450	474	312	252	218	132	3583
Wind (m/s)	4.7	4.8	6	4.8	4	4.5	3.9	3.5	4.2	4.5	5.7	4.8	4.6
HDD	745	649	428	230	63	4	0	0	39	256	414	673	3501
CDD	0	0	0	12	55	152	216	161	77	1	0	0	674

(Continued)

TABLE 7.16 (CONTINUED)
Climatic Data, Des Moines, Iowa

Average diurnal range (°K)	8.4
Annual mean range (°K)	38.6

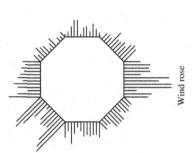

Wind rose

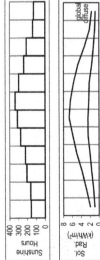
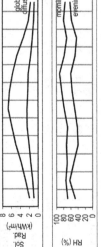
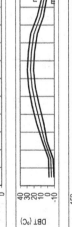
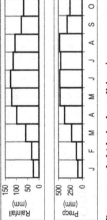

Recommended 'design' conditions[d]

Summer		
	DBT (°C)	32.3
	MCWB (°C)	23.9
	WBT (°C)	25.0
	MCDB (°C)	30.7
Winter		
	DBT (°C)	−18.7

Sources of data: https://energyplus.net/weather, [a]WMO (2010), [b]http://worldweather.wmo.int/en/city.html?cityId=748, [c]NOAA (2017), [d]ASHRAE (2009).

TABLE 7.17

Climatic Data, Dover, Delaware

Latitude	N 39° 7'				Longitude	W 75° 28'			Altitude			7 m	
Climate	Mixed Humid				ASHRAE	4A			Köppen			Cfa/Dfa	
Months	Jan	Feb	Mar	Apr	May	Jun	Jul	Aug	Sep	Oct	Nov	Dec	Year
Sunshine h[a]	155.7	154.7	202.8	217.0	245.1	271.2	275.6	260.1	219.3	204.5	154.7	137.7	2498.4
Cloud (%)	71.8	67.6	59.9	57.8	63.8	55.0	67.1	59.0	46.7	32.4	63.4	62.1	58.9
				Solar irradiation daily average (Wh/m²)									
Global	2065	2563	3624	4607	5193	5872	5097	5134	4363	3888	1826	1960	3849
Diffuse	881	1228	1699	2163	2815	2668	2795	2553	1929	1355	1155	755	1833
				Relative humidity (%)									
Morning	79	78	70	77	77	89	92	93	88	82	82	76	81.9
Evening	64	54	49	48	49	52	63	66	53	48	53	54	54.4
				Dry-bulb temperature (°C)									
Max	3.3	4.7	8.0	15.8	20.0	27.3	28.4	26.0	23.8	19.3	9.3	7.8	16.1
Min	−1.0	−0.5	0.8	6.8	11.1	18.3	21.6	18.8	15.5	10.3	2.7	1.2	8.8
Mean	1.2	2.1	4.4	11.3	15.6	22.8	25.0	22.4	19.7	14.8	6.0	4.5	12.5
Neutrality	20.9	20.9	20.9	21.3	22.6	24.9	25.6	24.7	23.9	22.4	20.9	20.9	21.7
Upper limit	23.4	23.4	23.4	23.8	25.1	27.4	28.1	27.2	26.4	24.9	23.4	23.4	24.2
Lower limit	18.4	18.4	18.4	18.8	20.1	22.4	23.1	22.2	21.4	19.9	18.4	18.4	19.2
Rain (mm)[a]	86.6	78.0	109.5	98.6	108.0	101.6	103.9	110.7	104.9	86.9	88.4	92.7	1169.8
Prec (mm)[b]	341	307	431	388	425	400	409	436	413	342	348	365	4605
Wind (m/s)	4.5	5.3	5.2	3.4	5.0	2.8	2.6	2.2	2.2	4.0	3.9	4.4	3.8
HDD	532	455	432	201	83	4	0	1	11	130	381	436	2666
CDD	0	0	0	4	16	147	213	132	50	15	2	0	579

(Continued)

TABLE 7.17 (CONTINUED)
Climatic Data, Dover, Delaware

Average diurnal range (°K)		7.3
Annual mean range (°K)		29.4

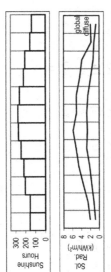
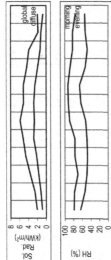
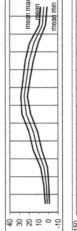
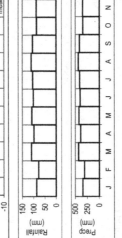

Wind rose

Sunshine Hours — 300, 200, 100, 0

Sol. Rad. (kWh/m²) — 8, 6, 4, 2, 0 — global, diffuse

RH (%) — 100, 80, 60, 40, 20, 0 — morning, evening

DBT (°C) — 40, 30, 20, 10, 0, -10 — mean max, mean, mean min

Rainfall (mm) — 150, 100, 50, 0

Precp (mm) — 500, 250, 0

J F M A M J J A S O N D

Recommended 'design' conditions

Summer		
	DBT (°C)	32.1
	MCWB (°C)	23.8
	WBT (°C)	25.3
	MCDB (°C)	29.6
Winter	DBT (°C)	-7.7

Sources of data: https://energyplus.net/weather, [a]WMO (2010), [b]http://worldweather.wmo.int/en/city.html?cityId=1975, [c]NOAA (2017), [d]ASHRAE (2009), Sunshine hours for Philadelphia International Airport (52 miles).

TABLE 7.18
Climatic Data, Fargo, North Dakota

	Latitude	N 46° 55'			Longitude		W 96° 49'			Altitude		274 m	
Climate		Very Cold Humid			ASHRAE		7A			Köppen		Dfb	
Months	Jan	Feb	Mar	Apr	May	Jun	Jul	Aug	Sep	Oct	Nov	Dec	Year
Sunshine h[a]	140.9	153.9	212.3	241.6	283.2	303.2	350.2	313.2	231.2	178.9	113.1	107.4	2629.1
Cloud (%)	60.6	49.9	57.5	41.8	53.9	41.6	32.2	45.8	46.4	42.0	62.2	61.0	49.58
Solar irradiation daily average (Wh/m^2)													
Global	1500	2506	3616	4774	5873	6313	6307	5451	4046	2630	1796	1317	3844
Diffuse	892	1185	1729	1904	2222	2570	2338	2062	1523	1257	972	669	1610
Relative humidity (%)													
Morning	71	72	84	73	74	83	88	88	86	78	81	81	79.9
Evening	67	65	63	41	44	48	53	46	46	44	72	75	55.3
Dry-bulb temperature (°C)													
Max	-11.4	-7.7	2.0	13.5	21.6	23.9	26.6	24.7	20.7	12.4	-0.9	-10.1	9.6
Min	-16.4	-14.6	-4.0	1.6	10.3	14.1	16.3	14.1	9.5	2.7	-5.5	-14.2	1.2
Mean	-13.9	-11.2	-1.0	7.6	16.0	19.0	21.5	19.4	15.1	7.6	-3.2	-12.2	5.4
Neutrality	20.9	20.9	20.9	20.9	22.7	23.7	24.4	23.8	22.5	20.9	20.9	20.9	19.5
Upper limit	23.4	23.4	23.4	23.4	25.2	26.2	27.0	26.3	25.0	23.4	23.4	23.4	22.0
Lower limit	18.4	18.4	18.4	18.4	20.2	21.2	22.0	21.3	20.0	18.4	18.4	18.4	17.0
Rain (mm)[a]	17.8	15.5	33.0	34.5	71.4	99.1	70.9	65.0	65.3	54.6	25.4	21.1	573.6
Prec (mm)[b]	70	61	130	136	281	390	279	256	257	215	100	83	2258
Wind (m/s)	6.3	5.6	6.0	5.5	4.7	5.0	4.2	4.5	3.8	5.0	6.1	3.5	5.0
HDD	1007	826	599	318	83	30	4	29	128	345	640	947	4956
CDD	0	0	0	3	20	69	112	70	34	0	0	0	308

(Continued)

TABLE 7.18 (CONTINUED)
Climatic Data, Fargo, North Dakota

| | Average diurnal range (°K) | 8.5 |
| | Annual mean range (°K) | 43.0 |

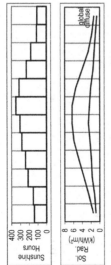
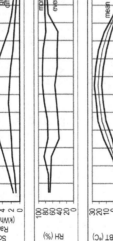
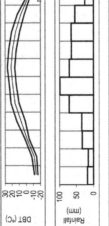
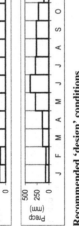

Wind rose

Recommended 'design' conditions

Summer		
	DBT (°C)	30.9
	MCWB (°C)	21.3
	WBT (°C)	23.0
	MCDB (°C)	28.7
Winter		
	DBT (°C)	−26.2

Sources of data: https://energyplus.net/weather, [a]WMO (2010), [b]http://worldweather.wmo.int/en/city.html?cityId=756, [c]NOAA (2017), [d]ASHRAE(2009).

TABLE 7.19

Climatic Data, Harrisburg, Pennsylvania

Latitude	N 40° 13'				Longitude			W 76° 50'			Altitude		104 m	
Climate	Cool Humid				ASHRAE			5A			Köppen		Dfb	
Months	Jan	Feb	Mar	Apr	May	Jun	Jul	Aug	Sep	Oct	Nov	Dec	Year
Sunshine h[a]	154.9	167.2	213.8	235.7	266.7	288.5	310.1	285.4	226.7	199.2	139.6	126	2613.8
Cloud (%)	55.0	64.2	54.4	50.9	53.2	54.2	56.4	61.9	57.1	46.0	60.7	56.1	55.83
Solar irradiation daily average (Wh/m²)													
Global	1948	2717	3741	4823	5561	6073	5697	5064	4118	3184	2111	1609	3887
Diffuse	971	1291	1685	2193	2676	2816	2882	2581	1988	1418	1110	829	1870
Relative humidity (%)													
Morning	70	72	64	78	77	82	90	90	96	88	84	69	80.0
Evening	63	54	41	50	54	55	56	58	61	53	58	50	54.4
Dry-bulb temperature (°C)													
Max	-0.7	2.7	9.7	16.4	20.9	26.3	28.2	26.7	22.9	17.5	10.6	2.7	15.3
Min	-4.0	-2.6	1.4	8.4	11.7	17.5	19.5	18.4	14.6	9.0	3.3	-1.8	8.0
Mean	-2.4	0.1	5.6	12.4	16.3	21.9	23.9	22.6	18.8	13.3	7.0	0.5	11.6
Neutrality	20.9	20.9	20.9	21.6	22.9	24.6	25.2	24.8	23.6	21.9	20.9	20.9	21.4
Upper limit	23.4	23.4	23.4	24.1	25.4	27.1	27.7	27.3	26.1	24.4	23.4	23.4	23.9
Lower limit	18.4	18.4	18.4	19.1	20.4	22.1	22.7	22.3	21.1	19.4	18.4	18.4	18.9
Rain (mm)[b]	74.4	77.7	89.9	88.6	105.4	102.6	108.2	91.7	110.7	86.4	87.6	84.8	1108
Prec (mm)[c]	288	239	337	310	379	360	461	320	407	327	323	323	4074
Wind (m/s)	4.0	4.2	4.0	3.6	3.0	3.1	2.9	2.4	2.4	2.6	3.5	3.8	3.3
HDD	636	510	391	189	74	3	2	3	24	181	344	558	2915
CDD	0	0	0	28	25	120	187	138	42	14	0	0	554

(Continued)

TABLE 7.19 (CONTINUED)
Climatic Data, Harrisburg, Pennsylvania

	Average diurnal range (°K)	7.4
	Annual mean range (°K)	32.2

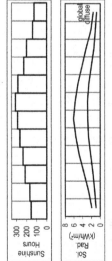

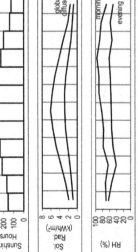

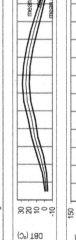

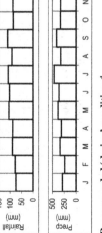

Wind rose

Recommended 'design' conditions[d]

Summer		
	DBT (°C)	32.0
	MCWB (°C)	22.5
	WBT (°C)	24.0
	MCDB (°C)	29.2
Winter	DBT (°C)	−10.4

Sources of data: https://energyplus.net/weather, [a]WMO (2010), [b]http://worldweather.wmo.int/en/city.html?cityId=768, [c]NOAA (2017), [d]ASHRAE (2009).

TABLE 7.20
Climatic Data, Hartford, Connecticut

	Jan	Feb	Mar	Apr	May	Jun	Jul	Aug	Sep	Oct	Nov	Dec	Year
Latitude N 41° 55'					**Longitude** W 72° 40'					**Altitude** 49 m			
Climate Cool Humid					**ASHRAE** 5A					**Köppen** Dfb			
Months	Jan	Feb	Mar	Apr	May	Jun	Jul	Aug	Sep	Oct	Nov	Dec	Year
Sunshine h[a]	169.8	176.1	213.9	228.2	258.6	273.4	293.1	269.6	223.6	199.4	139.4	139.5	2584.6
Cloud (%)	55.1	57.3	60.8	64.8	68.7	52.0	59.7	63.9	54.2	57.1	57.2	56.6	58.94
Solar irradiation daily average (Wh/m²)													
Global	1800	2550	3753	4576	5273	5819	5708	5042	4154	2830	2012	1494	3751
Diffuse	909	1244	1913	2330	2780	2738	2842	2432	2004	1432	970	851	1870
Relative humidity (%)													
Morning	61	72	69	69	75	91	90	89	85	84	67	77	77.4
Evening	43	50	46	38	43	55	52	54	51	57	45	54	49.0
Dry-bulb temperature (°C)													
Max	0.6	1.0	8.5	15.2	21.0	24.8	28.6	26.2	23.0	15.5	10.6	1.4	14.7
Min	-4.6	-3.8	1.1	5.0	10.6	14.5	18.5	17.0	12.4	7.3	3.0	-4.2	6.4
Mean	-2.0	-1.4	4.8	10.1	15.8	19.7	23.6	21.6	17.7	11.4	6.8	-1.4	10.6
Neutrality	20.9	20.9	20.9	20.9	22.7	23.9	25.1	24.5	23.3	21.3	20.9	20.9	21.1
Upper limit	23.4	23.4	23.4	23.4	25.2	26.4	27.6	27.0	25.8	23.8	23.4	23.4	23.6
Lower limit	18.4	18.4	18.4	18.4	20.2	21.4	22.6	22.0	20.8	18.8	18.4	18.4	18.6
Rain (mm)[b]	95.3	84.8	110.5	112.3	107.2	126.0	111.3	102.1	93.7	136.4	110.2	109.7	1299.5
Prec (mm)[c]	323	289	362	372	435	435	418	393	388	437	389	344	4585
Wind (m/s)	4.2	4.1	4.2	3.5	3.5	3.4	3.4	3.1	2.5	3.1	3.5	3.3	3.5
HDD	626	550	422	240	86	32	0	9	61	217	348	609	3200
CDD	0	0	0	1	18	81	166	119	40	0	0	0	425

(Continued)

TABLE 7.20 (CONTINUED)
Climatic Data, Hartford, Connecticut

Average diurnal range (°K)		8.3
Annual mean range (°K)		33.2

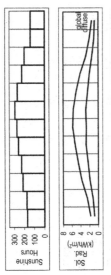

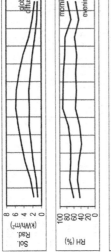

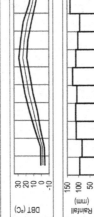

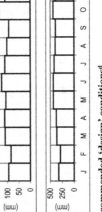

Wind rose

Recommended 'design' conditions[d]

Summer		
	DBT (°C)	31.0
	MCWB (°C)	22.6
	WBT (°C)	24.1
	MCDB (°C)	28.3
Winter	DBT (°C)	−11.9

Sources of data: https://energyplus.net/weather; [a]WMO (2010), [b]http://worldweather.wmo.int/en/city.html?cityId=2005, [c]NOAA (2017), [d]ASHRAE(2009)
Rainfall data for Middletown, Connecticut (16.7 miles).

TABLE 7.21

Climatic Data, Helena, Montana

	Latitude N 46° 35'			Longitude W 111° 58'						Altitude 1167 m			
	Climate Cold Dry			ASHRAE 6B						Köppen BSk			
Months	Jan	Feb	Mar	Apr	May	Jun	Jul	Aug	Sep	Oct	Nov	Dec	Year
Sunshine hª	119.4	149	225.8	243	282	308.7	370.3	324.1	254.6	202.9	118.6	99.9	2698.3
Cloud (%)	66.4	68.7	69.6	63.2	63.0	48.7	40.8	52.4	46.9	51.2	62.4	66.6	58.3
Solar irradiation daily average (Wh/m²)													
Global	1481	2379	3574	4951	5860	6720	7152	5779	4295	2662	1701	1310	3989
Diffuse	723	1084	1533	2082	2404	2398	1830	1843	1518	1100	823	636	1498
Relative humidity (%)													
Morning	79	72	75	75	68	70	62	66	73	81	75	87	73.6
Evening	51	54	40	36	31	31	26	30	35	44	49	63	40.8
Dry-bulb temperature (°C)													
Max	0.3	-0.5	7.2	11.9	17.4	22.9	27.4	25.0	21.2	11.8	6.4	-1.2	12.5
Min	-8.2	-6.4	-2.0	0.5	5.8	10.8	13.1	11.5	8.9	1.8	-1.9	-7.4	2.2
Mean	-4.0	-3.5	2.6	6.2	11.6	16.9	20.3	18.3	15.1	6.8	2.3	-4.3	7.3
Neutrality	20.9	20.9	20.9	20.9	21.4	23.0	24.1	23.5	22.5	20.9	20.9	20.9	20.1
Upper limit	23.4	23.4	23.4	23.4	23.9	25.5	26.6	26.0	25.0	23.4	23.4	23.4	22.6
Lower limit	18.4	18.4	18.4	18.4	18.9	20.5	21.6	21.0	20.0	18.4	18.4	18.4	17.6
Rain (mm)ᵇ	9.1	7.6	15.0	24.9	47.5	52.3	30.2	30.5	27.9	17.3	12.4	10.2	284.9
Prec (mm)ᶜ	36	30	59	98	187	206	119	120	110	68	49	40	1122
Wind (m/s)	2.6	2.6	3.8	4.1	4.2	3.6	3.5	3.3	3.1	2.4	3.1	2.8	3.3
HDD	704	607	491	354	192	62	7	35	114	364	507	714	4151
CDD	0	0	0	0	1	43	86	52	15	0	0	0	197

(Continued)

TABLE 7.21 (CONTINUED)
Climatic Data, Helena, Montana

	°K
Average diurnal range (°K)	10.3
Annual mean range (°K)	35.6

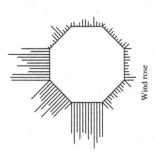

Wind rose

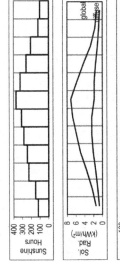
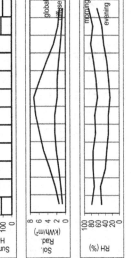
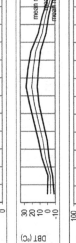
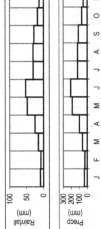
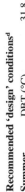

Sunshine Hours
Sol. Rad. (kWh/m²) — global, diffuse
RH (%) — morning, evening
DBT (°C) — mean max, mean, mean min.
Rainfall (mm)
Precip (mm)

J F M A M J J A S O N D

Recommended 'design' conditions[d]

Summer		
	DBT (°C)	31.8
	MCWB (°C)	15.9
	WBT (°C)	17.2
	MCDB (°C)	28.3
Winter	DBT (°C)	−22.3
Winter	DBT (°C)	−11.9

Sources of data: https://energyplus.net/weather, [a]WMO (2010), [b]http://worldweather.wmo.int/en/city.html?cityId=769, [c]NOAA (2017), [d]ASHRAE (2009).

TABLE 7.22

Climatic Data, Honolulu, Hawaii

	Jan	Feb	Mar	Apr	May	Jun	Jul	Aug	Sep	Oct	Nov	Dec	Year
Latitude	N 21° 19'				**Longitude**			W 155° 57'		**Altitude**			2 m
Climate	Very Hot Humid				ASHRAE			1A		Köppen			Aw
Sunshine h[a]	213.5	212.7	259.2	251.8	280.6	286.1	306.2	303.1	278.8	244.0	200.4	199.5	3035.9
Cloud (%)	46.8	44.1	46.8	51.0	57.4	47.8	49.4	44.0	54.2	40.9	54.7	49.0	48.83
Solar irradiation daily average (Wh/m²)													
Global	3984	4486	5244	5815	6308	6491	6524	6550	5819	5016	4149	3766	5346
Diffuse	1484	1792	2158	2336	2416	2427	2191	2043	2255	1837	1681	1398	2002
Relative humidity (%)													
Morning	80	89	82	76	80	76	75	78	79	83	80	80	80
Evening	55	63	55	51	52	53	52	49	56	52	56	58	55
Dry-bulb temperature (°C)													
Max	26.1	26.5	27.2	27.4	28.9	28.9	30.4	31.1	30.0	30.6	28.3	26.9	28.5
Min	19.6	19.6	19.9	20.9	22.0	23.3	24.4	23.9	24.4	22.4	22.0	21.2	22.0
Mean	22.9	23.1	23.6	24.2	25.5	26.1	27.4	27.5	27.2	26.5	25.2	24.1	25.2
Neutrality	24.9	24.9	25.1	25.3	25.7	25.9	26.3	26.3	26.2	26.0	25.6	25.3	25.6
Upper limit	27.4	27.4	27.6	27.8	28.2	28.4	28.8	28.8	28.7	28.5	28.1	27.8	28.1
Lower limit	22.4	22.5	22.6	22.8	23.2	23.4	23.8	23.8	23.7	23.5	23.1	22.8	23.1
Rain (mm)[a]	58.7	50.5	51.3	16.0	15.7	6.6	13.0	14.2	17.8	46.7	61.5	82.3	434.3
Prec (mm)[b]	231	199	202	63	62	26	51	56	70	184	242	324	1710
Wind (m/s)	4.1	3.7	4.2	4.8	4.4	5.9	6	5.2	5.4	3.8	5.2	4.4	4.8
HDD	0	0	0	0	0	0	0	0	0	0	0	0	0
CDD	139	135	160	173	224	236	276	281	266	255	203	176	2524

(Continued)

TABLE 7.22 (CONTINUED)
Climatic Data, Honolulu, Hawaii

	Average diurnal range (°K)	6.6
	Annual mean range (°K)	11.5

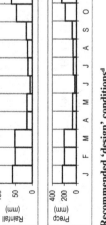

Wind rose

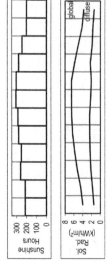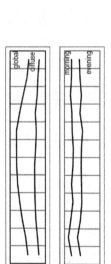

Recommended 'design' conditions[d]

Summer		
	DBT (°C)	31.7
	MCWB (°C)	23.1
	WBT (°C)	24.6
	MCDB (°C)	29.0
Winter		
	DBT (°C)	17.4

Sources of data: https://energyplus.net/weather, [a]WMO (2010), [b]http://worldweather.wmo.int/en/city.html?cityId=849, [c]NOAA (2017), [d]ASHRAE(2009).

TABLE 7.23

Climatic Data, Indianapolis, Indiana

Latitude	N 39° 43'				Longitude	W 86° 16'				Altitude	241 m		
Climate	Cool Humid					ASHRAE	5A			Köppen	Dfa		
Months	Jan	Feb	Mar	Apr	May	Jun	Jul	Aug	Sep	Oct	Nov	Dec	Year
Sunshine hª	132.1	145.7	178.3	214.8	264.7	287.2	295.2	273.7	232.6	196.6	117.1	102.4	2440.4
Cloud (%)	63.8	66.9	64.1	54.0	60.7	57.1	49.8	55.8	39.3	56.8	61.3	71.9	58.45
Solar irradiation daily average (Wh/m²)													
Global	1916	2863	3909	4986	5857	6330	6326	5427	4571	3346	2101	1626	4105
Diffuse	1099	1598	1906	2266	2919	2926	2561	2523	1928	1422	1127	979	1938
Relative humidity (%)													
Morning	86	80	76	80	79	87	92	88	93	75	82	81	83.3
Evening	72	64	54	50	45	56	53	54	54	49	61	73	57.1
Dry-bulb temperature (°C)													
Max	−2.0	0.2	10.5	16.3	23.2	27.2	29.4	28	23.3	16.5	8.4	0.3	15.1
Min	−6.0	−5.8	2.6	6.4	12.8	18.2	19.6	18.7	13.7	7.8	2.2	−3.7	7.2
Mean	−4.0	−2.8	6.6	11.4	18.0	22.7	24.5	23.4	18.5	12.2	5.3	−1.7	11.2
Neutrality	20.9	20.9	20.9	21.3	23.4	24.8	25.4	25.0	23.5	21.6	20.9	20.9	21.3
Upper limit	23.4	23.4	23.4	23.8	25.9	27.3	27.9	27.5	26.0	24.1	23.4	23.4	23.8
Lower limit	18.4	18.4	18.4	18.8	20.9	22.3	22.9	22.5	21.0	19.1	18.4	18.4	18.8
Rain (mm)ᵇ	56.1	54.1	83.8	99.1	133.9	111.5	108.2	81.0	57.2	79.0	94.0	78.2	1036.1
Prec (mm)ᶜ	221	213	330	390	527	439	426	319	225	311	370	308	4079
Wind (m/s)	5.1	4.6	5.2	4.5	3.7	4.0	3.5	3.4	3.9	4	4.8	4.8	4.3
HDD	696	588	363	203	45	2	0	1	47	203	387	619	3154
CDD	0	0	3	7	46	140	205	160	53	4	0	0	618

(Continued)

TABLE 7.23 (CONTINUED)
Climatic Data, Indianapolis, Indiana

Average diurnal range (°K)		7.9
Annual mean range (°K)		35.4

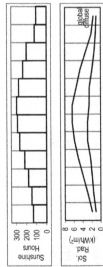
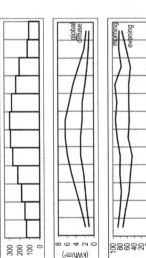
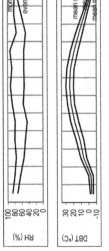
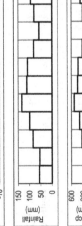

Wind rose

Recommended 'design' conditions[d]

Summer	DBT (°C)	31.5
	MCWB (°C)	23.5
	WBT (°C)	24.9
	MCDB (°C)	29.7
Winter	DBT (°C)	−14.2

Sources of data: https://energyplus.net/weather, [b]http://worldweather.wmo.int/en/city.html?cityId=772, [c]NOAA (2017), [d]ASHRAE(2009).

TABLE 7.24

Climatic Data, Jackson, Mississippi

	Jan	Feb	Mar	Apr	May	Jun	Jul	Aug	Sep	Oct	Nov	Dec	Year
Latitude N 32° 19'					Longitude W 90° 4'					Altitude 94 m			
Climate Hot Humid					ASHRAE 2A					Köppen Cfa			
Sunshine h[a]	154.5	165.3	223.5	251.1	276.2	298.5	283.4	273.1	232.7	235.2	174.0	152.1	2719.6
Cloud (%)	59.8	62.0	63.4	38.2	52.8	27.2	47.2	50.3	25.0	37.8	53.9	51.9	47.4
Solar irradiation daily average (Wh/m²)													
Global	2650	3423	4289	5744	5702	6305	6255	6016	4943	4256	2989	2626	4600
Diffuse	1158	1536	1926	1979	2666	2677	2905	2550	1997	1481	1197	1061	1928
Relative humidity (%)													
Morning	88	83	93	91	96	91	98	90	86	93	88	92	90.8
Evening	58	51	52	49	56	53	57	47	42	45	47	51	50.7
Dry-bulb temperature (°C)													
Max	11.5	15.0	20.4	23.8	26.3	30.9	32.5	33.2	29.8	24.7	19.9	13.5	23.5
Min	3.2	5.0	10.4	12.6	16.0	21.5	22.8	22.6	18.1	11.2	10.0	3.0	13.0
Mean	7.4	10.0	15.4	18.2	21.2	26.2	27.7	27.9	24.0	18.0	15.0	8.3	18.2
Neutrality	20.9	20.9	22.6	23.4	24.4	25.9	26.4	26.4	25.2	23.4	22.4	20.9	23.5
Upper limit	23.4	23.4	25.1	25.9	26.9	28.4	28.9	28.9	27.7	25.9	24.9	23.4	26.0
Lower limit	18.4	18.4	20.1	20.9	21.9	23.4	23.9	24.0	22.7	20.9	19.9	18.4	21.0
Rain (mm)[b]	127.3	111.5	129.5	129.8	123.2	104.9	115.6	101.9	81.8	101.3	121.2	131.8	1379.8
Prec (mm)[c]	407	417	475	486	569	484	449	300	338	377	490	535	5327
Wind (m/s)	3.6	3.4	3.7	2.7	2.6	2.1	2.8	2.9	2.5	2.8	3.7	2.4	2.9
HDD	354	249	122	40	9	0	0	0	4	66	147	329	1320
CDD	0	1	29	46	100	242	289	288	166	39	18	1	1219

(Continued)

TABLE 7.24 (CONTINUED)
Climatic Data, Jackson, Mississippi

| | Average diurnal range (°K) | 10.4 |
| | Annual mean range (°K) | 30.2 |

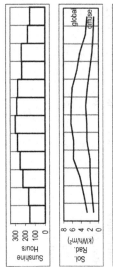
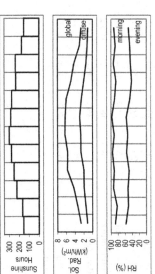
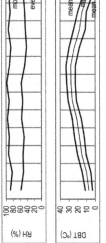
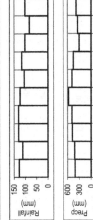
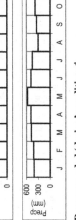
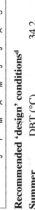

Wind rose

Recommended 'design' conditions[d]

Summer	DBT (°C)	34.2
	MCWB (°C)	24.6
	WBT (°C)	26.0
	MCDB (°C)	31.6
Winter	DBT (°C)	-3.5

Sources of data: https://energyplus.net/weather, [a]WMO (2010), [b]http://worldweather.wmo.int/en/city.html?cityId=773, [c]NOAA (2017), [d]ASHRAE(2009).

TABLE 7.25

Climatic Data, Jefferson City, Missouri

Latitude	N 38° 34'			Longitude		W 92° 9'			Altitude		167 m	
Climate	Mixed Humid			ASHRAE		4A			Köppen		Cfa	
Months	Jan	Feb	Mar	Apr	May	Jun	Jul	Aug	Sep	Oct	Nov	Dec	Year
Sunshine h[a]	161.5	154.3	193.5	226.9	264.1	294.1	313.4	288.5	229.1	210.7	150.6	140.3	2627
Cloud (%)	73.5	81.8	48.1	47.6	49.7	29.8	23.3	26.2	32.2	40.7	47.6	65.2	47.14
Solar irradiation daily average (Wh/m²)													
Global	2278	1916	4332	4757	5405	7012	6849	5926	4643	3510	2150	1458	4186
Diffuse	804	1292	1627	1950	2246	2418	2215	2134	1762	1284	979	805	1626
Relative humidity (%)													
Morning	83	83	83	84	90	90	91	94	90	89	85	82	87.0
Evening	66	68	49	53	55	51	50	58	53	51	53	66	56.1
Dry-bulb temperature (°C)													
Max	3.6	6.6	12.8	19.0	21.3	29.1	30.9	29.6	27.7	19.5	14.2	4.0	29.4
Min	−1.4	1.0	2.5	9.5	12.2	19.0	20.9	20.5	17.6	9.5	6.2	−1.8	17.4
Mean	1.1	3.8	7.7	14.3	16.8	24.1	25.9	25.1	22.7	14.5	10.2	1.1	13.9
Neutrality	20.9	20.9	20.9	22.2	23.0	25.3	25.8	25.6	24.8	22.3	21.0	20.9	22.1
Upper limit	23.4	23.4	23.4	24.7	25.5	27.8	28.3	28.1	27.3	24.8	23.5	23.4	24.6
Lower limit	18.4	18.4	18.4	19.7	20.5	22.8	23.3	23.1	22.3	19.8	18.46	18.4	19.6
Rain (mm)[b]	48.3	58.7	76.2	105.2	131.3	111.5	109.5	102.6	105.7	84.8	91.7	69.3	1094.8
Prec (mm)[c]	320	366	457	563	652	741	785	765	681	575	456	344	6705
Wind (m/s)	3.2	3.5	3.5	4.1	3.3	2.8	2.7	2.1	2.2	2.4	3.2	3.9	3.1
HDD	534	402	322	137	70	0	0	0	12	154	257	538	2426
CDD	0	0	1	26	31	191	247	214	129	28	1	0	868

(Continued)

TABLE 7.25 (CONTINUED)
Climatic Data, Jefferson City, Missouri

| | Average diurnal range (°K) | 8.6 |
| | Annual mean range (°K) | 32.7 |

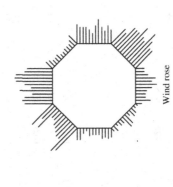

Wind rose

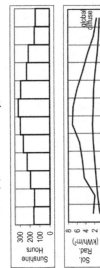

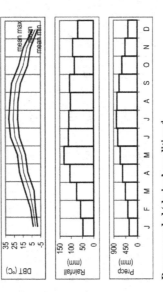

Recommended 'design' conditions[d]

Summer	DBT (°C)	33.0
	MCWB (°C)	24.1
	WBT (°C)	25.5
	MCDB (°C)	31.3
Winter	DBT (°C)	−11.0

Sources of data: https://energyplus.net/weather, [a]WMO (2010), [b]http://worldweather.wmo.int/en/city.html?cityId=1988, [c]NOAA (2017), [d]ASHRAE (2009), Sunshine hours for Columbia, Missouri (31.3 miles).

TABLE 7.26

Climatic Data, Juneau, Alaska

Latitude	N 58° 20′				Longitude			W 134° 34′			Altitude		4 m
Climate	Very Cold				ASHRAE			7			Köppen		Dfb
Months	**Jan**	**Feb**	**Mar**	**Apr**	**May**	**Jun**	**Jul**	**Aug**	**Sep**	**Oct**	**Nov**	**Dec**	**Year**
Sunshine hᵃ	80.9	89.2	137.3	182.3	231.7	189.3	182.9	161.6	109.6	66.2	58.5	41.2	1530.7
Cloud (%)	68.5	68.0	76.4	69.5	70.8	82.8	88.9	82.3	97.6	91.4	81.4	82.4	80.0
Solar irradiation daily average (Wh/m²)													
Global	390	1061	1828	3510	4926	4528	3528	3949	2049	1242	555	267	2319
Diffuse	265	599	1175	1819	2610	2714	2520	1805	1225	615	443	215	1334
Relative humidity (%)													
Morning	89	82	81	93	90	93	94	96	95	93	74	90	89.2
Evening	78	64	64	63	57	66	76	73	83	79	64	85	71.0
Dry-bulb temperature (°C)													
Max	−2	0.6	1.6	9.7	14.6	15.9	16.4	16.1	11.9	8.2	1.5	−1.1	7.8
Min	−5.5	−3.3	−1.7	3.2	7	9.3	11.3	9.4	9.2	4.9	−0.2	−2.9	3.4
Mean	−3.8	−1.4	0.0	6.5	10.8	12.6	13.9	12.8	10.6	6.6	0.7	−2.0	5.6
Neutrality	20.9	20.9	20.9	20.9	21.1	21.7	22.1	21.8	21.1	20.9	20.9	20.9	19.5
Upper limit	23.4	23.4	23.4	23.4	23.7	24.2	24.6	24.3	23.6	23.4	23.4	23.4	22.0
Lower limit	18.4	18.4	18.4	18.4	18.7	19.2	19.6	19.3	18.6	18.4	18.4	18.4	17.0
Rain (mm)ᵇ	202.7	170.4	159.8	117.9	126.0	112.3	138.2	207.3	323.1	336.0	214.4	234.4	2342.5
Prec (mm)ᶜ	535	413	378	294	340	324	460	573	864	863	599	584	6227
Wind (m/s)	2.9	2.9	4.5	3.3	3.3	3.4	3.5	2.9	4	3.2	3.8	3.2	3.4
HDD	683	549	566	351	217	157	134	161	231	370	527	623	4569
CDD	0	0	0	0	0	0	0	0	0	0	0	0	0

(Continued)

TABLE 7.26 (CONTINUED)
Climatic Data, Juneau, Alaska

Average diurnal range (°K)		4.5
Annual mean range (°K)		21.9

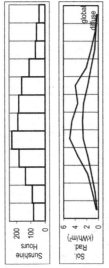
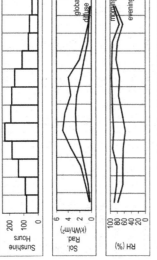
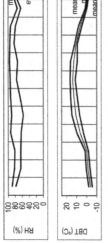
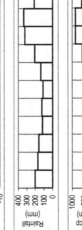

Wind rose

Recommended 'design' conditions[d]

Summer	DBT (°C)	21.1
	MCWB (°C)	14.6
	WBT (°C)	15.3
	MCDB (°C)	19.8
Winter	DBT (°C)	–13.2

Sources of data: https://energyplus.net/weather, [a]WMO (2010), [b]http://worldweather.wmo.int/en/city.html?cityId=775, [c]NOAA (2017), [d]ASHRAE(2009).

TABLE 7.27

Climatic Data, Lansing, Michigan

	Latitude	N 42° 46'			Longitude	W 84° 34'			Altitude	256 m			
	Climate	Cool Humid			ASHRAE	5A			Köppen	Dfa			
Months	Jan	Feb	Mar	Apr	May	Jun	Jul	Aug	Sep	Oct	Nov	Dec	Year
Sunshine h[a]	118.2	140.1	187.6	218.7	278.6	296.2	318.5	278.1	217.6	163.8	92.4	82.1	2391.9
Cloud (%)	76.1	74.1	58.9	63.1	55.0	55.4	55.2	62.5	52.5	61.4	71.8	85.4	64.3
Solar irradiation daily average (Wh/m²)													
Global	1582	2555	3459	4465	5694	6196	5943	5070	4278	2631	1710	1319	3742
Diffuse	998	1718	1674	1985	2594	2594	2579	2408	1802	1397	1029	966	1812
Relative humidity (%)													
Morning	81	85	84	74	83	90	93	90	92	94	88	84	86.5
Evening	68	76	65	46	49	48	54	57	58	64	74	77	61.3
Dry-bulb temperature (°C)													
Max	-3.3	-2.7	5.6	13.3	21.2	26.0	27.5	24.8	21.5	13.5	5.7	-1.1	12.7
Min	-7.5	-8.5	-1.0	4.4	10.2	13.8	17.0	16.6	12.7	5.5	1.6	-3.5	5.1
Mean	-5.4	-5.6	2.3	8.9	15.7	19.9	22.3	20.7	17.1	9.5	3.7	-2.3	8.9
Neutrality	20.9	20.9	20.9	20.9	22.7	24.0	24.7	24.2	23.1	20.9	20.9	20.9	20.9
Upper limit	23.4	23.4	23.4	23.4	25.2	26.5	27.2	26.7	25.6	23.4	23.4	23.4	23.4
Lower limit	18.4	18.4	18.4	18.4	20.2	21.5	22.2	21.7	20.6	18.4	18.4	18.4	18.4
Rain (mm)[b]	45.7	37.6	44.7	72.9	84.6	88.9	82.8	83.8	92.2	69.9	71.1	42.4	816.6
Prec (mm)[c]	165	147	206	303	336	345	284	323	350	253	278	187	3177
Wind (m/s)	5.9	5.0	5.7	5.5	4.7	3.8	3.8	3.9	3.6	4.2	4.6	5.6	4.7
HDD	736	672	497	282	91	24	2	19	65	287	440	636	3751
CDD	0	0	0	6	22	87	129	102	25	0	0	0	371

(Continued)

TABLE 7.27 (CONTINUED)
Climatic Data, Lansing, Michigan

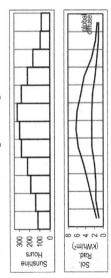

Average diurnal range (°K)		7.6
Annual mean range (°K)		36.0

Wind rose

Recommended 'design' conditions[d]

Summer		
	DBT (°C)	30.3
	MCWB (°C)	22.3
	WBT (°C)	23.7
	MCDB (°C)	28.3
Winter	DBT (°C)	−16.1

Sources of data: https://energyplus.net/weather, [a]WMO (2010), [b]http://worldweather.wmo.int/en/city.html?cityId=780, [c]NOAA (2017), [d]ASHRAE(2009).

TABLE 7.28

Climatic Data, Lincoln, Nebraska

	Latitude	N 40° 49'			Longitude		W 96° 46'			Altitude		357 m	
	Climate	Cool Humid				ASHRAE	5A				Köppen		Dfa
Months	Jan	Feb	Mar	Apr	May	Jun	Jul	Aug	Sep	Oct	Nov	Dec	Year
Sunshine h[a]	176.8	167.6	211.9	236.4	273.3	314.4	329.9	294.9	236.4	216.9	156.4	146.8	2761.7
Cloud (%)	54.2	37.0	52.5	60.5	51.2	56.0	34.0	59.6	33.8	31.6	53.5	44.8	47.38
Solar irradiation daily average (Wh/m²)													
Global	2009	2885	3784	4525	5237	6405	6645	5100	4503	3394	1946	1688	4010
Diffuse	1037	1052	1533	1995	2286	2479	2354	2352	1638	1295	799	787	1634
Relative humidity (%)													
Morning	75	76	79	86	82	86	86	92	91	83	85	81	83.5
Evening	49	48	45	52	51	50	52	58	49	44	55	61	51.2
Dry-bulb temperature (°C)													
Max	0.6	5.0	12.7	17.3	22.7	29.7	30.9	27.5	24.6	19.1	7.1	0.9	16.5
Min	−7.4	−4.6	1.7	7.6	13.1	19.5	20.2	17.0	12.7	7.4	−0.9	−6.6	6.6
Mean	−3.4	0.2	7.2	12.5	17.9	24.6	25.6	22.3	18.7	13.3	3.1	−2.9	11.6
Neutrality	20.9	20.9	20.9	21.7	23.3	25.4	25.7	24.7	23.6	21.9	20.9	20.9	21.4
Upper limit	23.4	23.4	23.4	24.2	25.8	27.9	28.2	27.2	26.1	24.4	23.4	23.4	23.9
Lower limit	18.4	18.4	18.4	19.2	20.9	22.9	23.2	22.2	21.1	19.4	18.4	18.4	18.9
Rain (mm)[b]	16.3	19.6	49.0	68.8	109.0	110.5	86.4	88.6	76.7	50.0	36.3	24.1	735.3
Prec (mm)[c]	64	77	193	271	429	435	340	349	302	197	143	95	2895
Wind (m/s)	4.3	5.1	5.4	5.3	5.6	4.6	4.2	3.1	3.7	4.3	4.2	4.0	4.5
HDD	687	509	352	183	60	0	2	3	44	198	478	670	3186
CDD	0	0	2	11	55	203	231	133	55	29	0	0	719

(*Continued*)

TABLE 7.28 (CONTINUED)
Climatic Data, Lincoln, Nebraska

Average diurnal range (°K) 9.9
Annual mean range (°K) 38.3

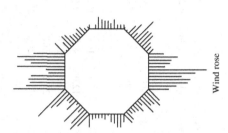

Wind rose

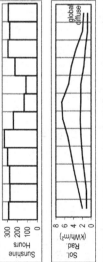
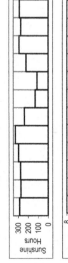
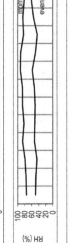
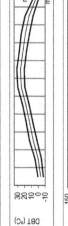
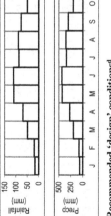
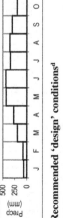

Sunshine Hours — 300 200 100 0

Sol. Rad. (kWh/m²) — 8 6 4 2 0 — global, diffuse

RH (%) — 100 80 60 40 20 — morning, evening

DBT (°C) — 30 20 10 0 -10 — mean max, mean, mean min

Rainfall (mm) — 150 100 50 0

Precp (mm) — 500 250 0

J F M A M J J A S O N D

Recommended 'design' conditions[d]

Summer		
DBT (°C)	34.3
MCWB (°C)	23.5
WBT (°C)	25.0
MCDB (°C)	31.7

Winter		
DBT (°C)	−17.6

Sources of data: https://energyplus.net/weather, [a]WMO (2010), [b]http://worldweather.wmo.int/en/city.html?cityId=783, [c]NOAA (2017), [d]ASHRAE(2009).

TABLE 7.29

Climatic Data, Little Rock, Arkansas

Latitude	N 34° 45'				Longitude	W 92° 13'				Altitude	78 m			
Climate	Warm Humid				ASHRAE	3A				Köppen	Cfa			
Months	Jan	Feb	Mar	Apr	May	Jun	Jul	Aug	Sep	Oct	Nov	Dec	Year
Sunshine hᵃ	180.9	188.2	244.5	276.7	325.3	346.2	351	323	271.9	251	176.9	166.2	3101.8
Cloud (%)	53.7	53.2	61.8	53.2	58.9	44.1	47.7	39.2	47.6	43.0	56.0	52.1	50.9
Solar irradiation daily average (Wh/m²)													
Global	2467	2904	4104	5317	5796	6497	6569	6196	4848	4021	2747	2258	4477
Diffuse	1109	1227	1938	2147	2764	2815	2577	2161	1924	1448	1234	1025	1864
Relative humidity (%)													
Morning	79	69	80	85	91	93	93	90	90	82	79	75	83.8
Evening	58	43	57	52	61	54	52	50	53	45	54	53	52.7
Dry-bulb temperature (°C)													
Max	6.8	12.7	16.0	21.9	25.5	30.9	32.3	32.5	29.1	22.6	16.9	10.0	21.4
Min	-0.6	3.8	7.2	12.0	17.1	21.0	22.7	22.4	19.3	12.0	8.6	2.5	12.3
Mean	3.1	8.3	11.6	17.0	21.3	26.0	27.5	27.5	24.2	17.3	12.8	6.3	16.9
Neutrality	20.9	20.9	21.4	23.1	24.4	25.8	26.3	26.3	25.3	23.2	21.8	20.9	23.0
Upper limit	23.4	23.4	23.9	25.6	26.9	28.3	28.8	28.8	27.8	25.7	24.3	23.4	25.5
Lower limit	18.4	18.4	18.9	20.6	21.9	23.3	23.8	23.8	22.8	20.7	19.3	18.4	20.5
Rain (mm)ᵇ	86.9	93.0	121.2	121.7	126.0	84.1	96.8	71.1	84.1	121.9	136.9	127.3	1271
Prec (mm)ᶜ	355	366	468	514	487	365	327	259	318	491	528	497	4975
Wind (m/s)	4.1	3.5	4.0	3.7	2.9	2.7	2.7	2.4	2.7	2.8	2.4	4.1	3.2
HDD	475	284	219	63	8	0	0	0	1	62	186	385	1683
CDD	0	0	8	28	103	239	295	289	174	21	3	0	1160

(Continued)

TABLE 7.29 (CONTINUED)
Climatic Data, Little Rock, Arkansas

	Average diurnal range (°K)	9.1
	Annual mean range (°K)	33.3

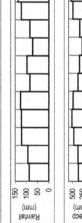
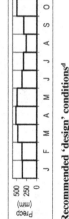

Wind rose

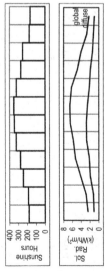
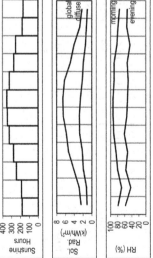
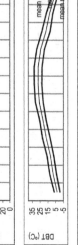

Recommended 'design' conditions[d]

Summer	DBT (°C)	35.2
	MCWB (°C)	25.1
	WBT (°C)	26.2
	MCDB (°C)	32.6
Winter	DBT (°C)	−5.4

Sources of data: https://energyplus.net/weather, [a]WMO (2010), [b]http://worldweather.wmo.int/en/city.html?cityId=784, [c]NOAA (2017), [d]ASHRAE (2009).

TABLE 7.30

Climatic Data, Louisville, Kentucky

Latitude	N 38° 10'				Longitude			W 85° 43'		Altitude		147 m	
Climate	Mixed Humid				ASHRAE			4A		Köppen		Dfa	
Months	Jan	Feb	Mar	Apr	May	Jun	Jul	Aug	Sep	Oct	Nov	Dec	Year
Sunshine h[a]	140.5	148.9	188.6	221.1	263.4	288.9	293.6	272.6	234.3	208.5	135.7	118.3	2514.4
Cloud (%)	59.6	68.3	69.1	54.9	57.0	52.1	52.2	55.0	54.2	52.2	59.3	70.6	58.7
Solar irradiation daily average (Wh/m²)													
Global	2018	2842	3858	5079	5717	6614	6296	5526	4534	3380	2303	1848	4168
Diffuse	1053	1306	1838	2179	2570	2670	2683	2357	1862	1464	1099	1021	1842
Relative humidity (%)													
Morning	79	77	80	77	81	81	94	90	92	90	84	69	82.8
Evening	59	58	52	47	49	53	57	55	59	56	57	52	54.5
Dry-bulb temperature (°C)													
Max	3.2	3.8	13.2	17.9	24.0	28.6	30.0	28.9	24.6	19.8	11.9	6.5	17.7
Min	-2.5	-1.4	4.2	8.3	14.1	19.3	21.2	19.8	16.4	10.6	4.6	1.5	9.7
Mean	0.4	1.2	8.7	13.1	19.1	24.0	25.6	24.4	20.5	15.2	8.3	4.0	13.7
Neutrality	20.9	20.9	20.9	21.9	23.7	25.2	25.7	25.3	24.2	22.5	20.9	20.9	22.0
Upper limit	23.4	23.4	23.4	24.4	26.2	27.7	28.2	27.8	26.7	25.0	23.4	23.4	24.5
Lower limit	18.4	18.4	18.4	19.4	21.2	22.7	23.2	22.9	21.7	20.0	18.4	18.4	19.5
Rain (mm)[b]	82.3	80.8	105.9	101.9	133.9	96.3	107.4	84.6	77.5	81.8	91.2	97.3	1140.9
Prec (mm)[c]	338	318	416	408	512	414	417	331	316	322	361	401	4554
Wind (m/s)	4.9	4.5	4.4	4	3.4	3.2	3.3	2.8	2.5	3.0	3.8	4.2	3.7
HDD	563	479	290	170	33	0	0	0	27	122	308	450	2442
CDD	0	0	5	28	72	186	236	189	95	25	0	0	836

(Continued)

TABLE 7.30 (CONTINUED)
Climatic Data, Louisville, Kentucky

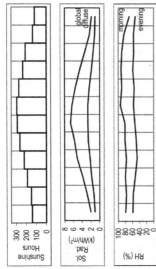

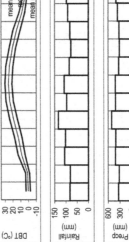

| Average diurnal range (°K) | 8.0 |
| Annual mean range (°K) | 32.5 |

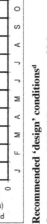

Wind rose

Recommended 'design' conditions[d]

Summer	DBT (°C)	32.9
	MCWB (°C)	24.0
	WBT (°C)	25.3
	MCDB (°C)	30.9
Winter	DBT (°C)	−9.7

Sources of data: https://energyplus.net/weather, [a]WMO (2010), [b]http://worldweather.wmo.int/en/city.html?cityId=785, [c]NOAA (2017), [d]ASHRAE(2009).

TABLE 7.31
Climatic Data, Madison Dane, Wisconsin

| | | N 43° 7' | | | | Longitude | | W 89° 19' | | Altitude | | 262 m | |
| | | Cold Humid | | | | ASHRAE | | 6A | | Köppen | | Dfa | |
Months	Jan	Feb	Mar	Apr	May	Jun	Jul	Aug	Sep	Oct	Nov	Dec	Year
Sunshine h[a]	143.0	152.3	187.3	206.7	263.1	293.1	304.9	270.2	213.8	172.5	111.4	109.5	2427.8
Cloud (%)	60.8	63.3	65.0	68.6	62.5	56.9	55.0	55.1	45.3	56.9	54.7	76.4	60.04
					Solar irradiation daily average (Wh/m²)								
Global	1922	2694	3627	4687	5822	6487	6173	5435	4030	2699	1708	1567	3904
Diffuse	1140	1533	2082	2340	2732	2615	2784	2785	1540	1253	892	980	1890
					Relative humidity (%)								
Morning	81	79	78	83	86	91	94	94	94	93	80	86	86.6
Evening	68	66	54	50	54	53	63	58	61	57	54	75	59.4
					Dry-bulb temperature (°C)								
Max	-7.1	-3.1	7.0	12.9	21.3	25.0	27.4	26.0	19.7	13.5	4.7	-4.6	11.9
Min	-12.2	-10.1	0.1	2.4	10.9	13.4	18.1	16.6	10.9	4.5	-1.3	-8.3	3.8
Mean	-9.7	-6.6	3.6	7.7	16.1	19.2	22.8	21.3	15.3	9.0	1.7	-6.5	7.8
Neutrality	20.9	20.9	20.9	20.9	22.8	23.8	24.9	24.4	22.5	20.9	20.9	20.9	20.9
Upper limit	23.4	23.4	23.4	23.4	25.3	26.3	27.4	26.9	25.0	23.4	23.4	23.4	23.4
Lower limit	18.4	18.4	18.4	18.4	20.3	21.3	22.4	21.9	20.0	18.4	18.4	18.4	18.4
Rain (mm)[b]	31.2	36.8	55.9	86.4	90.2	115.3	106.2	108.5	79.5	61.0	60.7	44.2	875.9
Prec (mm)[c]	123	145	220	340	355	454	418	427	313	240	239	174	3448
Wind (m/s)	5.2	4.8	5.3	4.3	4.4	3.7	4.0	3.8	2.7	3.9	3.1	4.8	4.2
HDD	574	442	314	144	53	3	0	0	11	172	273	494	2480
CDD	0	0	4	0	63	151	228	195	84	6	0	0	731

(Continued)

TABLE 7.31 (CONTINUED)
Climatic Data, Madison Dane, Wisconsin

	Average diurnal range (°K)	8.1
	Annual mean range (°K)	39.6

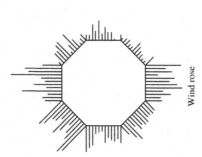

Wind rose

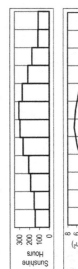

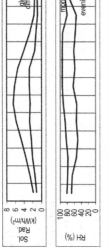

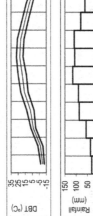

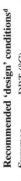

Recommended 'design' conditions[d]

Summer		
	DBT (°C)	30.5
	MCWB (°C)	22.7
	WBT (°C)	23.9
	MCDB (°C)	28.6
Winter	DBT (°C)	−19.4

Sources of data: https://energyplus.net/weather, [a]WMO (2010), [b]http://worldweather.wmo.int/en/city.html?cityId=788, [c]NOAA (2017), [d]ASHRAE(2009).

TABLE 7.32

Climatic Data, Minneapolis, Minnesota

Latitude	N 44° 52'				Longitude	W 93° 13'				Altitude		254 m	
Climate	Cold Humid				ASHRAE	6A				Köppen		Dfa	
Months	Jan	Feb	Mar	Apr	May	Jun	Jul	Aug	Sep	Oct	Nov	Dec	Year
Sunshine h[a]	156.7	178.3	217.5	242.1	295.2	321.9	350.5	307.2	233.2	181.0	112.8	114.3	2710.7
Cloud (%)	65.8	58.2	64.0	63.8	58.1	57.1	51.9	55.5	55.0	52.8	64.9	65.9	59.4
Solar irradiation daily average (Wh/m²)													
Global	1724	2590	3691	4763	5902	6418	6020	5167	4025	2814	1593	1209	3826
Diffuse	785	1149	1513	1968	2339	2578	2417	2291	1656	1235	755	544	1603
Relative humidity (%)													
Morning	71	70	76	76	77	78	80	86	87	87	76	73	78.1
Evening	57	55	50	53	45	43	55	58	50	55	61	58	53.3
Dry-bulb temperature (°C)													
Max	−9.0	−1.4	5.7	11.0	21.2	25.2	27.9	24.2	20.6	12.9	2.7	−4.7	11.4
Min	−13.3	−8.0	−1.4	4.0	11.1	15.2	19.1	16.8	11.4	4.3	−1.3	−8.6	4.1
Mean	−11.2	−4.7	2.2	7.5	16.2	20.2	23.5	20.5	16.0	8.6	0.7	−6.7	7.7
Neutrality	20.9	20.9	20.9	20.9	22.8	24.1	25.1	24.2	22.8	20.9	20.9	20.9	20.9
Upper limit	23.4	23.4	23.4	23.4	25.3	26.6	27.6	26.7	25.3	23.4	23.4	23.4	23.4
Lower limit	18.4	18.4	18.4	18.4	20.3	21.6	22.6	21.7	20.3	18.4	18.4	18.4	18.4
Rain (mm)[b]	22.9	19.6	48.0	67.6	85.3	108.0	102.6	109.2	78.2	61.7	45.0	29.5	777.6
Prec (mm)[c]	90	77	189	266	336	425	404	430	308	243	177	116	3061
Wind (m/s)	4.6	4.7	5.1	5.0	4.8	4.2	4.4	4.0	4.5	4.4	4.6	4.8	4.6
HDD	911	639	495	326	82	22	0	14	103	304	531	774	4201
CDD	0	0	0	12	36	93	179	96	38	0	0	0	454

(Continued)

TABLE 7.32 (CONTINUED)
Climatic Data, Minneapolis, Minnesota

| | Average diurnal range (°K) | 7.3 |
| | Annual mean range (°K) | 41.2 |

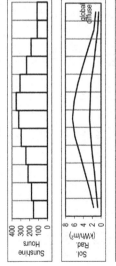
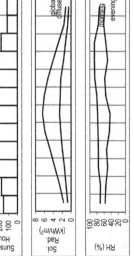
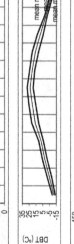
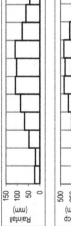
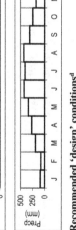

Wind rose

Recommended 'design' conditions[d]

Summer		
	DBT (°C)	31.0
	MCWB (°C)	22.4
	WBT (°C)	23.9
	MCDB (°C)	29.0
Winter		
	DBT (°C)	−22.0

Sources of data: https://energyplus.net/weather, [a]WMO (2010), [b]http://worldweather.wmo.int/en/city.html?cityId=276, [c]NOAA (2017), [d]ASHRAE(2009).

TABLE 7.33
Climatic Data, Montgomery, Alabama

Latitude	N 32° 17'			Longitude	W 86° 24'			Altitude	62 m		
Climate	Warm Humid			ASHRAE	3A			Köppen	Cfa		
Months	Jan	Feb	Mar	Apr	May	Jun	Jul	Aug	Sep	Oct	Nov	Dec	Year
Sunshine h[a]	153.1	166.0	219.4	250.8	267.4	261.8	262.1	251.9	226.4	228.3	171.4	153.1	2611.7
Cloud (%)	58.9	56.6	61.6	34.2	63.6	53.6	32.2	50.8	48.4	39.2	48.0	53.1	50.02
Solar irradiation daily average (Wh/m²)													
Global	2779	3603	4584	5686	6017	6235	5855	5777	5109	4142	3162	2517	4622
Diffuse	1251	1604	2022	2183	2890	3122	2891	2812	2286	1675	1231	998	2080
Relative humidity (%)													
Morning	85	84	90	94	94	94	95	87	95	78	90	73	88.3
Evening	52	52	48	50	52	53	59	57	46	45	49	45	50.7
Dry-bulb temperature (°C)													
Max	12.0	14.8	21.7	24.5	27.3	31.1	31.6	31.4	30.4	23.8	19.3	15.3	23.6
Min	3.3	5.3	9.7	13.3	16.5	20.8	22.1	22.9	18.3	12.7	7.8	5.1	13.2
Mean	7.7	10.1	15.7	18.9	21.9	26.0	26.9	27.2	24.4	18.3	13.6	10.2	18.4
Neutrality	20.9	20.9	22.7	23.7	24.6	25.8	26.1	26.2	25.3	23.5	22.0	21.0	23.5
Upper limit	23.4	23.4	25.2	26.2	27.1	28.3	28.6	28.7	27.9	26.0	24.5	23.5	26.0
Lower limit	18.4	18.4	20.2	21.2	22.1	23.3	23.6	23.7	22.9	21.0	19.5	18.5	21.0
Rain (mm)[b]	118.1	134.1	151.1	102.1	89.9	103.4	133.1	100.6	100.8	74.2	117.1	123.4	1347.9
Prec (mm)[c]	465	528	595	402	354	407	524	396	397	292	461	486	5307
Wind (m/s)	3.4	3.9	3.1	2.8	3.0	2.3	2.3	1.7	2.2	2.7	2.7	2.9	2.8
HDD	339	239	94	34	4	0	0	0	0	48	165	269	1192
CDD	0	0	12	57	126	235	272	268	177	36	5	0	1188

(*Continued*)

TABLE 7.33 (CONTINUED)
Climatic Data, Montgomery, Alabama

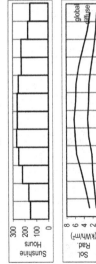
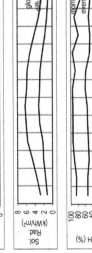
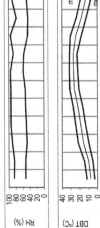
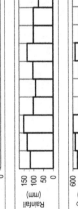

| | Average diurnal range (°K) | 10.5 |
| | Annual mean range (°K) | 28.3 |

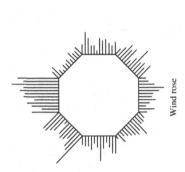

Wind rose

Recommended 'design' conditions[d]

Summer	DBT (°C)	34.5
	MCWB (°C)	24.5
	WBT (°C)	25.9
	MCDB (°C)	31.8
Winter	DBT (°C)	−2.6

Sources of data: https://energyplus.net/weather; [a]WMO (2010), [b]http://worldweather.wmo.int/en/city.html?cityId=793, [c]NOAA (2017), [d]ASHRAE(2009).

TABLE 7.34

Climatic Data, Montpelier, Vermont

		Latitude N 44° 12'			Longitude W 72° 34'					Altitude 343 m			
		Climate Cold Humid			ASHRAE 6A					Köppen Dfb			
Months	Jan	Feb	Mar	Apr	May	Jun	Jul	Aug	Sep	Oct	Nov	Dec	Year
Sunshine h[a]	126.9	146.8	190.7	206.2	251.4	270.1	301.9	258.2	201.0	159.2	91.1	91.6	2295.1
Cloud (%)	81.2	63.6	73.0	72.2	61.8	74.3	55.7	43.0	52.8	70.4	74.5	72.6	66.3
Solar irradiation daily average (Wh/m²)													
Global	1680	2348	2893	3756	4839	4721	5435	5548	3682	2178	1474	1105	3305
Diffuse	663	1101	1395	1983	2487	2801	2471	1881	1643	1147	846	638	1588
Relative humidity (%)													
Morning	82	73	83	78	81	90	89	96	94	88	80	77	84.3
Evening	67	47	57	46	46	64	48	60	59	65	61	63	56.9
Dry-bulb temperature (°C)													
Max	-1.1	-5.1	3.8	12.3	16.2	20.2	24.3	24.1	20.5	12.6	6.4	-3.1	10.9
Min	-5.2	-10.8	-3.4	3.8	6.2	13.5	13.7	13.9	11.7	5.9	2.0	-7.0	3.7
Mean	-3.2	-8.0	0.2	8.1	11.2	16.9	19.0	19.0	16.1	9.3	4.2	-5.1	7.3
Neutrality	20.9	20.9	20.9	20.9	21.3	23.0	23.7	23.7	22.8	20.9	20.9	20.9	20.9
Upper limit	23.4	23.4	23.4	23.4	23.8	25.5	26.2	26.2	25.3	23.4	23.4	23.4	23.4
Lower limit	18.4	18.4	18.4	18.4	18.8	20.5	21.2	21.2	20.3	18.4	18.4	18.4	18.4
Rain (mm)[b]	75.7	68.3	72.4	80.3	98.0	104.6	112.3	103.1	87.9	99.6	90.2	85.1	1077.5
Prec (mm)[c]	245	204	239	266	337	380	408	401	312	344	317	274	3727
Wind (m/s)	2.6	4	3.5	3.8	3.8	2.7	3.2	1.7	2.7	2.2	3.5	2.1	3.0
HDD	660	737	563	293	212	74	17	18	93	284	422	724	4097
CDD	0	0	0	0	11	33	60	36	25	2	0	0	167

(Continued)

TABLE 7.34 (CONTINUED)
Climatic Data, Montpelier, Vermont

Average diurnal range (°K)	7.2
Annual mean range (°K)	35.1

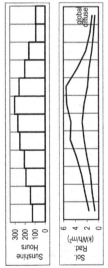

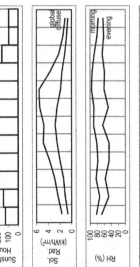

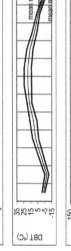

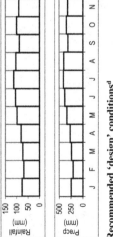

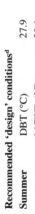

Wind rose

Recommended 'design' conditions[d]

Summer		
DBT (°C)	27.9
MCWB (°C)	20.1
WBT (°C)	21.5
MCDB (°C)	26.2

Winter		
DBT (°C)	−21.1

Sources of data: https://energyplus.net/weather; [a]WMO (2010), [b]http://worldweather.wmo.int/en/city.html?cityId=2008, [c]NOAA (2017), [d]ASHRAE(2009), Sunshine hours for Burlington, Vermont (38.9 miles).

TABLE 7.35

Climatic Data, Nashville, Tennessee

Latitude	N 36° 07'				Longitude	W 86° 40'			Altitude	177 m			
Climate	Mixed Humid				ASHRAE	4A			Köppen	Cfa			
Months	Jan	Feb	Mar	Apr	May	Jun	Jul	Aug	Sep	Oct	Nov	Dec	Year
Sunshine h[a]	139.6	145.2	191.3	231.5	261.8	277.7	279.0	262.1	226.4	216.8	148.1	130.6	2510.1
Cloud (%)	69.9	69.0	61.5	55.5	61.1	44.3	39.8	41.1	44.8	37.8	49.0	50.9	52.1
Solar irradiation daily average (Wh/m²)													
Global	2154	3218	4131	5293	5694	6603	6190	5739	4460	3812	2518	1903	4310
Diffuse	1252	1659	1899	2274	2772	2972	2886	2395	1967	1360	1129	618	1932
Relative humidity (%)													
Morning	83	81	78	78	89	91	88	90	88	92	68	81	83.9
Evening	59	56	49	42	51	50	53	53	51	51	40	55	50.8
Dry-bulb temperature (°C)													
Max	5.8	6.3	16.7	20.1	25.5	29.9	31.3	31.1	27.2	20.9	16.0	10.0	20.1
Min	-0.4	-0.6	7.8	9.9	15.2	19.1	22.2	21.9	17.3	10.9	7.4	2.0	11.1
Mean	2.7	2.9	12.3	15.0	20.4	24.5	26.8	26.5	22.3	15.9	11.7	6.0	15.6
Neutrality	20.9	20.9	21.6	22.5	24.1	25.4	26.1	26.0	24.7	22.7	21.4	20.9	22.6
Upper limit	23.4	23.4	24.1	25.0	26.6	27.9	28.6	28.5	27.2	25.2	23.9	23.4	25.1
Lower limit	18.4	18.4	19.1	20.0	21.6	22.9	23.6	23.5	22.2	20.2	18.9	18.4	20.1
Rain (mm)[b]	95.3	100.1	104.4	101.6	139.7	105.2	92.5	80.5	86.6	77.2	109.5	107.7	1200.3
Prec (mm)[c]	375	394	411	400	550	414	364	317	341	304	431	424	4725
Wind (m/s)	3.9	4.2	4.7	3.1	2.3	3.0	3.1	2.4	2.6	3.3	3.6	4.1	3.4
HDD	483	434	200	126	27	0	0	0	0	92	213	397	1972
CDD	0	1	13	34	96	195	273	257	124	20	12	0	1025

(Continued)

TABLE 7.35 (CONTINUED)
Climatic Data, Nashville, Tennessee

Average diurnal range (°K)		9.0
Annual mean range (°K)		31.9

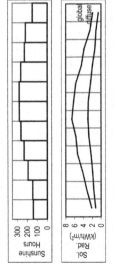

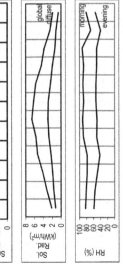

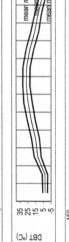

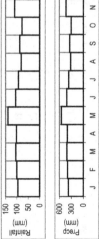

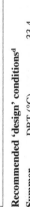

Wind rose

Recommended 'design' conditions[d]

Summer		
	DBT (°C)	33.4
	MCWB (°C)	23.8
	WBT (°C)	25.1
	MCDB (°C)	31.0
Winter	DBT (°C)	−7.6

Sources of data: https://energyplus.net/weather, [a]WMO (2010), [b]http://worldweather.wmo.int/en/city.html?cityId=794, [c]NOAA (2017), [d]ASHRAE(2009), Sunshine hours for Burlington, Vermont (38.9 miles).

TABLE 7.36
Climatic Data, Oklahoma City, Oklahoma

	Jan	Feb	Mar	Apr	May	Jun	Jul	Aug	Sep	Oct	Nov	Dec	Year
Latitude	N 35° 22'				Longitude		W 97° 35'			Altitude		398 m	
Climate	Warm Humid				ASHRAE		3A			Köppen		Cfa	
Months	Jan	Feb	Mar	Apr	May	Jun	Jul	Aug	Sep	Oct	Nov	Dec	Year
Sunshine h[a]	200.8	189.7	244.2	271.3	295.2	326.1	356.6	329.3	263.7	245.1	186.5	180.9	3089.4
Cloud (%)	51.3	61.9	50.5	66.0	58.0	46.5	31.4	41.2	36.9	37.9	47.2	46.8	47.98
Solar irradiation daily average (Wh/m²)													
Global	2779	3444	4560	5479	6044	6800	7121	6244	5299	4022	2909	2505	4767
Diffuse	1001	1583	1716	2027	2427	2381	2149	2134	1663	1385	1111	871	1704
Relative humidity (%)													
Morning	72	80	80	81	84	88	80	77	83	82	88	80	81.3
Evening	41	51	47	50	52	51	40	39	46	43	52	52	47.0
Dry-bulb temperature (°C)													
Max	6.7	11.1	15.5	21.1	25.2	29.4	33.2	33.8	28.7	22.9	15.6	8.7	21.0
Min	-3.3	3.2	6.0	11.8	16.4	20.3	22.6	22.0	18.4	12.3	5.5	0.8	11.3
Mean	1.7	7.2	10.8	16.5	20.8	24.9	27.9	27.9	23.6	17.6	10.6	4.8	16.2
Neutrality	20.9	20.9	21.1	22.9	24.2	25.5	26.4	26.4	25.1	23.3	21.1	20.9	22.8
Upper limit	23.4	23.4	23.6	25.4	26.8	28.0	29.0	29.0	27.6	25.8	23.6	23.4	25.3
Lower limit	18.4	18.4	18.6	20.4	21.8	23.0	24.0	24.0	22.6	20.8	18.6	18.4	20.3
Rain (mm)[b]	17.3	31.0	70.6	74.4	103.4	119.6	68.3	83.6	86.6	82.8	46.7	38.4	822.7
Prec (mm)[c]	139	158	306	307	465	493	293	328	406	371	198	188	3652
Wind (m/s)	6.2	6.0	6.7	6.0	5.9	4.3	4.6	4.6	4.7	4.7	4.7	5.0	5.3
HDD	530	314	239	74	23	0	0	0	11	83	247	446	1967
CDD	0	0	4	21	104	210	302	307	162	51	0	0	1161

(Continued)

TABLE 7.36 (CONTINUED)
Climatic Data, Oklahoma City, Oklahoma

	Average diurnal range (°K)	9.7
	Annual mean range (°K)	37.1

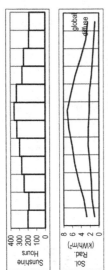

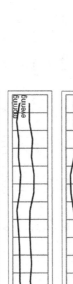

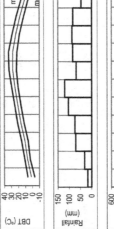

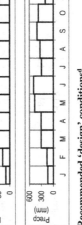

Wind rose

Recommended 'design' conditions[d]

Summer	DBT (°C)	36.0
	MCWB (°C)	23.4
	WBT (°C)	24.9
	MCDB (°C)	32.2
Winter	DBT (°C)	−8.1

Sources of data: https://energyplus.net/weather, [a]WMO (2010), [b]http://worldweather.wmo.int/en/city.html?cityId=799, [c]NOAA (2017), [d]ASHRAE(2009).

TABLE 7.37

Climatic Data, Olympia, Washington

Latitude	N 46° 58'				Longitude		W 122° 54'			Altitude		63 m	
Climate	Mixed Marine				ASHRAE		4C			Köppen		Cfb	
Months	Jan	Feb	Mar	Apr	May	Jun	Jul	Aug	Sep	Oct	Nov	Dec	Year
Sunshine h[a]	69.8	108.8	178.4	207.3	253.7	268.4	312.0	281.4	221.7	142.6	72.7	52.9	2169.7
Cloud (%)	88.5	80.5	67.5	69.9	67.9	66.3	48.6	39.0	58.9	71.4	87.1	83.4	69.1
Solar irradiation daily average (Wh/m^2)													
Global	1030	1532	2800	4063	4917	5553	6003	5099	3664	2246	1149	874	3244
Diffuse	817	1055	1581	2066	2618	2791	2538	1995	1786	1266	862	656	1669
Relative humidity (%)													
Morning	94	89	97	98	94	95	88	92	93	98	97	97	94.3
Evening	84	67	64	55	57	52	45	44	60	68	81	82	63.3
Dry-bulb temperature (°C)													
Max	6.7	8.0	11.0	14.6	17.0	21.0	25.0	24.7	20.7	16.0	9.7	6.8	15.1
Min	3.4	2.5	1.9	4.0	6.9	9.1	11.6	11.0	11.1	7.4	5.5	2.1	6.4
Mean	5.1	5.3	6.5	9.3	12.0	15.1	18.3	17.9	15.9	11.7	7.6	4.5	10.7
Neutrality	20.9	20.9	20.9	20.9	21.5	22.5	23.5	23.3	22.7	21.4	20.9	20.9	21.1
Upper limit	23.4	23.4	23.4	23.4	24.0	25.0	26.0	25.8	25.2	23.9	23.4	23.4	23.6
Lower limit	18.4	18.4	18.4	18.4	19.0	20.0	21.0	20.8	20.2	18.9	18.4	18.4	18.6
Rain (mm)[b]	199.1	133.9	134.4	89.9	59.2	44.7	16.0	23.9	43.4	116.8	219.2	189.5	1270
Prec (mm)[c]	784	527	529	354	233	176	63	94	171	460	863	746	5000
Wind (m/s)	3.4	3.6	2.6	3.3	3.0	3.0	2.6	2.4	2.5	2.1	3.4	2.3	2.9
HDD	415	369	369	271	183	90	37	36	92	236	332	441	2871
CDD	0	0	0	0	0	1	36	26	0	0	0	0	63

(Continued)

TABLE 7.37 (CONTINUED)
Climatic Data, Olympia, Washington

Average diurnal range (°K)	8.7
Annual mean range (°K)	23.1

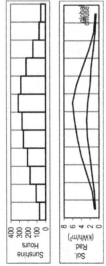
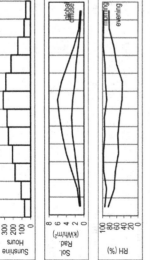
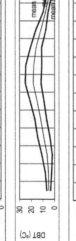

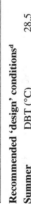

Wind rose

Recommended 'design' conditions[d]

Summer	DBT (°C)	28.5
	MCWB (°C)	18.2
	WBT (°C)	18.8
	MCDB (°C)	27.3
Winter	DBT (°C)	−4.4

Sources of data: https://energyplus.net/weather, [a]WMO (2010), [b]http://worldweather.wmo.int/en/city.html?cityId=2019, [c]NOAA (2017), [d]ASHRAE(2009), Sunshine hours for Seattle–Tacoma International airport, Washington (39.78 miles).

TABLE 7.38

Climatic Data, Phoenix, Arizona

Latitude	N 33° 27'				Longitude			W 111° 58'		Altitude		337 m	
Climate	Hot Dry				ASHRAE			2B		Köppen		BWh	
Months	Jan	Feb	Mar	Apr	May	Jun	Jul	Aug	Sep	Oct	Nov	Dec	Year
Sunshine h[a]	256.0	257.2	318.4	353.6	401.0	407.8	378.5	360.8	328.6	308.9	256.0	244.8	3871.6
Cloud (%)	26.4	40.8	30.7	25.5	24.7	26.7	43.1	26.1	29.4	28.1	25.8	40.0	30.61
Solar irradiation daily average (Wh/m²)													
Global	3287	4162	5336	7086	7844	8318	7616	7127	6340	4823	3773	3070	5732
Diffuse	999	1282	1653	1779	1996	1946	2207	1965	1580	1271	977	874	1544
Relative humidity (%)													
Morning	50	71	57	36	39	30	42	45	42	47	52	67	48.2
Evening	23	32	28	13	14	12	19	21	18	22	24	32	21.5
Dry-bulb temperature (°C)													
Max	18.8	21.4	22.4	30.6	33.5	40.1	41.4	39.2	36.5	31.3	24.2	17.2	29.7
Min	8.2	10.5	12.1	15.8	19.9	27.1	29.9	28.1	24.6	18.8	12.5	7.1	17.9
Mean	13.5	16.0	17.3	23.2	26.7	33.6	35.7	33.7	30.6	25.1	18.4	12.2	23.8
Neutrality	22.0	22.7	23.1	25.0	26.1	28.2	28.2	28.2	27.3	25.6	23.5	21.6	25.2
Upper limit	24.5	25.2	25.7	27.5	28.6	30.7	30.7	30.7	29.8	28.1	26.0	24.1	27.7
Lower limit	19.5	20.2	20.7	22.5	23.6	25.7	25.7	25.7	24.8	23.1	21.0	19.1	22.7
Rain (mm)[b]	24.9	24.6	26.9	6.6	2.0	0.5	27.2	24.6	15.5	16.5	16.8	21.6	207.7
Prec (mm)[c]	91	92	99	28	11	2	105	100	64	58	65	88	803
Wind (m/s)	2.3	2.4	2.4	3.8	3.9	3	3.6	2.5	2.1	2.7	2.1	2.3	2.8
HDD	155	93	59	15	0	0	0	0	0	1	27	195	545
CDD	0	0	38	184	288	480	544	490	372	212	26	0	2661

(Continued)

TABLE 7.38 (CONTINUED)
Climatic Data, Phoenix, Arizona

Average diurnal range (°K)		11.8
Annual mean range (°K)		34.3

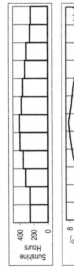
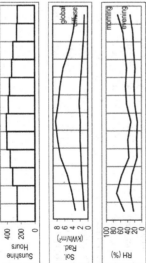
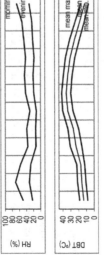
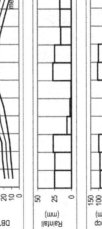
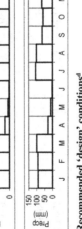

Wind rose

Recommended 'design' conditions[d]

Summer		
	DBT (°C)	42.3
	MCWB (°C)	21.0
	WBT (°C)	24.0
	MCDB (°C)	35.4
Winter	DBT (°C)	5.2

Sources of data: https://energyplus.net/weather, [a]WMO (2010), [b]http://worldweather.wmo.int/en/city.html?cityId=806, [c]NOAA (2017), [d]ASHRAE(2009).

TABLE 7.39

Climatic Data, Pierre, South Dakota

Latitude	N 44° 22'				Longitude		W 100° 16'			Altitude		528 m		
Climate	Cold Humid				ASHRAE		6A			Köppen		Dfa		
Months	Jan	Feb	Mar	Apr	May	Jun	Jul	Aug	Sep	Oct	Nov	Dec	Year
Sunshine h[a]	179.7	182.8	229.9	251.7	307.0	332.8	362.4	329.3	258.9	215.9	152.1	144.1	2946.6
Cloud (%)	62.7	57.0	62.0	58.7	48.3	54.8	48.8	50.0	37.7	31.0	71.3	55.0	53.1
Solar irradiation daily average (Wh/m²)													
Global	1786	2535	3940	4790	5817	6768	6822	5691	4664	3143	1896	1507	4113
Diffuse	807	1138	1810	2007	2353	2306	2315	1924	1499	1273	1021	784	1603
Relative humidity (%)													
Morning	80	76	76	82	77	87	79	81	83	75	84	68	79.0
Evening	62	52	56	45	40	47	35	42	36	35	56	53	46.6
Dry-bulb temperature (°C)													
Max	-5.3	0.6	6.8	14.6	21.2	26.8	30.8	28.9	24.6	16.2	6.1	-0.9	14.2
Min	-10.7	-7.8	-1.2	3.3	10	14.9	17.7	17.3	10.2	4.8	-1.8	-7.8	4.1
Mean	-8.0	-3.6	2.8	9.0	15.6	20.9	24.3	23.1	17.4	10.5	2.2	-4.4	9.4
Neutrality	20.9	20.9	20.9	20.9	22.6	24.3	25.3	25.0	23.2	21.1	20.9	20.9	20.9
Upper limit	23.4	23.4	23.4	23.4	25.1	26.8	27.8	27.5	25.7	23.6	23.4	23.4	23.4
Lower limit	18.4	18.4	18.4	18.4	20.1	21.8	22.8	22.5	20.7	18.6	18.4	18.4	18.4
Rain (mm)[b]	10.7	15.0	31.2	46.0	80.0	90.7	66.3	45.7	47.5	41.9	19.3	14.0	508.3
Prec (mm)[c]	42	59	123	181	315	357	261	180	187	165	76	55	2001
Wind (m/s)	4.8	5.7	6.1	5.9	5.4	4.7	4.7	4.8	4.4	5.5	5.1	5.4	5.2
HDD	832	622	482	280	94	17	0	0	58	266	500	731	3882
CDD	0	0	0	0	18	102	191	153	29	0	0	0	493

(Continued)

TABLE 7.39 (CONTINUED)
Climatic Data, Pierre, South Dakota

| | Average diurnal range (°K) | 10.1 |
| | Annual mean range (°K) | 41.5 |

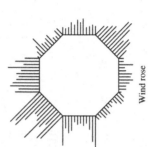

Wind rose

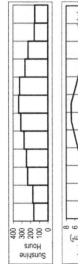
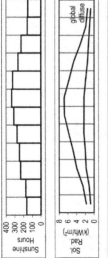
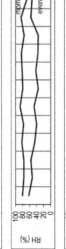
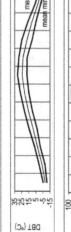
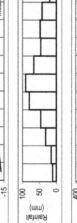
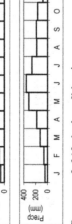

Recommended 'design' conditions[d]

Summer	DBT (°C)	35.0
	MCWB (°C)	21.4
	WBT (°C)	23.0
	MCDB (°C)	31.1
Winter	DBT (°C)	−21.4

Sources of data: https://energyplus.net/weather, [a]WMO (2010), [b]http://worldweather.wmo.int/en/city.html?cityId=2023, [c]NOAA (2017), [d]ASHRAE (2009), Sunshine hours for Huron, South Dakota (105.54 miles).

TABLE 7.40

Climatic Data, Providence, Rhode Island

		Latitude N 41° 43'				Longitude W 71° 25'					Altitude 16 m		
Climate		Cool Humid				ASHRAE 5A					Köppen Cfb		
Months	Jan	Feb	Mar	Apr	May	Jun	Jul	Aug	Sep	Oct	Nov	Dec	Year
Sunshine h[a]	171.7	172.6	215.6	225.1	254.9	274.1	290.6	262.8	233.0	208.7	148.0	148.6	2605.7
Cloud (%)	61.3	58.0	72.3	65.5	69.6	59.5	61.8	60.2	55.8	55.0	62.6	57.5	61.6
					Solar irradiation daily average (Wh/m²)								
Global	1863	2611	3581	4424	5553	6085	5628	5089	4137	3012	2040	1607	3803
Diffuse	1004	1351	1563	2098	2747	2694	2718	2510	1904	1340	1027	798	1813
					Relative humidity (%)								
Morning	73	68	77	70	82	83	85	90	90	85	85	76	80.3
Evening	59	46	55	43	51	53	54	60	57	53	64	49	53.7
					Dry-bulb temperature (°C)								
Max	1.2	3.8	6.9	12.8	19.4	23.4	27.9	25.2	23.0	16.3	9.0	2.7	14.3
Min	−4.0	−2.5	1.5	4.7	10.7	15.1	19.4	17.9	14.5	8.5	3.2	−2.5	7.2
Mean	−1.4	0.7	4.2	8.8	15.1	19.3	23.7	21.6	18.8	12.4	6.1	0.1	10.8
Neutrality	20.9	20.9	20.9	20.9	22.5	23.8	25.1	24.5	23.6	21.6	20.9	20.9	21.1
Upper limit	23.4	23.4	23.4	23.4	25.0	26.3	27.6	27.0	26.1	24.1	23.4	23.4	23.6
Lower limit	18.4	18.4	18.4	18.4	20.0	21.3	22.6	22.0	21.1	19.1	18.4	18.4	18.6
Rain (mm)[b]	98.0	83.6	127.3	110.7	90.2	92.5	83.6	91.4	99.6	99.8	114.6	107.2	1198.5
Prec (mm)[c]	386	329	501	436	355	364	329	360	392	393	451	422	4718
Wind (m/s)	4.8	4.9	4.7	5.1	4.4	3.8	3.8	4.0	4.1	4.1	4.2	4.4	4.4
HDD	607	500	440	286	124	27	4	4	38	197	365	573	3165
CDD	0	0	0	0	35	64	167	115	47	9	0	0	437

(Continued)

TABLE 7.40 (CONTINUED)
Climatic Data, Providence, Rhode Island

| | Average diurnal range (°K) | 7.1 |
| | Annual mean range (°K) | 31.9 |

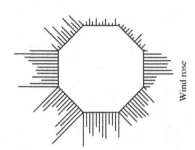

Wind rose

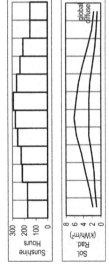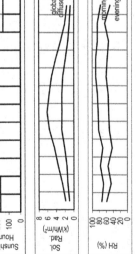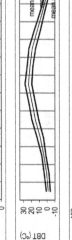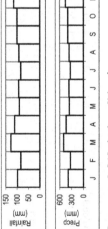

Recommended 'design' conditions[d]

Summer	DBT (°C)	30.4
	MCWB (°C)	22.1
	WBT (°C)	23.8
	MCDB (°C)	27.8
Winter	DBT (°C)	−11.2

Sources of data: https://energyplus.net/weather, [a]WMO (2010), [b]http://worldweather.wmo.int/en/city.html?cityId=811, [c]NOAA (2017), [d]ASHRAE (2009).

TABLE 7.41

Climatic Data, Raleigh, North Carolina

Latitude	N 35° 52'			Longitude	W 78°46'			Altitude	127 m			
Climate	Mixed Humid			ASHRAE	4A			Köppen	Cfa			
Months	Jan	Feb	Mar	Apr	May	Jun	Jul	Aug	Sep	Oct	Nov	Dec	Year
Sunshine h[a]	163.8	173.1	228.9	250.7	258.4	267.7	259.5	239.6	217.6	215.4	174.0	157.6	2606.3
Cloud (%)	54.8	59.2	46.2	50.4	60.1	59.9	54.3	45.8	53.0	39.3	49.9	53.6	52.21
Solar irradiation daily average (Wh/m²)													
Global	2394	3211	4340	5609	5767	6393	6064	5586	4493	3664	2438	2232	4349
Diffuse	1082	1494	1788	2147	2504	2880	2753	2891	2321	1446	894	917	1926
Relative humidity (%)													
Morning	86	74	81	86	94	93	93	97	94	96	82	83	88.3
Evening	55	44	40	40	55	51	53	56	58	56	48	54	50.8
Dry-bulb temperature (°C)													
Max	8.6	9.8	16.3	21.9	25.1	29.2	30.9	29.8	27.0	20.4	15.8	10.4	20.4
Min	1.1	1.3	5.4	9.9	15.2	19.1	21.3	19.8	17.9	9.9	7.2	2.1	10.9
Mean	4.9	5.6	10.9	15.9	20.2	24.2	26.1	24.8	22.5	15.2	11.5	6.3	15.6
Neutrality	20.9	20.9	21.2	22.7	24.0	25.3	25.9	25.5	24.8	22.5	21.4	20.9	22.6
Upper limit	23.4	23.4	23.7	25.2	26.5	27.8	28.4	28.0	27.3	25.0	23.9	23.4	25.1
Lower limit	18.4	18.4	18.7	20.2	21.5	22.8	23.4	23.0	22.3	20.0	18.9	18.4	20.1
Rain (mm)[b]	88.9	82.0	104.4	74.2	83.1	89.4	120.1	108.2	110.7	82.6	79.2	78.0	1100.8
Prec (mm)[c]	350	323	411	292	327	352	473	426	436	325	312	307	4334
Wind (m/s)	3.9	4.0	3.9	3.8	3.2	2.9	2.3	2.7	2.8	1.9	3.8	3.0	3.2
HDD	429	359	232	97	22	0	0	0	6	116	219	385	1865
CDD	0	1	2	29	86	178	246	199	123	11	1	0	876

(Continued)

TABLE 7.41 (CONTINUED)
Climatic Data, Raleigh, North Carolina

| | Average diurnal range (°K) | 9.6 |
| | Annual mean range (°K) | 29.8 |

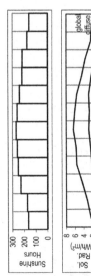
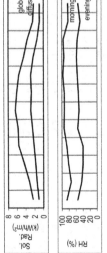
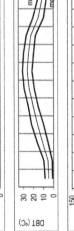
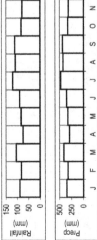
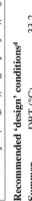

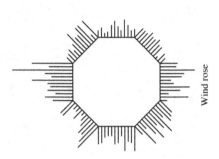

Wind rose

Recommended 'design' conditions[d]

Summer	DBT (°C)	33.2
	MCWB (°C)	24.2
	WBT (°C)	25.2
	MCDB (°C)	31.0
Winter	DBT (°C)	−5.0

Sources of data: https://energyplus.net/weather, [a]WMO (2010), [b]http://worldweather.wmo.int/en/city.html?cityId=813, [c]NOAA (2017), [d]ASHRAE(2009).

TABLE 7.42
Climatic Data, Reno Tahoe, Nevada

	Latitude	N 39° 28'			Longitude		W 119° 46'			Altitude		1342 m	
Climate		Cool Dry			ASHRAE		5B			Köppen		BSk	
Months	Jan	Feb	Mar	Apr	May	Jun	Jul	Aug	Sep	Oct	Nov	Dec	Year
Sunshine h[a]	195.6	204.2	291.0	332.1	375.8	393.8	424.0	390.8	343.9	295.2	212.0	187.5	3645.9
Cloud (%)	60.7	55.1	48.6	55.5	49.0	35.4	18.8	20.5	23.0	36.0	45.8	36.0	40.4
Solar irradiation daily average (Wh/m²)													
Global	2263	3285	4267	5815	6978	7828	7743	7247	5738	4071	2793	2123	5013
Diffuse	1000	1158	1622	2079	2155	1794	1575	1421	1176	1066	872	797	1393
Relative humidity (%)													
Morning	79	77	77	67	66	66	60	57	73	75	72	79	70.7
Evening	44	36	36	26	22	21	14	13	22	27	33	42	28.0
Dry-bulb temperature (°C)													
Max	7.1	9.4	11.5	16.9	22.0	26.7	32.8	32.2	25.4	18.6	12.6	5.4	18.4
Min	−3.9	−2.1	0.3	3.3	6.3	10.9	13.9	11.7	7.7	2.4	−2.3	−4.6	3.6
Mean	1.6	3.7	5.9	10.1	14.2	18.8	23.4	22.0	16.6	10.5	5.2	0.4	11.0
Neutrality	20.9	20.9	20.9	20.9	22.2	23.6	25.0	24.6	22.9	20.9	20.9	20.9	21.2
Upper limit	23.4	23.4	23.4	23.4	24.7	26.1	27.5	27.1	25.4	23.4	23.4	23.4	23.7
Lower limit	18.4	18.4	18.4	18.4	19.7	21.1	22.5	22.1	20.4	18.4	18.4	18.4	18.7
Rain (mm)[b]	25.9	26.9	19.8	13.5	17.0	14.2	5.6	5.8	9.7	15.5	25.9	30.0	209.8
Prec (mm)[c]	103	102	76	47	49	51	18	23	35	51	82	103	740
Wind (m/s)	2.4	3.3	2.4	3.4	4.3	3.5	3.4	3.2	2.7	2.6	2.4	1.9	3.0
HDD	536	410	385	234	135	10	0	0	49	248	409	572	2988
CDD	0	0	0	0	41	60	194	153	13	0	0	0	461

(Continued)

TABLE 7.42 (CONTINUED)
Climatic Data, Reno Tahoe, Nevada

| | Average diurnal range (°K) | 14.8 |
| | Annual mean range (°K) | 37.4 |

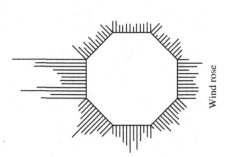

Wind rose

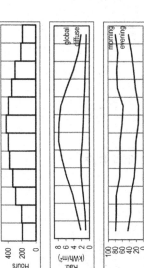

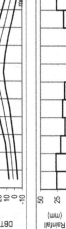

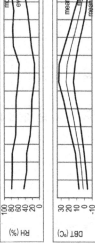

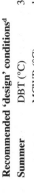

Recommended 'design' conditions[d]

Summer	DBT (°C)	33.9
	MCWB (°C)	15.9
	WBT (°C)	16.9
	MCDB (°C)	30.9
Winter	DBT (°C)	−8.3

Sources of data: https://energyplus.net/weather, [a]WMO (2010), [b]http://worldweather.wmo.int/en/city.html?cityId=815, [c]NOAA (2017), [d]ASHRAE(2009).

TABLE 7.43

Climatic Data, Richmond, Virginia

	Jan	Feb	Mar	Apr	May	Jun	Jul	Aug	Sep	Oct	Nov	Dec	Year
Latitude N 37° 31'					Longitude W 77° 19'					Altitude 50 m			
Climate Mixed Humid					ASHRAE 4A					Köppen Cfa			
Months	Jan	Feb	Mar	Apr	May	Jun	Jul	Aug	Sep	Oct	Nov	Dec	Year
Sunshine h[a]	172.5	179.7	233.3	261.6	288.0	306.4	301.4	278.9	237.9	222.8	183.5	163.0	2829
Cloud (%)	60.2	55.8	54.8	54.0	59.6	63.4	67.5	45.2	51.1	52.1	49.3	48.4	55.1
Solar irradiation daily average (Wh/m²)													
Global	2345	3131	4177	5321	5798	6158	6031	5672	4384	3439	2497	2066	4252
Diffuse	1127	1245	1674	2231	2590	2905	2578	2485	1888	1452	1053	833	1838
Relative humidity (%)													
Morning	79	77	72	84	85	81	91	93	98	94	81	85	85.0
Evening	52	48	39	48	47	49	54	47	66	55	48	52	50.4
Dry-bulb temperature (°C)													
Max	5.6	8.7	15.3	20.1	25.7	28.5	30.8	30.9	25.0	19.8	14.5	10.8	19.6
Min	-1.5	0.1	5.1	9.1	14.1	18.4	22.0	20.2	17.2	10.0	6.4	2.4	10.3
Mean	2.1	4.4	10.2	14.6	19.9	23.5	26.4	25.6	21.1	14.9	10.5	6.6	15.0
Neutrality	20.9	20.9	21.0	22.3	24.0	25.1	26.0	25.7	24.3	22.4	21.0	20.9	22.4
Upper limit	23.4	23.4	23.5	24.8	26.5	27.6	28.5	28.2	26.8	24.9	23.5	23.4	24.9
Lower limit	18.4	18.4	18.5	19.8	21.5	22.6	23.5	23.2	21.8	19.9	18.5	18.4	19.9
Rain (mm)[b]	77.2	70.1	102.6	83.1	96.0	99.8	114.6	118.4	104.9	75.7	82.3	82.8	1107.5
Prec (mm)[c]	304	276	404	327	378	393	451	466	413	298	324	326	4360
Wind (m/s)	4.3	4.6	4.2	4.1	3.6	3.3	3.1	3.1	3.4	3.0	4.4	3.3	3.7
HDD	513	395	255	138	18	0	0	0	8	119	247	377	2070
CDD	0	0	8	29	77	164	257	223	92	4	2	2	858

(*Continued*)

TABLE 7.43 (CONTINUED)
Climatic Data, Richmond, Virginia

	Average diurnal range (°K)	9.4
	Annual mean range (°K)	32.4

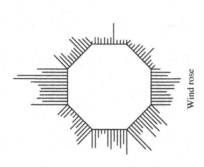

Wind rose

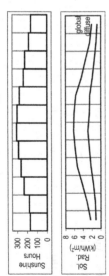
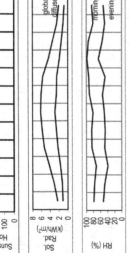
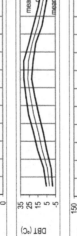
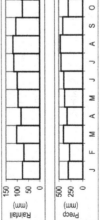
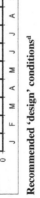

Sunshine Hours: 300 200 100 0

Sol. Rad. (kWh/m²): 8 6 4 2 0 — global, diffuse

RH (%): 100 80 60 40 20 — morning, evening

DBT (°C): 35 25 15 5 -5 — mean max., mean, mean min

Rainfall (mm): 150 100 50 0

Precp (mm): 500 250 0

J F M A M J J A S O N D

Recommended 'design' conditions[d]

Summer		
	DBT (°C)	33.5
	MCWB (°C)	24.1
	WBT (°C)	25.3
	MCDB (°C)	31.0
Winter		
	DBT (°C)	−6.2

Sources of data: https://energyplus.net/weather, [a]WMO (2010), [b]http://worldweather.wmo.int/en/city.html?cityId=816, [c]NOAA (2017), [d]ASHRAE(2009).

TABLE 7.44
Climatic Data, Sacramento, California

	Jan	Feb	Mar	Apr	May	Jun	Jul	Aug	Sep	Oct	Nov	Dec	Year
Latitude	N 38° 42'				Longitude	W 121° 24'				Altitude	7 m		
Climate	Warm Dry				ASHRAE	3B				Köppen	Csa		
Months	**Jan**	**Feb**	**Mar**	**Apr**	**May**	**Jun**	**Jul**	**Aug**	**Sep**	**Oct**	**Nov**	**Dec**	**Year**
Sunshine h[a]	145.5	201.3	278.0	329.6	406.3	419.5	440.2	406.9	347.8	296.7	194.9	141.1	3607.8
Cloud (%)	64.9	61.5	28.9	22.0	13.9	7.2	6.0	2.0	13.0	26.5	42.5	55.9	28.69
Solar irradiation daily average (Wh/m²)													
Global	1734	2635	4732	6394	7345	8093	7872	7050	5417	3650	2512	1882	4943
Diffuse	908	1380	1726	2001	2041	2078	1999	1743	1663	1380	1196	1022	1595
Relative humidity (%)													
Morning	96	95	89	84	82	78	81	87	83	86	89	95	87.1
Evening	65	65	47	35	33	29	28	28	31	44	51	66	43.5
Dry-bulb temperature (°C)													
Max	13.3	12.6	20.4	23.8	26.2	32.0	33.9	33.2	29.7	23.0	18.2	12.9	23.3
Min	3.9	5.0	8.1	9.1	10.9	15.0	15.5	14.3	13.7	10.3	6.4	4.2	9.7
Mean	8.6	8.8	14.3	16.5	18.6	23.5	24.7	23.8	21.7	16.7	12.3	8.6	16.5
Neutrality	20.9	20.9	22.2	22.9	23.6	25.1	25.5	25.2	24.5	23.0	21.6	20.9	22.9
Upper limit	23.4	23.4	24.7	25.4	26.1	27.6	28.0	27.7	27.0	25.5	24.1	23.4	25.4
Lower limit	18.4	18.4	19.7	20.4	21.1	22.6	23.0	22.7	22.0	20.5	19.1	18.4	20.4
Rain (mm)[b]	92.5	88.1	69.9	29.2	17.3	5.3	0	1.3	7.4	24.1	52.8	82.6	470.5
Prec (mm)[c]	364	347	275	115	68	21	0	5	29	95	208	325	1852
Wind (m/s)	2.4	3.9	3.2	4.1	3.9	4.1	4	4.1	2.9	3.4	2.9	2.4	3.4
HDD	321	265	132	76	41	1	0	0	5	79	197	319	1436
CDD	0	0	2	25	48	151	178	158	97	11	0	0	670

(Continued)

TABLE 7.44 (CONTINUED)
Climatic Data, Sacramento, California

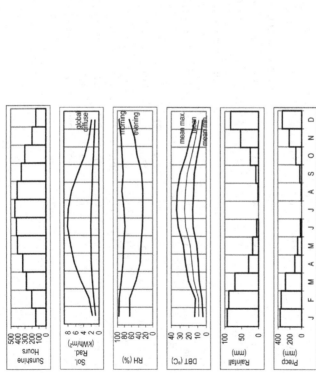

| | Average diurnal range (°K) | 13.6 |
| | Annual mean range (°K) | 30.0 |

Wind rose

Recommended 'design' conditions[d]

Summer	DBT (°C)	36.6
	MCWB (°C)	20.9
	WBT (°C)	21.9
	MCDB (°C)	34.5
Winter	DBT (°C)	1.1

Sources of data: https://energyplus.net/weather, [a]WMO (2010), [b]http://worldweather.wmo.int/en/city.html?cityId=820, [c]NOAA (2017), [d]ASHRAE(2009).

TABLE 7.45
Climatic Data, Salem, Oregon

Latitude	N 44° 54'				**Longitude**		W 123° 0'			**Altitude**		60 m	
Climate	Mixed Marine				ASHRAE		4C			Köppen		Cfb	
Months	**Jan**	**Feb**	**Mar**	**Apr**	**May**	**Jun**	**Jul**	**Aug**	**Sep**	**Oct**	**Nov**	**Dec**	**Year**
Sunshine h[a]	85.6	116.4	191.1	221.1	276.1	290.2	331.9	298.1	235.7	151.7	79.3	63.7	2340.9
Cloud (%)	72.2	71.6	66.3	75.1	67.0	50.9	32.4	34.8	43.3	59.6	77.9	75.0	60.51
Solar irradiation daily average (Wh/m²)													
Global	1332	1986	3108	4328	5445	6260	6787	5569	4561	2556	1361	1013	3692
Diffuse	857	1185	1656	2347	2721	2556	2217	1859	1655	1151	953	656	1651
Relative humidity (%)													
Morning	89	88	93	91	92	88	84	89	90	90	91	89	89.5
Evening	80	71	61	59	51	43	37	38	42	56	75	76	57.4
Dry-bulb temperature (°C)													
Max	7.4	8.5	12.5	15.2	18.8	23.8	27.8	26.7	24.3	17.0	11.2	6.6	16.7
Min	3.1	3.0	5.2	6.5	7.9	11.2	12.4	12.4	10.9	7.1	6.9	3.2	7.5
Mean	5.3	5.8	8.9	10.9	13.4	17.5	20.1	19.6	17.6	12.1	9.1	4.9	12.1
Neutrality	20.9	20.9	20.5	21.2	21.9	23.2	24.0	23.9	23.3	21.5	20.6	20.9	21.5
Upper limit	23.4	23.4	23.0	23.7	24.4	25.7	26.5	26.4	25.8	24.0	23.1	23.4	24.0
Lower limit	18.4	18.4	18.0	18.7	19.4	20.7	21.5	21.4	20.8	19.0	18.1	18.4	19.0
Rain (mm)[b]	151.4	115.8	101.3	71.4	56.4	39.4	11.7	11.4	32.5	77.0	165.1	174.2	1007.6
Prec (mm)[c]	596	456	399	281	222	155	46	45	128	303	650	686	3967
Wind (m/s)	3.3	4.8	2.6	3.5	3.3	3.7	3.2	3	2.6	2.8	4	4.4	3.4
HDD	406	358	297	228	145	49	8	1	48	209	285	417	2451
CDD	0	0	0	0	3	31	74	46	8	0	0	0	162

(Continued)

TABLE 7.45 (CONTINUED)
Climatic Data, Salem, Oregon

	Average diurnal range (°K)	9.2
	Annual mean range (°K)	24.8

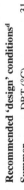

Wind rose

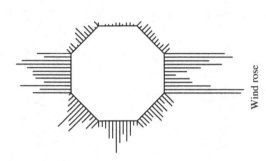

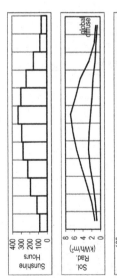
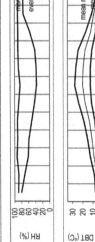
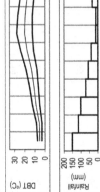
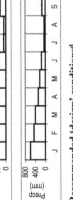

Recommended 'design' conditions[d]

Summer	DBT (°C)	31.1
	MCWB (°C)	18.8
	WBT (°C)	19.5
	MCDB (°C)	29.5
Winter	DBT (°C)	−3.2

Sources of data: https://energyplus.net/weather, [a]WMO (2010), [b]http://worldweather.wmo.int/en/city.html?cityId=822, [c]NOAA (2017), [d]ASHRAE(2009), Sunshine hours for Portland, Oregon (43.73 miles).

TABLE 7.46

Climatic Data, Salt Lake City, Utah

Latitude	N 40° 46′				Longitude		W 111° 58′			Altitude		1288 m	
Climate	Cool Dry				ASHRAE		5B			Köppen		BSk	
Months	Jan	Feb	Mar	Apr	May	Jun	Jul	Aug	Sep	Oct	Nov	Dec	Year
Sunshine h[a]	127.4	163.1	241.9	269.1	321.7	360.5	380.5	352.5	301.1	248.1	150.4	113.1	3029.4
Cloud (%)	69.7	70.8	64.2	64.1	62.6	39.1	33.2	59.2	43.1	48.2	57.6	64.5	56.4
Solar irradiation daily average (Wh/m²)													
Global	1886	2920	3983	5385	6321	7602	7278	6374	5330	3690	2295	1555	4552
Diffuse	1102	1481	1665	2148	2411	2156	1924	1941	1495	1229	1411	784	1646
Relative humidity (%)													
Morning	78	78	73	68	70	54	52	51	58	69	76	86	67.8
Evening	58	49	42	36	39	21	21	19	27	40	50	64	38.8
Dry-bulb temperature (°C)													
Max	3.8	4.9	11.3	15.4	19	27.7	32.9	31.1	25.4	17.3	8.8	3.2	16.7
Min	-1.8	-1.1	3.1	6.0	10.2	13.7	19.6	19.1	13.7	7.8	2	-2.5	7.5
Mean	1.0	1.9	7.2	10.7	14.6	20.7	26.3	25.1	19.6	12.6	5.4	0.4	12.1
Neutrality	20.9	20.9	20.9	21.1	22.3	24.2	25.9	25.6	23.9	21.7	20.9	20.9	21.6
Upper limit	23.4	23.4	23.4	23.6	24.8	26.7	28.4	28.1	26.4	24.2	23.4	23.4	24.1
Lower limit	18.4	18.4	18.4	18.6	19.8	21.7	23.4	23.1	21.4	19.2	18.4	18.4	19.1
Rain (mm)[b]	31.8	31.8	45.5	50.5	49.5	24.9	15.5	17.5	30.7	38.6	36.8	35.8	
Prec (mm)[c]	125	125	179	199	195	98	61	69	121	152	145	141	1610
Wind (m/s)	4.0	3.1	4.7	4.2	4.5	4.4	4.2	3.9	4.7	4.0	4.0	3.3	4.1
HDD	547	464	347	224	111	0	0	0	42	194	401	567	2908
CDD	0	0	0	2	10	106	263	217	71	0	0	0	669

(Continued)

TABLE 7.46 (CONTINUED)
Climatic Data, Salt Lake City, Utah

	Average diurnal range (°K)	9.3
	Annual mean range (°K)	35.4

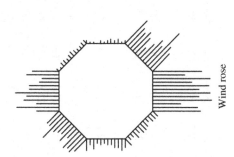

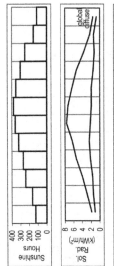
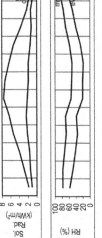
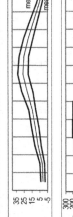
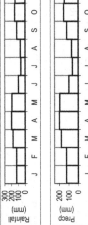
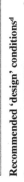

Wind rose

Recommended 'design' conditions[d]

Summer	DBT (°C)	34.9
	MCWB (°C)	17.0
	WBT (°C)	18.7
	MCDB (°C)	30.1
Winter	DBT (°C)	−9.9

Sources of data: https://energyplus.net/weather, [a]WMO (2010), [b]http://worldweather.wmo.int/en/city.html?cityId=823, [c]NOAA (2017), [d]ASHRAE (2009).

TABLE 7.47

Climatic Data, Sante Fe, New Mexico

Latitude	N 35° 37'					Longitude		W 106° 4'			Altitude		1934 m	
Climate	Cool Dry					ASHRAE		5B			Köppen		BSk	
Months	Jan	Feb	Mar	Apr	May	Jun	Jul	Aug	Sep	Oct	Nov	Dec	Year
Sunshine h[a]	234.2	225.3	270.2	304.6	347.4	359.3	335.0	314.2	286.7	281.4	233.8	223.3	3415.4
Cloud (%)	37.1	34.5	28.1	35.7	33.8	24.3	45.7	30.9	22.0	26.5	30.7	26.1	31.3
Solar irradiation daily average (Wh/m²)													
Global	2996	4265	5687	6523	7586	7757	7037	6614	6008	4529	3341	2900	5437
Diffuse	667	832	1276	1694	1780	1646	2011	1838	1297	904	871	590	1284
Relative humidity (%)													
Morning	64	75	60	58	67	48	76	79	78	72	70	72	68.3
Evening	31	37	28	20	26	18	35	32	37	27	34	35	30.0
Dry-bulb temperature (°C)													
Max	5.7	8.2	15.9	18.1	22.9	28.6	27.5	27.6	22.4	18.5	9.8	5.4	17.6
Min	−4.9	−2.6	1.8	3.1	8.1	12.4	15.5	14.5	10.3	4.1	−1.2	−5.9	4.6
Mean	0.4	2.8	8.9	10.6	15.5	20.5	21.5	21.1	16.4	11.3	4.3	−0.3	11.1
Neutrality	20.9	20.9	20.9	21.1	22.6	24.2	24.5	24.3	22.9	21.3	20.9	20.9	21.2
Upper limit	23.4	23.4	23.4	23.6	25.1	26.7	27.0	26.8	25.4	23.8	23.4	23.4	23.7
Lower limit	18.4	18.4	18.4	18.6	20.1	21.7	22.0	21.8	20.4	18.8	18.4	18.4	18.7
Rain(mm)[b]	17.5	18.5	28.7	22.9	24.4	35.3	70.4	57.7	41.9	35.6	22.9	22.9	398.7
Prec (mm)[c]	61	52	81	72	94	111	176	260	163	135	68	77	1350
Wind (m/s)	3.9	3.9	4.1	4.7	4.7	4.7	3.6	3.5	3.6	4.4	4	3.8	4.1
HDD	573	443	297	217	88	7	0	1	85	232	440	599	2982
CDD	0	0	0	0	26	103	109	94	28	0	0	0	360

(Continued)

TABLE 7.47 (CONTINUED)
Climatic Data, Sante Fe, New Mexico

Average diurnal range (°K)	13.0
Annual mean range (°K)	34.5

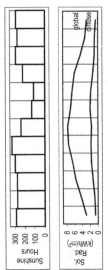
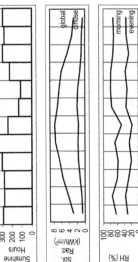
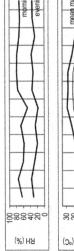
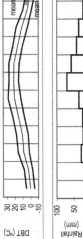

Sunshine Hours — 300 200 100 0

Sol. Rad. (kWh/m²) — 8 6 4 2 0 — global, diffuse

RH (%) — 100 80 60 40 20 0 — morning, evening

DBT (°C) — 30 20 10 0 -10 — mean max, mean, mean min

Rainfall (mm) — 100 50 0

Precp (mm) — 300 150 0

J F M A M J J A S O N D

Wind rose

Recommended 'design' conditions[d]

Summer		
DBT (°C)	33.8
MCWB (°C)	15.6
WBT (°C)	18.0
MCDB (°C)	27.4
Winter	
DBT (°C)	-6.0

Sources of data: https://energyplus.net/weather, [a]WMO (2010), [b]http://worldweather.wmo.int/en/city.html?cityId=2035, [c]NOAA (2017), [d]ASHRAE(2009),
Recommended design conditions and Sunshine hours Albuquerque International airport, New Mexico (64.4 miles).

TABLE 7.48
Climatic Data, Springfield, Illinois

| | Latitude N 39° 50' | | | | Longitude W 89° 40' | | | | Altitude 179 m | | | |
| | Climate Cool Humid | | | | ASHRAE 5A | | | | Köppen Dfa | | | |
Months	Jan	Feb	Mar	Apr	May	Jun	Jul	Aug	Sep	Oct	Nov	Dec	Year
Sunshine h[a]	160.7	158.7	186.5	225.8	281.2	308.0	320.7	291.0	248.4	214.0	140.2	129.3	2664.5
Cloud (%)	58.2	59.4	58.4	48.9	40.5	59.1	53.4	32.0	43.8	45.6	48.6	55.3	50.3
Solar irradiation daily average (Wh/m²)													
Global	1884	2905	3777	4951	5794	6548	6284	5649	4727	3520	2093	1749	4157
Diffuse	912	1321	1745	2062	2549	2622	2666	2287	1729	1350	986	846	1756
Relative humidity (%)													
Morning	75	80	87	83	82	79	82	94	87	80	72	78	81.6
Evening	57	67	58	57	50	47	56	57	51	43	50	56	54.1
Dry-bulb temperature (°C)													
Max	-1.8	-0.3	11.5	16.7	23.1	28.7	29.2	28.6	25	17.9	11.1	4.4	16.2
Min	-7.3	-6.4	3.6	8.3	13.4	19.1	20.9	18.5	14.2	7.2	4.4	-2.4	7.8
Mean	-4.6	-3.4	7.6	12.5	18.3	23.9	25.1	23.6	19.6	12.6	7.8	1.0	12.0
Neutrality	20.9	20.9	20.9	21.7	23.5	25.2	25.6	25.1	23.9	21.7	20.9	20.9	21.5
Upper limit	23.4	23.4	23.4	24.2	26.0	27.7	28.1	27.6	26.4	24.2	23.4	23.4	24.0
Lower limit	18.4	18.4	18.4	19.2	21.0	22.7	23.1	22.6	21.4	19.2	18.4	18.4	19.0
Rain (mm)[b]	46.2	46.0	66.8	89.2	107.7	113.3	100.1	82.3	73.7	80.0	81.5	64.0	950.8
Prec (mm)[c]	182	181	263	351	424	446	394	324	290	315	321	252	3743
Wind (m/s)	4.6	5.2	5.4	5.4	4.5	4.7	4.0	2.7	4.6	4.6	5.5	4.6	4.7
HDD	712	603	339	196	35	2	0	0	28	193	321	540	2969
CDD	0	0	6	29	45	171	209	165	64	3	1	0	693

(Continued)

TABLE 7.48 (CONTINUED)
Climatic Data, Springfield, Illinois

Average diurnal range (°K)		8.4
Annual mean range (°K)		36.5

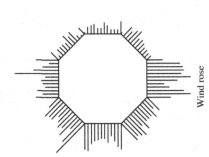

Wind rose

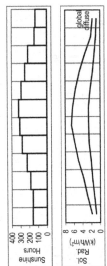
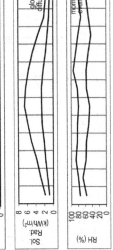
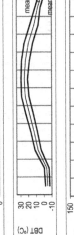
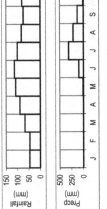
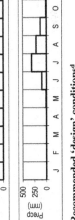

Recommended 'design' conditions[d]

Summer		
	DBT (°C)	32.5
	MCWB (°C)	24.2
	WBT (°C)	25.5
	MCDB (°C)	30.9
Winter		
	DBT (°C)	−15.2

Sources of data: https://energyplus.net/weather, [a]WMO (2010), [b]http://worldweather.wmo.int/en/city.html?cityId=834, [c]NOAA (2017), [d]ASHRAE(2009).

TABLE 7.49
Climatic Data, Tallahassee, Florida

	Jan	Feb	Mar	Apr	May	Jun	Jul	Aug	Sep	Oct	Nov	Dec	Year
Latitude	N 30° 22'				Longitude			W 82°22'		Altitude		21 m	
Climate	Hot Humid				ASHRAE			2A		Köppen		Cfa	
Months	Jan	Feb	Mar	Apr	May	Jun	Jul	Aug	Sep	Oct	Nov	Dec	Year
Sunshine h[a]	187.7	188.1	250.8	296.8	327.9	304.8	278.6	262.6	251.8	261.2	212.8	187.8	3010.9
Cloud (%)	59.1	57.1	63.4	48.0	50.5	59.6	26.9	54.5	57.5	45.3	50.9	55.9	52.4
Solar irradiation daily average (Wh/m²)													
Global	2851	3621	4597	6010	6129	5726	5711	5471	4837	4213	3222	2573	4580
Diffuse	1178	1213	1734	2277	2283	2667	2711	2248	2257	1662	1314	987	1878
Relative humidity (%)													
Morning	86	82	89	92	98	97	97	93	95	92	93	88	91.8
Evening	50	47	47	41	58	61	61	55	55	52	52	53	52.7
Dry-bulb temperature (°C)													
Max	16.1	18.7	23.0	27.0	27.8	30.7	31.5	31.5	29.7	25.1	23.1	17.3	25.1
Min	6.6	7.4	12.0	13.4	16.7	21.3	23.3	22.6	19.8	14.1	12.4	6.3	14.7
Mean	11.4	13.1	17.5	20.2	22.3	26.0	27.4	27.1	24.8	19.6	17.8	11.8	19.9
Neutrality	21.3	21.8	23.2	24.1	24.7	25.9	26.3	26.2	25.5	23.9	23.3	21.5	24.0
Upper limit	23.8	24.4	25.7	26.6	27.2	28.4	28.8	28.7	28.0	26.4	25.8	24.0	26.5
Lower limit	18.8	19.4	20.7	21.6	22.2	23.4	23.8	23.7	23.0	21.4	20.8	19.0	21.5
Rain (mm)[b]	110.2	123.2	150.9	77.7	88.1	196.3	182.1	186.7	119.1	82.0	88.9	99.1	1540.3
Prec (mm)[c]	434	485	594	306	347	773	717	735	469	323	350	390	5923
Wind (m/s)	2.4	3.2	3.8	2.8	2.8	2.4	1.9	2.3	2.1	2.1	2.5	2.3	2.6
HDD	242	156	71	16	1	0	0	0	0	24	75	231	816
CDD	1	4	36	57	143	236	284	268	190	60	28	1	1308

(*Continued*)

TABLE 7.49 (CONTINUED)
Climatic Data, Tallahassee, Florida

| | Average diurnal range (°K) | 10.5 |
| | Annual mean range (°K) | 25.2 |

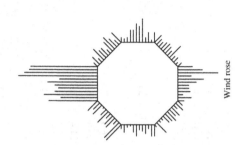

Wind rose

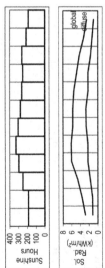

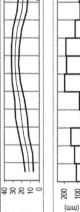

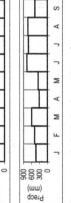

Recommended 'design' conditions[d]

Summer		
	DBT (°C)	34.2
	MCWB (°C)	24.5
	WBT (°C)	26.1
	MCDB (°C)	31.2
	DBT (°C)	−1.7
Winter		

Sources of data: https://energyplus.net/weather, [a]WMO (2010), [b]http://worldweather.wmo.int/en/city.html?cityId=837, [c]NOAA (2017), [d]ASHRAE (2009), Sunshine hours for Apalachicola, Florida (76 miles).

TABLE 7.50

Climatic Data, Topeka, Kansas

Latitude	N 39° 04'				Longitude		W 95° 37'			Altitude		269 m	
Climate	Mixed Humid				ASHRAE		4A			Köppen		Dfa	
Months	Jan	Feb	Mar	Apr	May	Jun	Jul	Aug	Sep	Oct	Nov	Dec	Year
Sunshine hᵃ	177.4	168.8	212.6	231.7	268.5	293.0	326.9	291.7	233.4	212.4	157.8	150.5	2724.7
Cloud (%)	51.8	43.0	57.1	49.4	63.8	37.2	45.8	44.6	36.2	48.5	45.2	51.3	47.8
Solar irradiation daily average (Wh/m²)													
Global	2289	2912	4147	4946	5792	6374	6327	5865	4842	3427	2483	2013	4285
Diffuse	902	1031	1666	1889	2603	2387	2455	2067	1548	1254	931	789	1627
Relative humidity (%)													
Morning	80	77	83	81	89	84	86	88	90	88	78	78	83.5
Evening	52	46	51	50	54	51	53	54	51	54	46	52	51.2
Dry-bulb temperature (°C)													
Max	2.2	7.9	11.5	19.0	23.6	28.6	30.0	29.5	26.9	19.1	13.3	3.9	18.0
Min	-5.6	-1.2	1.4	9.4	14.2	18.9	20.6	19.6	15.8	8.9	2.9	-4.1	8.4
Mean	-1.7	3.4	6.5	14.2	18.9	23.8	25.3	24.6	21.4	14.0	8.1	-0.1	13.2
Neutrality	20.9	20.9	20.9	22.2	23.7	25.2	25.6	25.4	24.4	22.1	20.9	20.9	21.9
Upper limit	23.4	23.4	23.4	24.7	26.2	27.7	28.1	27.9	26.9	24.6	23.4	23.4	24.4
Lower limit	18.4	18.4	18.4	19.7	21.2	22.7	23.1	22.9	21.9	19.6	18.4	18.4	19.4
Rain (mm)ᵇ	22.9	32.3	65.5	95.5	130.3	146.6	104.6	121.7	104.4	79.0	52.6	34.0	989.4
Prec (mm)ᵇ	86	132	249	353	491	540	382	424	366	303	185	135	3646
Wind (m/s)	4.7	3.8	4.6	4.4	3.6	4.4	4.1	3.4	2.9	3.7	5	4.1	4.1
HDD	627	426	379	146	25	12	1	2	12	146	324	590	2690
CDD	0	0	0	28	50	180	229	195	96	6	0	0	784

(Continued)

TABLE 7.50 (CONTINUED)
Climatic Data, Topeka, Kansas

Average diurnal range (°K)	9.6	
Annual mean range (°K)	35.6	

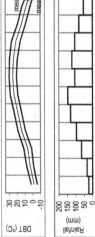
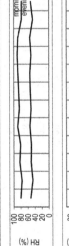

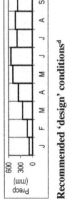
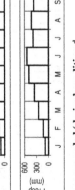
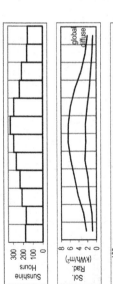

Wind rose

Recommended 'design' conditions[d]

Summer		
	DBT (°C)	34.5
	MCWB (°C)	24.3
	WBT (°C)	25.4
	MCDB (°C)	32.2
Winter		
	DBT (°C)	−14.0

Sources of data: https://energyplus.net/weather, [a]WMO (2010), [b]http://worldweather.wmo.int/en/city.html?cityId=839, [c]NOAA (2017), [d]ASHRAE(2009).

TABLE 7.51
Climatic Data, Trenton, New Jersey

	Jan	Feb	Mar	Apr	May	Jun	Jul	Aug	Sep	Oct	Nov	Dec	Year
Latitude	N 40° 16'				Longitude			W 74° 49'		Altitude			65 m
Climate	Cool Humid				ASHRAE			5A		Köppen			Dfa
Months	Jan	Feb	Mar	Apr	May	Jun	Jul	Aug	Sep	Oct	Nov	Dec	Year
Sunshine h[a]	163.1	169.7	207.4	227.2	248.1	262.8	269.2	252.5	215.0	201.5	149.3	140.1	2505.9
Cloud (%)	56.6	68.5	56.5	48.8	49.3	44.5	32.8	45.5	66.8	27.1	45.9	54.1	49.7
Solar irradiation daily average (Wh/m²)													
Global	1778	2267	3659	4980	5965	6088	6317	4936	3684	3613	2141	1601	3919
Diffuse	889	1153	1697	2176	2619	2898	2681	2401	1954	1402	1022	785	1806
Relative humidity (%)													
Morning	72	73	77	81	86	87	82	85	94	85	82	72	81.3
Evening	51	59	50	53	51	49	44	53	62	44	58	53	52.3
Dry-bulb temperature (°C)													
Max	1.8	3.1	10.8	17.1	20.5	26.5	30.4	29.1	22.5	19.1	12.7	5.7	16.6
Min	−3.2	−1.8	3.1	8.8	10.5	16.5	20.2	21.3	15.2	8.8	5.5	−0.1	8.7
Mean	−0.7	0.7	7.0	13.0	15.5	21.5	25.3	25.2	18.9	14.0	9.1	2.8	12.7
Neutrality	20.9	20.9	20.9	21.8	22.6	24.5	25.6	25.6	23.6	22.1	20.9	20.9	21.7
Upper limit	23.4	23.4	23.4	24.3	25.1	27.0	28.1	28.1	26.1	24.6	23.4	23.4	24.2
Lower limit	18.4	18.4	18.4	19.3	20.1	22.0	23.1	23.1	21.1	19.6	18.4	18.4	19.2
Rain (mm)[b]	80.3	58.7	105.2	89.9	111.0	112.0	125.7	104.1	108.5	106.2	84.1	94.0	1179.7
Prec (mm)[c]	316	231	414	354	437	441	495	410	427	418	331	370	4644
Wind (m/s)	4.9	4.6	3.9	3.6	3.3	3.1	2.8	2.3	3.4	3.4	3.4	3.5	3.5
HDD	591	488	346	188	103	5	0	0	38	151	273	487	2670
CDD	0	0	0	34	26	118	225	216	60	9	0	0	688

(Continued)

TABLE 7.51 (CONTINUED)
Climatic Data, Trenton, New Jersey

Average diurnal range (°K)	7.9
Annual mean range (°K)	33.6

Wind rose

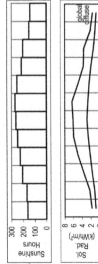
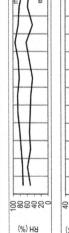
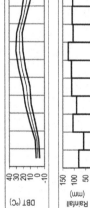
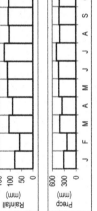

Sunshine Hours — 300, 200, 100, 0

Sol. Rad. (kWh/m²) — 8, 6, 4, 2, 0 — global, diffuse

RH (%) — 100, 80, 60, 40, 20 — morning, evening

DBT (°C) — 40, 30, 20, 10, 0, -10 — mean max., mean, mean min.

Rainfall (mm) — 150, 100, 50, 0

Precp (mm) — 600, 300, 0

J F M A M J J A S O N D

Recommended 'design' conditions[d]

Summer		
DBT (°C)	32.1
MCWB (°C)	23.2
WBT (°C)	24.4
MCDB (°C)	29.8
Winter	
DBT (°C)	-9.9

Sources of data: https://energyplus.net/weather, [a]WMO (2010), [b]http://worldweather.wmo.int/en/city.html?cityId=2041, [c]NOAA (2017), [d]ASHRAE(2009).

REFERENCES

ASHRAE (2009) ASHRAE Climatic Design Conditions 2009/2013/2017, available at http://ashrae-meteo. info/places.php?continent=North%20America.

DOE (n.d.) All Regions - North and Central America WMO Region 4 - USA, Department of Energy, US Government, available at https://energyplus.net/weather-region/north_and_central_america_wmo_region_4/USA%20 %20.

DOE (n.d.) EnergyPlus Weather Data, Department of Energy, US Government, available at https://energyplus. net/weather.

Kabre C (1999) WINSHADE: A Computer Design Tool for Solar Control. *Building and Environment*, 34(3): 263–274.

NOAA (2017) Climate Normals, National Oceanic and Atmospheric Administration, US Government, available at https://www.ncdc.noaa.gov/data-access/land-based-station-data/land-based-datasets/climate-normals.

Wikimedia Commons Contributors (2017) File: US Map - States and Capitals.png. October 12, Wikimedia Commons, the Free Media Repository, retrieved 16:56, April 16, 2020 from https://commons.wikimedia. org/w/index.php?title=File:US_map_-_states_and_capitals.png&oldid=262675659.

Wilcox S and Marion W (2008) *User's Manual for TMY3 Data Sets*, NREL/TP-581-43156, April, National Renewable Energy Laboratory, Golden, CO.

WMO (2010) World Meteorological Organization Standard Normal, United Nations Statistics Division, available at http://data.un.org/Explorer.aspx.

WMO (2020) World Weather Information Service, World Meteorological Organization, available at http:// worldweather.wmo.int/en/home.html.

Index

Printed in the United States
By Bookmasters